TABLE A.3 Critical Values for the t Distribution

One-tailed α = .10		.05	.025	.01	.005
Two-tailed α = .20		.10	.05	.02	.01
df = 1	3.078	6.314	12.706	31.821	63.657
2	1.886	2.920	4.303	6.965	9.925
3	1.638	2.353	3.182	4.541	5.841
4	1.533	2.132	2.776	3.747	4.604
5	1.476	2.015	2.571	3.365	4.032
6	1.440	1.943	2.447	3.143	3.707
7	1.415	1.895	2.365	2.998	3.499
8	1.397	1.860	2.306	2.896	3.355
9	1.383	1.833	2.262	2.821	3.250
10	1.372	1.812	2.228	2.764	3.169
11	1.363	1.796	2.201	2.718	3.106
12	1.356	1.782	2.179	2.681	3.055
13	1.350	1.771	2.160	2.650	3.012
14	1.345	1.761	2.145	2.624	2.977
15	1.341	1.753	2.131	2.602	2.947
16	1.337	1.746	2.120	2.583	2.921
17	1.333	1.740	2.110	2.567	2.898
18	1.330	1.734	2.101	2.552	2.878
19	1.328	1.729	2.093	2.539	2.861
20	1.325	1.725	2.086	2.528	2.845
21	1.323	1.721	2.080	2.518	2.831
22	1.321	1.717	2.074	2.508	2.819
23	1.319	1.714	2.069	2.500	2.807
24	1.318	1.711	2.064	2.492	2.797
25	1.316	1.708	2.060	2.485	2.787
26	1.315	1.706	2.056	2.479	2.779
27	1.314	1.703	2.052	2.473	2.771
28	1.313	1.701	2.048	2.467	2.763
29	1.311	1.699	2.045	2.462	2.756
30	1.310	1.697	2.042	2.457	2.750
40	1.303	1.684	2.021	2.423	2.704
50	1.299	1.676	2.009	2.403	2.678
60	1.296	1.671	2.000	2.390	2.660
70	1.294	1.667	1.994	2.381	2.648
80	1.292	1.664	1.990	2.374	2.639
90	1.291	1.662	1.987	2.368	2.632
100	1.290	1.660	1.984	2.364	2.626
125	1.288	1.657	1.979	2.357	2.616
150	1.287	1.655	1.976	2.351	2.609
200	1.286	1.653	1.972	2.345	2.601
∞	1.282	1.645	1.960	2.326	2.576

Note: Table entry gives t^c corresponding to Pr $(t \geq t^c) = \alpha$ for one-tailed tests and Pr $(|t| \geq t^c) = \alpha$ for two-tailed tests.

Source: Computed using Fortran subroutines from the IMSL Library.

Economic Statistics and Econometrics

Economic Statistics and Econometrics

SECOND EDITION

State University of New York at Albany

Macmillan Publishing Company
New York

Collier Macmillan Publishers
London

PRINTED IN THE UNITED STATES OF AMERICA

Macmillan Publishing Company
866 Third Avenue, New York, New York 10022

Collier Macmillan Canada, Inc.

Library of Congress Cataloging in Publication Data

Mirer, Thad W.
Economic statistics and econometrics.

Bibliography: p.
Includes index.
1. Economics—Statistical methods. 2. Econometrics.
I. Title.
HB137.M57 1988 330'.028 87-24027
ISBN 0-02-381821-2

Printing: 1 2 3 4 5 6 7 8 Year: 8 9 0 1 2 3 4 5 6 7

For IRENE
and PAUL
and DANNY

Preface

This text is designed for courses in economic statistics and introductory econometrics that aim to mix the development of technique with its application to real economic analysis. No background in statistics or calculus is required, and no matrix algebra is used. Nonetheless, the subject matter is technical, and it is presented with careful explanation rather than oversimplification.

Most of the work of revision has gone into reorganizing and rewriting Chapters 1 through 13, which constitute the core of the book. The number of problems in these chapters has been increased by one third. The major structural changes are noted in the following summary outline.

Part I deals with data and descriptive statistics. A discussion of elasticity has been added to the introduction. A special chapter on the nature of economic data presents two data sets—one cross-section and one time-series—that serve as the basis for most of the quantitative examples in the book. In the chapter on descriptive statistics, the section on bivariate statistics has been moved to the end in order to permit it to be deferred to later coverage.

Part II deals with the specification and estimation of regression models. In order to focus on the specification and interpretation of models, questions of inference and the properties of estimators are treated later, in Part IV. The chapter on the theory of simple regression received a lot of rewriting, and a new section exploring the distinction between true and estimated regressions has been added. The section on linear transformations has been simplified by focusing first on the question of units adjustment. A chapter on the application of simple regression introduces a variety of specification forms, including a new use of first differences. Throughout the text, regression applications based on the data in Chapter 2 serve as mini case studies of the use of econometrics. The chapter on multiple regression now begins with a treatment of the case of two regressors, and the special applications sections have been reorganized. The section on general specification questions has moved to the end of the chapter, and parts of it now look forward to new treatments in Part IV.

Part III deals with random variables and probability distributions. It begins with an optional chapter on probability theory. The basic chapter on random variables received a lot of rewriting, but covers the same material. The normal and t distributions are explained with an eye toward their later use, and an appendix on chi-square and F distributions provides background for special topics appearing later in the book.

Part IV develops sampling theory and the methods of statistical inference wholly in the context of regression. This novel presentation focuses attention where it is most critical: on the regression coefficients. In the chapter on sampling theory, the basic sections have been reorganized, and a new section on sample selection has been added. Hypothesis testing for single coefficients is developed in a series of special cases using t tests, and confidence intervals are developed for estimation and prediction. Treatment of the statistical properties of estimators has been upgraded, and it now leads into discussions of multicollinearity and misspecification that are based on comparisons of alternative sampling distributions.

Part V covers a series of topics in econometrics that extend the methods presented earlier: autocorrelation, heteroscedasticity, F tests, dummy dependent variables, distributed lags, time-series use of regression, and simultaneous-equation models. The changes include brief discussions of Durbin's two-step method in estimating models with autocorrelation and the Breusch-Pagan test for heteroscedasticity; also, there is an expanded discussion of the Chow test in time-series models.

Part VI covers some standard topics in statistics: sampling and inference for the mean and variance of a random variable, chi-square tests, and a comparison of analysis of variance with regression.

The back matter includes a set of carefully designed statistical tables, the answers to selected problems, and a bibliography. (A solutions manual for all the problems in the book is available from the publisher, for instructors using the text.)

The core of the book has been designed for one-term courses in economic statistics or introductory econometrics that have no statistics prerequisites. Most texts that have a title like "Statistics for Business and Economics" give detailed attention to topics in basic statistics but give short shrift to the use of regression in economics. This leaves students with very little to carry on to the rest of their studies. This book is designed for a course with the opposite priorities: only those statistical topics that constitute a foundation for basic econometrics are covered, and regression models are treated in detail.

Within the first four parts of the book, the more difficult material involving derivations and logarithms can be used selectively or not at all, without interfering with subsequent discussions of basic material. These sections are identified through a series of footnotes. The order of Parts II and III can be switched, yielding the traditional separation between statistics and econometrics.

Longer courses, or those serving students with stronger backgrounds, can include some or all of the material from Part V. Instructors who include computer use as part of the course will find the data sets in Chapter 2 helpful: students can replicate and modify the examples presented throughout the text, and new relations can be explored.

Special care has been taken in the production of the book to make it useful. All major drawings of normal, t, chi-square, and F distributions were computer generated (including several three-dimensional figures); thus they are technically correct. The basic statistical reference tables were also computer generated, permitting the construction of a table of P-values for the t distribution. The technical notation has been kept as consistent and simple as possible, and all equation displays are numbered for easy classroom reference. The problems at the end of each chapter are separated by section, and a star marks those for which answers appear at the back of the book.

I am grateful to many people for their help in making this second edition. Various instructors who have used the text in their classes made very useful suggestions, and I am sure that they will recognize their impact. Weihua Fu and Daeshik Kang provided helpful assistance.

At Macmillan, Jack Repcheck and Ken MacLeod oversaw this project, and John Travis guided the book through production. Throughout two editions, my wife and children have borne the side effects of my work. I used to think it was merely tradition or writers' cant to thank one's family, but I have learned better.

Thad W. Mirer

Contents

Introduction

When the Nobel Memorial Prize in Economic Science was first awarded, in 1969, it was given to Ragnar Frisch of Norway and Jan Tinbergen of The Netherlands for their pioneering work in econometrics. At the time, few people had heard of the subject and even fewer knew much about it. Today econometrics is widely recognized as the primary tool of empirical economic analysis.

Put simply, ***econometrics*** involves the development and use of special statistical methods within a framework that is consistent with the ways of economic inquiry. It is an extension of the field of ***statistics,*** which deals with techniques for collecting and analyzing data that arise in many different contexts. ***Economic statistics*** involves the application of these general techniques to economic questions.

1.1 The Nature of the Subjects

A time-honored example will illustrate the nature of the subjects and preview the topics covered in this book. In his *General Theory,* John Maynard Keynes developed the concept of an aggregate consumption function as a stable relation between consumer expenditures and aggregate income. Although the consumption function was only one part of his macroeconomic theory, its elaboration and testing were crucial in the validation and use of Keynes' other ideas. This was one of the first problems on which the then-young field of econometrics cut its teeth.

Econometric analysis starts from a statement of economic theory, whether it be derived from some sophisticated mathematical optimization technique or from some plain reasoning, and develops a ***structural equation*** that specifies how the value of one variable is determined by the values of other variables. In the case of the Keynesian consumption function, the simplest specification is

$$C = \beta_0 + \beta_1 Y + u \qquad (1.1)$$

where C stands for consumption and Y stands for income. The parameters β_0 and β_1 are specified to be unknown constants and are referred to as the ***structural coefficients*** in the equation (β is lowercase "beta," the Greek "b"). The ***disturbance*** u may be thought of as a variable reflecting all the factors in addition to income that help determine consumption. These factors may include variables that are relatively unimportant, and therefore not specifically mentioned as determinants of consumption, as well as pure chance and error in the measurement of C.

The structural equation serves as a ***model*** of the economic process determining the level of consumption in the economy. As a model, it necessarily abstracts from reality by simplifying the complexity of the true economic process under consideration. It seeks to get down to the essentials. However, many of the techniques of statistics and econometrics are based on the premise that the specified model is an accurate representation of the way the world works. If the model is wrong in its essentials, then all the quantitative and qualitative conclusions that are drawn from the analysis of data using these techniques may be far off the mark. The challenge of econometrics is to blend knowledge of economic theory and behavior with knowledge of statistical techniques in order to produce well-specified models.

Not all equations in empirical economics are structural equations describing economic behavior. For example, the familiar GNP accounting identity

$$Y = C + I + G \qquad (1.2)$$

has no unknown parameters and involves no disturbance term. The equation perfectly describes a relation among variables that holds true because of the way the variables are defined and measured. Here, there is no econometric problem.

Returning to the specification (1.1), the main task is to estimate the value of β_0 and β_1 using available data. Applying a technique known as ***regression,*** in Chapter 6 we estimate β_0 and β_1 to be 0.568 and 0.907, respectively, based on data presented in Chapter 2. We write the estimated model as

$$\hat{C} = 0.568 + 0.907Y \qquad (1.3)$$

where $\hat{C}$ stands for the value of C that is predicted to occur in conjunction with any given value for Y. The result is illustrated schematically in Figure 1.1,

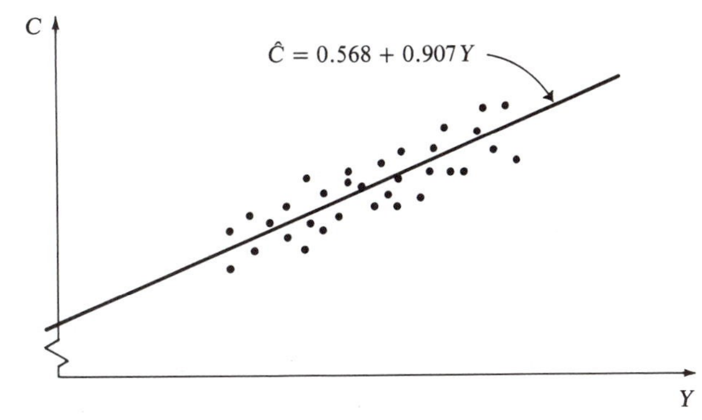

FIGURE 1.1 Each of the plotted points represents the paired values of consumption (C) and income (Y) in a particular year. The graphed line represents the estimated model of the aggregate consumption function, given as Equation (1.3). (The plotted points are merely illustrative; they are not the actual data points used in the estimation.)

where the estimated model is graphed through a scatter plot of data on C and Y. Since the original specification includes a disturbance term, we recognize that the predictions made with this estimated model are always subject to some error. Also, since the numbers 0.568 and 0.907 are only estimates of the true β_0 and β_1, we recognize that there is a further source of error in our predictions.

In addition to, or instead of, making predictions we might have special interest in the values of the parameters themselves. For example, in modern technical terms the essence of Keynes' consumption theory is that the marginal propensity to consume is less than 1; this implies that $\beta_1 < 1$. Treating this as a hypothesis to be tested, we note that since our estimated value for β_1 is 0.907, it appears that the hypothesis is confirmed. However, remembering that 0.907 is only an estimate of β_1, and therefore subject to estimation error, should we really consider the hypothesis to be confirmed? To answer this question, which is related to questions regarding the errors associated with our estimates and predictions, we must gain a firm understanding of the statistical foundations on which econometrics is built.

The field of statistics is divided into two parts, descriptive statistics and statistical inference. ***Descriptive statistics*** is concerned with summarizing the information in data on one or more variables, and it provides the methods for estimating the values of various parameters, including the coefficients of an econometric model. ***Statistical inference*** is concerned with the relation between these estimates and the true values of the parameters, and it provides the basis for testing hypotheses and for assessing the errors that are always present in estimation.

As noted earlier, statistical analysis of economic questions need not be econometric in nature. For example, to estimate the extent of poverty in the United States, the U.S. Bureau of the Census takes an annual survey of more than

50,000 households, asking questions about income and related items. Taking a representative survey involves sampling, and using the results to estimate the proportion of households that are poor involves inference. These techniques are statistical, but not econometric.

1.2 The Plan of This Book

Most readers of this book are students in courses in economic statistics or econometrics, and it has been designed for them. (Naturally, others are welcome, too.) These subjects get taught in a wide variety of ways at different levels, and the book aims to be as flexible as possible. Parts I through IV, with possible deletions, provide the basis for a course in economic statistics. Parts II, IV, and V provide the basis for an introductory course in econometrics.

The field of statistics is quite broad and it has a wide variety of applications. This book covers only those fundamental statistical ideas that make up the foundations of econometrics. This coverage is satisfactory at the introductory level, but students planning advanced work in econometrics are well advised to learn statistics on its home ground (i.e., in departments of mathematics and statistics).

Econometrics is a broad field also. This book gives a thorough treatment of the fundamental principles of specification and estimation of single-equation regression models and of the corresponding procedures of statistical inference. The standard "advanced topics" in econometrics are treated in a straightforward way, aiming for appreciation and understanding rather than rigorous detail.

The subjects of statistics and econometrics are inherently technical, but this book is not for technicians. Instead, it aims to explain and apply the basic material in as careful a manner as possible. The only mathematical background needed is algebra. However, this does not make the book easy. It is necessary to have a certain level of technical appreciation and experience in order to apply the methods correctly. Most sophomores can easily learn to put numbers into a computer and get back statistical results. To do this wisely—to make the statistical analysis valid—requires a real understanding of the techniques involved.

The reader should bear in mind that each chapter is a self-contained unit, building on most of the previous chapters. Within each chapter, however, the importance and meaning of some material in early sections may not be clear until later sections are covered. Given the difficulty of some of the material, the chapters demand reading and then rereading, perhaps several times. In Chapters 2 through 13, sections and appendixes that are especially difficult are marked with an asterisk (*).

Each chapter has a set of problems at the end, separated according to the section on which they draw. These problems are meant to be learning aids, not simply chores to be done, and they can lead one through a review of the chapter after it has been read thoroughly. Those problems for which answers appear at the back of the book are marked by a star (⋆).

Problems

Section 1.1

1.1 What specific economic factors other than income are likely to affect consumption? What noneconomic factors may affect consumption?

⋆ **1.2** Based on the estimated model in Equation (1.3), determine the predicted value of consumption resulting from an income level of 1000 (the units have not been specified).

1.3 Based on Equation (1.3), determine the impact on predicted consumption of an increase in income of 100.

1.4 Write down an equation that may serve as a simple model representing how the quantity demanded of a certain good depends on its price. What do you expect the signs of each of the structural coefficients to be? In your model is price the only factor that determines the quantity demanded?

⋆ **1.5** Write down an equation that may serve as a simple model representing how the quantity supplied of a certain good depends on its price. What differences are there between this and the demand model?

1.6 The equation $S = Y - C$ expresses the macroeconomic statement that saving equals income minus consumption. It this a model that needs econometric analysis?

⋆ **1.7** It is important for the U.S. Department of Labor to obtain a fairly precise estimate of the nation's unemployment rate each month. Is this a task of econometrics?

1.8 It is important to know how the rate of inflation will be affected if the rate of growth of the money supply is increased. Is this a task of econometrics?

Appendix

For Problems 1.9–1.13, consider the linear equation $y = 1.0 + 0.5x$.

1.9 Determine the slope and the intercept, and graph this equation for the values of x from 0 to 10.

1.10 Let point A have $x = 3$ and point B have $x = 8$. Determine the y values of A and B algebraically and graphically.

⋆ Answers for problems marked with a star are given at the end of the book.

★ **1.11** Using points A and B from Problem 1.10, determine the slope of the line.

★ **1.12** If x increases by 4, what is the associated change in y? If x decreases by 1, what is the associated change in y?

1.13 Determine the elasticity of this linear function when $x = 3$. Determine the elasticity when $x = 8$.

★ **1.14** The function $y = x^2$ has a slope equal to $2x$ at any point. Determine the slope and the elasticity when $x = 2$. Determine the slope and elasticity when $x = 4$.

1.15 For the function in Problem 1.14, determine the ratio $\Delta y/\Delta x$ for the movement between the point where $x = 2$ to the point where $x = 4$.

★ **1.16** For the function in Problem 1.14, determine the arc elasticity for the movement between the point where $x = 2$ to the point where $x = 4$.

APPENDIX: Functions and Graphs

A mathematical function is a relation that associates a single value of the variable y with any particular value of the variable x. In very general form, we may write

$$y = f(x) \tag{1.4}$$

In the graph of a function, the value of x being considered is measured horizontally and the associated value of y is given by the height of the curve.

Figure 1.2 shows the graph of a straight line, which is associated with a function whose equation is

$$y = b_0 + b_1 x \tag{1.5}$$

Such an equation is called a ***linear equation.*** If $x = 0$, then $y = b_0$, and therefore b_0 is the ***intercept*** of the line with the y axis drawn vertically through $x = 0$. The ***slope*** of a straight line is defined as the ratio of the change in y (i.e., Δy) to the change in x (i.e., Δx) resulting from moving from one point to another along the line. In other words,

$$\text{slope} = \frac{\Delta y}{\Delta x} \tag{1.6}$$

In (1.5) the slope is equal to b_1, the coefficient multiplying x. To see that this is true, let p' correspond to the y and x values y' and x', and let p'' correspond to y'' and x''. Since

$$y'' = b_0 + b_1 x'' \quad \text{and} \quad y' = b_0 + b_1 x' \tag{1.7}$$

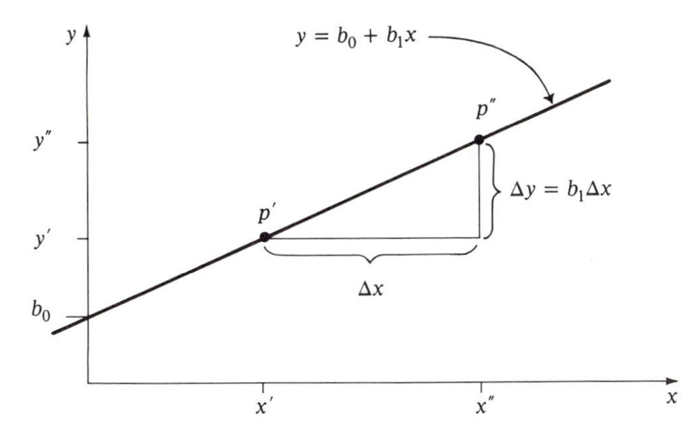

FIGURE 1.2 The graph of a linear equation is a straight line, whose distinguishing features are its intercept and its slope. The intercept is the value of y corresponding to $x = 0$. The slope is given by the ratio $\Delta y/\Delta x$ associated with moving along the line from any one point to another.

then

$$y'' - y' = (b_0 - b_0) + b_1(x'' - x') \quad \text{or} \quad \Delta y = b_1\,\Delta x \tag{1.8}$$

Thus $b_1 = \Delta y/\Delta x$, the slope. If b_1 is positive, the line slopes upward from left to right; if b_1 is zero, it is horizontal; and if b_1 is negative, the line slopes downward from left to right. If the absolute value of the slope is large, the line is steep, and if the absolute value is small, the line is relatively flat. In the special case where $\Delta x = 1$, we see that $b_1 = \Delta y$. Thus the coefficient b_1 gives the change in y that is associated with a unit change in x (i.e., $\Delta x = 1$). This interpretation is basic for understanding regression models, which are formulated as linear equations.

For example, if we start with a particular linear equation,

$$y = f(x) = 2.5 + 0.9x \tag{1.9}$$

we see immediately that its slope is 0.9 and its intercept is 2.5. To graph the straight line corresponding to this equation, all we need to do is determine any two points on the line and then draw the line through them. In this example, we could use the intercept point ($x = 0$, $y = 2.5$) and the point corresponding to $x = 10$ and

$$y = 2.5 + (0.9)(10) = 2.5 + 9 = 11.5 \tag{1.10}$$

that is, the point ($x = 10$, $y = 11.5$).

Figure 1.3a shows the graph of a function that is not a straight line. In Chapter 6 we see several different equation forms whose graphs might resemble this figure. The slope of such a graph is defined at each and every point on the curve

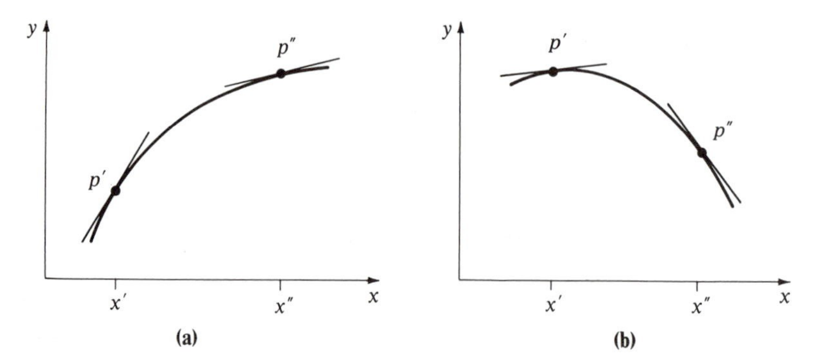

FIGURE 1.3 The slope of a nonlinear function usually varies from point to point along the curve. The slope at a particular point is defined as equal to the slope of the straight line that is tangent to the curve at that point. In (a), the slope is greater at p' than at p''. In (b), the slope at p' is algebraically greater than at p'', but the negative slope at p'' is greater in absolute value than that at p'.

as being the slope of the straight line that can be drawn tangent to the curve at that point. The slope of the function is greater at p' than at p''. In Figure 1.3b the slope at p' is positive, whereas that at p'' is negative but greater in absolute value.

The slope is a particular way of characterizing how the change in y is related to the change in x as we move along a function: the slope is the ratio of the two algebraic changes. Another way of characterizing how these changes are related is by the ***elasticity,*** which often finds its way into economists' discussions. For example, when we talk about the elasticity of demand with respect to price, we are thinking of the quantity demanded as being a function of price. In this case, the elasticity tells us something about how quantity changes are related to price changes as we move along the demand curve (function).

The elasticity of a function tells us how the proportional change in y is related to the proportional change in x as we move along the function. Specifically, the elasticity is the ratio of the proportional changes:

$$\text{elasticity} = \frac{\Delta y/y}{\Delta x/x} \tag{1.11}$$

Note that a proportional change, such as $\Delta y/y$, equals the algebraic change (Δy) divided by the level or value of the variable (y). Simple algebra allows us to see that the elasticity definition can be reexpressed in various ways:

$$\frac{\Delta y/y}{\Delta x/x} = \frac{\Delta y/\Delta x}{y/x} = \frac{\Delta y}{\Delta x} \cdot \frac{x}{y} \tag{1.12}$$

Notice that the elasticity is equal to the slope divided by the ratio of y to x (or multiplied by the ratio of x to y).

Many students find the concept of elasticity difficult to understand, even though they find the concept of slope quite easy. To help develop an understanding, it is important to realize that elasticity and slope play the same role: they characterize how changes in y are related to changes in x as we move along the function $y = f(x)$. In some cases it is natural to talk about the slope, while in other cases the elasticity is a more useful concept.

Two special cases provide an interesting contrast. Along a linear function, like (1.5), the slope takes on the same value (b_1) everywhere. However, unless $b_0 = 0$ the elasticity takes on a different value at each point. To see that this is true, note that the ratio y/x changes as we move along the line. By contrast, there are functions that have the opposite property: the value of the elasticity is the same at every point, but the slope changes as we move along the function. For this type of function, which is explored in Chapter 6, the elasticity provides a simple way to characterize how changes in y are related to changes in x; the slope provides a more cumbersome characterization in this case because it takes on a different value everywhere.

In general, as we move along a function both the slope and the elasticity values may change. For example, in Figure 1.3b we see that the slope is continuously decreasing in algebraic value, taking on a positive value at p' and a negative value at p''. In the range where the slope is negative, the elasticity is clearly changing: both the slope and the ratio x/y increase in absolute value as x increases. In Figure 1.3a, however, we cannot be sure at a quick glance. At p'' the slope is smaller than at p', but the ratio x/y is greater. In fact, this figure is similar to the graph of a function that has a constant elasticity taking on some specific value between zero and 1.

Both the slope and the elasticity are clearly defined at any single point along a function. However, these basic concepts need to be broadened if they are to be applied to movements from one particular point to another. For example, suppose that we are thinking of moving from p' to p'' along the function graphed in Figure 1.3a. If we draw a straight line between p' and p'', its slope is the ratio between the Δy and Δx in this case. Note, however, that this slope is smaller than the slope of the function at p' and greater than the slope at p''. Thus these two well-defined slopes do not correctly describe the ratio $\Delta y/\Delta x$ for the movement between the two points. Similarly, the well-defined elasticities at p' and p'' do not exactly describe the ratio of the proportional change of y to the proportional change of x.

If we compute the ratio of these proportional changes, we obtain the ***arc elasticity*** between p' and p''. A practical question in computing the arc elasticity involves deciding what value of y (such as y', y'', or some other) and of x to use as the base for the proportional changes. This decision depends on the particular application involved; usually the y value will be either the average of y' and y'' or it will be the initial value y', and similarly for x.

Data and Description

Economic Data

In this chapter we consider some general notions regarding data and present two sets of real economic data that serve as the bases for many of the examples presented later in the book. The appendix discusses the nature and use of natural logarithms, and in the last section of the chapter these are applied to the problem of measuring growth.

2.1 General Considerations

Data (the plural of *datum*) are the quantitative facts or pieces of information that we deal with in any statistical analysis. In common parlance, data are usually referred to as "statistics," but we reserve that term for a very different technical meaning. Taken together, all the data we consider in any particular analysis constitute a ***data set.***

In any economic process that we are studying empirically, the data are a set of cases or instances of the occurrence of that process. These cases or instances are called ***observations,*** and the nature of these cases or instances defines the ***unit of observation.*** For example, in macroeconomics the study of the consumption function focuses on an economy in aggregate as the unit of observation, and the observations may be different years' experiences of a single economy or they may be the experiences of different economies. At the microeconomic level, the study of consumption behavior usually focuses on the family as the unit of observation, because it is theorized that persons in a family make collective economic decisions. The observations may consist of data for the same family during different periods of time or for different families.

A data ***variable*** is a measurable characteristic about which we collect data. For example, in a microeconomic study of consumption behavior the relevant variables might be consumption, income, family size, the number of children, and so on. Variables often are given symbolic names, such as X, Y, and Z. However, in reporting the results of data analysis it often makes sense to use variable names that are mnemonics for the corresponding characteristics (e.g., *CONSUM, INCOME, FAMSIZE*) or even to use plain words.

Generally, variables are typed as being either discrete or continuous. The distinction between these types is precisely drawn in the probability theory dealing with random variables in Chapter 9, but in dealing with data variables the distinction is sometimes imprecise. A ***discrete*** data variable usually results from counting, and its values are usually integers. The number of distinct values taken on by the variable is often rather limited, even when the number of observations in the data set is quite large. For example, the number of persons in a family is clearly a discrete variable; family size takes on only integer values, and even in a survey of many families its values may range only from 2 to 15 or so. By contrast, a ***continuous*** data variable usually results from measuring, possibly with great precision, and in principle its values can be any of the real numbers (perhaps within a specified interval). In practice, however, measurement problems often serve to impose some practical limit on the number of distinct values that the variable can take on. For example, the income of families might be measured very precisely (down to the penny, perhaps), and in a survey of 100 families there might well be 100 different values for the variable.

In some cases the type of a variable is unclear. For example, workers might report the exact number of days that they worked in the preceding five years. In such a case, where the number of different values occurring for a discrete variable is very large, the data tend to resemble those for a continuous variable. The number of days worked might be considered practically continuous, and it could be treated as such for some purposes. In Chapter 4 we will explore methods of organizing and summarizing data that make use of the distinction between the two types of variables.

A data set usually consists of measurements on more than one variable. For our purposes, it is presumed that all the variables are measured for each and every observation that is in the data. In other words, there are no missing data.

These basic concepts fit neatly together to make up the notion of a ***rectangular data set.*** Suppose that we have n observations on the variables Y, X, and Z. The data can be thought of as being in a rectangular array, with each row constituting an observation and each column containing the values of a variable, as in Table 2.1. Each element in the array is a value that is symbolized by the variable name with a subscript indicating the observation number. A typical observation, which we denote as the ith, consists of the values Y_i, X_i, and Z_i.

TABLE 2.1 Rectangular Data Set for Y, X, and Z

Observation Number	Y	X	Z
1	Y_1	X_1	Z_1
2	Y_2	X_2	Z_2
3	Y_3	X_3	Z_3
⋮			
i	Y_i	X_i	Z_i
⋮			
n	Y_n	X_n	Z_n

An actual data set is a set of numbers that can be thought of as being arranged in this format.

Usually, a given set of observations are thought of as being a ***sample*** drawn from some larger ***population.*** For example, the people interviewed in a survey are usually chosen to be representative of all people, or perhaps all people in a certain social or economic category. For some statistical purposes, such as estimating the nation's unemployment rate, it is essential that the sample be representative of the corresponding population. The techniques for carring out proper representative sampling are based on important statistical considerations, but we do not deal with them.

Somewhat surprisingly, for purposes of econometric regression estimation it is *not* important that the sample constituting our data set be statistically representative of the corresponding population, in the ordinary sense. Thus our selection of data is not governed by standard sampling techniques. *The essential criterion for selecting our data is that all the observations result from the same economic process, that is, that they follow the same behavioral pattern.*

The reason for this criterion is straightforward: our use of data is to estimate the parameters of an equation that specifies a particular structural or behavioral relation in some economic process. The data must result from that process in order to be relevant; if they result from some other process they would give misleading information about the process being considered. For example, in Chapter 1 we looked at the aggregate consumption function. Suppose now that the true marginal propensity to consume was different during the 1970s from what it was in the 1960s, and that during each decade it was a genuinely fixed parameter. It would be appropriate to use each decade's data separately to estimate a consumption function describing behavior during those years. But to use the data from the 1970s to help estimate the parameters of behavior during the 1960s would not make sense.

Of course, it is not possible to know a priori whether a given set of observations are in fact results of the same process, so good judgment is essential in selecting data. Subject to this criterion, it is useful to have observations that display a wide range of results, that is, to have variables that take on a wide range of values. If there is not much variation among the observations, it is difficult to determine how one variable affects another.

In comparison with researchers in other disciplines, economists usually are not involved in the collection of original data. In macroeconomics, most of the relevant data are prepared by government agencies. In microeconomics, governments are also a major source of data because of their administrative records and the information received through regulatory functions. In addition, governments carry out extensive surveys of economic behavior and conditions. Beyond government, there are business firms and research organizations that collect and disseminate data of various types. With this wealth of information on so many economic activities being easily available, it is rare to find economists or their students out in the field interviewing people or probing the records of economic units. Although this permits a greater share of research time and effort to be devoted to the analysis of data rather than its collection, it makes it more difficult for the researcher to judge the basic quality of the data that he or she is using. It should be noted, however, that economists are becoming more involved with survey techniques and the collection of data; the negative income tax experiments and similar projects are major examples of this activity.

2.2 Cross-Section Data

Cross-section data arise when the observations are data for different entities (such as persons, firms, or nations) for which a common set of variables is measured at about the same point in time. For example, the results of a survey in which various people are interviewed regarding their labor force activity and earnings constitute a cross-section data set, as do the revenue and expenditure figures for the 50 state governments in the United States in any particular year. Although the data relate to different entities and display different values for the measured characteristics, in order to make them useful for estimating the coefficients of a particular structural relation we must presume that the observations follow the same behavioral pattern.

For use throughout this book, a small cross-section data set has been selected from the responses to the *Survey of Financial Characteristics of Consumers* and its sequel *Survey of Changes in Family Finances,* which were carried out for the Board of Governors of the Federal Reserve System in 1963 and 1964. In addition to being quite suitable for our purposes, these data are chosen because they were subjected to an especially interesting validation, which is discussed below.

The data set consists of 100 observations on 12 variables. These observations were selected from among those families headed by a male aged 25–54 who was not predominantly self-employed, in order to produce data relevant for estimating an earnings function. Also, in order to be able to examine normal behavior, families with wealth greater than $100,000 (in 1962 dollars) were not included. For convenience, the number of observations was reduced to 100 by a random selection. Normally, one would use as many observations as feasible, as long as they are appropriate for the estimation at hand.

The data are displayed in Table 2.2, arranged as a rectangular data set. The 12 variables refer to behavior or status during 1963, except as noted. Given mnemonic names, they are:

1. *SIZE* — The number of persons in the family.
2. *ED* — The number of years of education received by the head.
3. *AGE* — The age of the head, in years.
4. *EXP* — The labor market experience of the head, in years, calculated as $EXP = AGE - ED - 5$.
5. *MONTHS* — The number of months during which the head worked.
6. *RACE* — The race of the head, coded 1 for whites and 2 for blacks.
7. *REG* — The region of residence, coded 1 for Northeast, 2 for North Central, 3 for South, and 4 for West.
8. *EARNS* — The wage and salary earnings of the head, expressed in thousands of dollars.
9. *INCOME* — The total income of the family members, expressed in thousands of dollars.
10. *WEALTH* — The wealth of the family on December 31, 1962, expressed in thousands of dollars.
11. *SAVING* — The saving (flow) of the family, expressed in thousands of dollars.
12. *ID* — The observation number, repeated as a variable for convenience.

The variable *EXP* is constructed from *AGE* and *ED*, and it is approximately equal to the number of years since a person left school. This is intended to serve as a proxy for true labor market experience. Since a proxy is not a true measure, the use of such a variable introduces potential for error and distortion. Variables *RACE* and *REG* contain coded information, which must be treated in special ways. The other variables are measured in natural units. Note that the financial variables are expressed in thousands of dollars, for convenience; as we shall see, this does not distort any of the statistical procedures. Each observation constitutes one row in the table, and the values of a single variable are contained in a single column. There are no missing values, because survey respondents who did not provide complete information were excluded from consideration.

TABLE 2.2 Cross-Section Data Set

Obs.	(1) *SIZE*	(2) *ED*	(3) *AGE*	(4) *EXP*	(5) *MONTHS*	(6) *RACE*
1	4	2	40	33	12	2
2	4	9	33	19	12	1
3	2	17	31	9	12	1
4	3	9	50	36	12	1
5	4	12	28	11	12	1
6	4	13	33	15	12	1
7	5	17	36	14	12	1
8	5	16	44	23	12	1
9	5	9	48	34	12	2
10	5	16	31	10	12	1
11	10	9	41	27	12	1
12	4	10	41	26	12	1
13	7	11	36	20	12	1
14	5	14	31	12	12	1
15	5	7	27	15	12	1
16	5	8	42	29	12	1
17	4	12	28	11	11	1
18	2	6	46	35	12	2
19	3	12	47	30	12	1
20	7	8	35	22	12	1
21	3	9	41	27	9	1
22	4	17	30	8	12	1
23	6	12	38	21	12	1
24	3	11	48	32	12	1
25	3	10	36	21	12	1
26	3	12	45	28	12	1
27	6	8	44	31	6	1
28	4	10	44	29	12	1
29	3	3	46	38	12	1
30	4	12	26	9	12	1
31	5	12	50	33	12	1
32	4	8	46	33	11	1
33	5	8	33	20	12	1
34	4	12	41	24	12	1
35	5	17	33	11	12	1
36	4	12	41	24	12	1
37	3	12	29	12	11	2
38	9	11	27	11	12	1
39	5	12	42	25	12	1
40	5	16	39	18	12	1
41	6	12	36	19	12	1
42	4	8	34	21	12	1
43	4	12	40	23	12	1
44	4	12	37	20	12	1
45	5	17	44	22	12	1

(7) *REG*	(8) *EARNS*	(9) *INCOME*	(10) *WEALTH*	(11) *SAVING*	(12) *ID*
3	1.920	1.920	0.470	0.030	1
1	12.403	12.403	3.035	0.874	2
4	5.926	6.396	2.200	0.370	3
2	7.000	7.005	11.600	1.200	4
3	6.990	6.990	0.300	0.275	5
1	6.500	6.500	2.200	1.400	6
3	26.000	26.007	11.991	31.599	7
1	15.000	15.363	17.341	1.766	8
3	5.699	14.999	9.852	3.984	9
3	8.820	9.185	8.722	1.017	10
4	7.000	10.600	0.616	1.004	11
1	6.176	12.089	23.418	0.687	12
2	6.200	6.254	7.600	−0.034	13
3	5.800	9.010	0.358	−1.389	14
2	6.217	6.217	0.108	1.000	15
2	5.500	5.912	5.560	1.831	16
1	4.800	4.800	0.970	0.613	17
3	1.820	2.340	2.600	0.050	18
4	4.558	7.832	31.867	0.013	19
2	7.468	9.563	1.704	1.389	20
1	6.600	7.600	4.820	0.602	21
1	12.850	13.858	32.807	2.221	22
1	5.800	5.802	10.305	1.588	23
3	7.479	19.362	12.652	5.082	24
1	5.700	8.000	7.631	1.846	25
1	12.000	17.200	14.392	0.914	26
1	3.578	4.091	6.649	2.483	27
3	9.600	9.600	6.995	0.837	28
3	3.686	10.425	9.138	1.274	29
3	6.480	6.512	2.933	−0.275	30
4	6.383	7.675	38.260	1.092	31
1	5.610	12.418	12.661	1.157	32
1	6.000	6.079	0.820	0.340	33
2	6.300	6.979	21.286	0.373	34
1	10.513	10.517	9.723	3.307	35
2	30.000	30.996	95.187	10.668	36
1	3.427	5.283	0.171	1.105	37
2	8.500	8.511	3.105	3.500	38
1	11.300	12.700	7.385	0.541	39
3	16.960	16.770	16.049	3.020	40
1	8.300	8.300	0.050	0.650	41
2	5.375	5.375	4.464	0.989	42
4	4.770	6.265	7.203	2.532	43
2	4.320	8.520	9.145	6.120	44
4	10.720	24.226	54.524	−2.749	45

(*continued*)

TABLE 2.2 Cross-Section Data Set (*continued*)

Obs.	(1) *SIZE*	(2) *ED*	(3) *AGE*	(4) *EXP*	(5) *MONTHS*	(6) *RACE*
46	2	4	49	40	12	1
47	5	12	33	16	12	1
48	6	14	36	17	12	1
49	4	15	51	31	12	1
50	5	12	37	20	12	1
51	4	19	33	9	12	1
52	4	14	39	20	12	1
53	3	12	44	27	12	1
54	4	7	50	38	12	1
55	4	12	39	22	12	1
56	6	7	46	34	12	1
57	4	12	43	26	12	1
58	6	11	40	24	12	2
59	2	9	40	26	12	1
60	8	7	39	27	12	1
61	6	10	34	19	6	2
62	4	10	32	17	12	2
63	3	16	42	21	12	1
64	2	8	52	39	12	1
65	6	12	29	12	12	1
66	2	12	27	10	12	1
67	5	10	37	22	12	1
68	2	12	52	35	12	1
69	3	12	32	15	12	1
70	4	12	35	18	12	1
71	3	13	31	13	12	1
72	5	9	36	22	10	1
73	6	16	34	13	12	1
74	3	12	54	37	12	1
75	4	12	52	35	10	1
76	6	9	28	14	12	1
77	6	12	44	27	12	1
78	4	17	29	7	12	1
79	4	9	50	36	7	1
80	4	8	50	37	12	1
81	4	16	44	23	12	1
82	4	9	34	20	9	1
83	7	10	39	24	12	1
84	5	12	39	22	12	1
85	4	14	29	10	12	1
86	3	8	38	25	12	1
87	5	10	30	15	12	1
88	3	10	50	35	12	1
89	2	8	33	20	12	1
90	4	9	35	21	12	1

(7) *REG*	(8) *EARNS*	(9) *INCOME*	(10) *WEALTH*	(11) *SAVING*	(12) *ID*
3	0.750	0.750	4.000	0.000	46
4	7.310	7.356	6.800	−1.036	47
3	9.000	9.000	6.890	1.351	48
1	14.000	14.660	13.500	−1.150	49
2	3.900	5.593	9.837	−0.248	50
2	10.000	11.841	10.384	0.388	51
3	7.200	7.700	6.842	1.157	52
3	6.500	10.550	4.929	1.656	53
2	8.000	13.700	34.124	3.959	54
2	9.500	12.242	11.731	5.369	55
2	6.000	7.803	5.695	1.405	56
3	6.400	9.879	25.029	0.220	57
3	5.190	9.154	0.600	−0.298	58
3	4.548	7.067	45.105	−0.276	59
2	4.860	4.496	8.511	−0.578	60
4	2.736	4.636	20.205	−1.360	61
4	6.000	9.003	4.727	5.277	62
1	7.800	13.820	2.270	0.980	63
4	6.163	8.891	18.916	2.637	64
1	8.600	8.632	14.194	0.984	65
3	7.899	8.385	13.662	−0.076	66
4	5.048	5.403	0.159	0.902	67
2	4.133	8.573	21.700	10.733	68
3	6.500	6.516	1.180	0.716	69
2	6.000	6.000	5.900	0.200	70
4	10.116	16.778	2.531	0.006	71
1	6.000	9.504	44.461	1.464	72
4	8.950	8.953	4.863	0.948	73
4	4.952	8.703	8.534	0.835	74
1	8.681	12.667	26.085	−2.883	75
2	6.500	6.504	3.775	0.298	76
4	7.668	8.180	3.032	0.481	77
2	11.600	11.600	2.167	5.033	78
3	3.100	5.602	5.072	−0.111	79
3	4.586	10.390	4.100	0.000	80
2	27.000	30.610	51.892	4.115	81
1	1.500	3.941	1.260	2.575	82
3	1.789	2.936	17.128	−0.112	83
4	11.068	11.068	11.542	−5.577	84
4	8.338	8.338	2.272	2.750	85
3	2.943	6.683	6.100	0.095	86
1	7.212	7.212	0.857	1.348	87
1	7.500	10.411	3.678	0.178	88
3	5.250	8.850	1.650	−0.695	89
1	5.066	8.334	2.143	0.787	90

(*continued*)

TABLE 2.2 Cross-Section Data Set (*continued*)

Obs.	(1) *SIZE*	(2) *ED*	(3) *AGE*	(4) *EXP*	(5) *MONTHS*	(6) *RACE*
91	3	16	36	15	12	1
92	4	12	33	16	12	1
93	6	20	38	13	12	1
94	4	12	46	29	12	1
95	4	16	50	29	12	1
96	2	16	54	33	12	1
97	5	12	31	14	12	1
98	2	18	27	4	12	1
99	5	12	40	23	12	1
100	6	18	34	11	12	1

Just looking at the data in the table conveys very little insight about how the economy works, although an experienced researcher can "see" things in the table that another person might not. Statistical analysis, including econometrics, provides methods for learning about economic behavior from seemingly unfathomable masses of numbers such as this.

Errors in Data

Before any data are analyzed by statistical methods, the researcher should have a good understanding of exactly what characteristics the data variables purport to measure and of how good these measures are. In conjunction with the *Survey of Financial Characteristics of Consumers* (SFC) and with the cooperation of the Federal Reserve Board, a group of researchers undertook a validation study to assess the responses of interviewees to questions in the SFC regarding the ownership of savings accounts. The first step was to obtain bank records on savings accounts and their owners. Next, the owners were interviewed, following the same procedure that was used in the original SFC. Finally, the responses of the interviewees were matched with the corresponding bank records. The differences between the responses and the bank records are errors in the data, most of which are considered to be the fault of the respondents.

The findings of the study are quite interesting. The major source of error was the failure to report an existing savings account. Among the respondents to the validation survey, almost half failed to report an account that they owned, thus tending to underreport the amount of their wealth. (The amounts in savings accounts are a component of *WEALTH*, and are very different from *SAVING*.) A second type of error was misreporting the size of an account. Among the accounts whose existence was reported, the size of small accounts was overreported and the size of large accounts was underreported. Among accounts that were actually smaller than $1000, the average reported value was $245 greater than the true value, and among the accounts larger than $10,000, the average reported value was $795 less than the true value.

(7) *REG*	(8) *EARNS*	(9) *INCOME*	(10) *WEALTH*	(11) *SAVING*	(12) *ID*
2	12.848	13.923	18.182	4.642	91
2	6.214	6.214	0.275	1.260	92
1	12.202	12.323	28.953	2.687	93
2	8.190	14.963	11.230	0.720	94
2	7.200	10.060	25.462	5.109	95
1	30.000	32.080	98.033	1.800	96
2	9.190	9.260	5.539	1.684	97
2	7.500	10.450	2.860	1.475	98
3	7.852	9.138	11.197	0.566	99
1	12.000	12.350	30.906	25.405	100

These types of errors in data should alert us to be careful in applying and interpreting statistical techniques. Unfortunately, there is not much that can be done to improve the quality of data once it is obtained. Some of the judgment involved in econometrics is deciding which variables in a given set of data are sufficiently good to work with. Also, it makes sense to examine each observation to see if there are any obvious errors, and then either correct them or delete the observation from further consideration. This is an inappropriate procedure, however, if there is any chance that the researcher would correct or delete observations that do not conform to his beliefs about the way the world works.

2.3 Time-Series Data

Time-series data arise when the observations are data for the same entity in different periods of time. For example, records of a person's employment and earnings in each year of his life constitute a time-series data set, as do the official National Income Accounts compiled by the Bureau of Economic Analysis in the U.S. Department of Commerce. The data are essentially historical, and the factor of time is usually very important.

Although the economy changes over time, to make time-series data useful for estimating the coefficients of a particular structural relation we must believe that all the observations are subjected to the same, unchanging economic process. This assumption can be relaxed somewhat by explicitly including the factor of time in the specification of the structural relation; however, essentially the same assumption is still being made.

One of the major uses of time-series econometric models is to make predictions about future events. In doing this, we assume that the economic process represented in data from the past will continue unchanged in the future. The observations in our data set, then, can be viewed as a sample from the results of a continuing economic process.

For use throughout this book, a small time-series data set has been selected from the series (variables) compiled by the U.S. President's Council of Economic Advisers and published in the *Economic Report of the President 1982*. The unit of observation is the aggregate American economy in a given year. The sample period spans the years 1956 through 1980, yielding a total of 25 observations. During this time period, which starts after the Korean War, the basic structure of the economy is presumed not to have changed. However, at least with regard to inflation and price determination, this assumption may be untenable for the last 5 or 10 years and deserves to be investigated further.

The data are displayed in Table 2.3, arranged as a rectangular data set. The 12 variables, given mnemonic names, are:

1. *PGNP* The implicit price deflator for gross national product, expressed as index numbers with 1972 = 100.0.
2. *GNP* (Real) gross national product, measured in billions of 1972 dollars.
3. *INV* (Real) gross private domestic investment, measured in billions of 1972 dollars.

TABLE 2.3 Time-Series Data Set

Obs.	(Year)	(1) *PGNP*	(2) *GNP*	(3) *INV*	(4) *CON*	(5) *DPI*
1	(1956)	62.79	671.6	102.6	405.4	446.2
2	(1957)	64.93	683.8	97.0	413.8	455.5
3	(1958)	66.04	680.9	87.5	418.0	460.7
4	(1959)	67.60	721.7	108.0	440.4	479.7
5	(1960)	68.70	737.2	104.7	452.0	489.7
6	(1961)	69.33	756.6	103.9	461.4	503.8
7	(1962)	70.61	800.3	117.6	482.0	542.9
8	(1963)	71.67	832.5	125.1	500.5	542.3
9	(1964)	72.77	876.4	133.0	528.0	580.8
10	(1965)	74.36	929.3	151.9	557.5	616.3
11	(1966)	76.76	984.8	163.0	585.7	646.8
12	(1967)	79.06	1011.4	154.9	602.7	673.5
13	(1968)	82.54	1058.1	161.6	634.4	701.3
14	(1969)	86.79	1087.6	171.4	657.9	722.5
15	(1970)	91.45	1085.6	158.5	672.1	751.6
16	(1971)	96.01	1122.4	173.9	696.8	779.2
17	(1972)	100.00	1185.9	195.0	737.1	810.3
18	(1973)	105.69	1255.0	217.5	768.5	865.3
19	(1974)	114.92	1248.0	195.5	763.6	858.4
20	(1975)	125.56	1233.9	154.8	780.2	875.8
21	(1976)	132.11	1300.4	184.5	823.7	907.4
22	(1977)	139.83	1371.7	213.5	863.9	939.8
23	(1978)	150.05	1436.9	229.7	904.8	981.5
24	(1979)	162.77	1483.0	232.6	930.9	1011.5
25	(1980)	177.36	1480.7	203.6	935.1	1018.4

4. *CON* (Real) personal consumption expenditures, measured in billions of 1972 dollars.

5. *DPI* (Real) disposable personal income, measured in billions of 1972 dollars.

6. *RINF1* The annual rate of inflation in the GNP deflator, expressed as percent per annum.

7. *RINF2* The annual rate of inflation in the Consumer Price Index, expressed as percent per annum.

8. *UPCT* The unemployment rate, measured as a percent of the civilian labor force.

9. *M1* Money supply (M1), measured in billions of current dollars as the average of daily figures during December (the observations for 1956–1958 were obtained by splicing and adjusting comparable data from the 1980 *Report*).

10. *RTB* The interest rate on new issues of U.S. Treasury bills, expressed as percent per annum.

11. *RAAA* The interest yield of corporate bonds rated Aaa by Moody's Investors Service, expressed as percent per annum.

(6) *RINF1*	(7) *RINF2*	(8) *UPCT*	(9) *M1*	(10) *RTB*	(11) *RAAA*	(12) *T*
3.205	1.496	4.1	135.0	2.658	3.36	1
3.408	3.563	4.3	133.8	3.267	3.89	2
1.710	2.728	6.8	138.9	1.839	3.79	3
2.362	0.808	5.5	141.2	3.405	4.38	4
1.627	1.604	5.5	142.2	2.928	4.41	5
0.917	1.015	6.7	146.7	2.378	4.35	6
1.846	1.116	5.5	149.4	2.778	4.33	7
1.501	1.214	5.7	154.9	3.157	4.26	8
1.535	1.309	5.2	162.0	3.549	4.40	9
2.185	1.722	4.5	169.6	3.954	4.49	10
3.228	2.857	3.8	173.8	4.881	5.13	11
2.996	2.881	3.8	185.2	4.321	5.51	12
4.402	4.200	3.6	199.5	5.339	6.18	13
5.149	5.374	3.5	205.9	6.677	7.03	14
5.369	5.920	4.9	216.8	6.458	8.04	15
4.986	4.299	5.9	231.0	4.348	7.39	16
4.156	3.298	5.6	252.4	4.071	7.21	17
5.690	6.225	4.9	266.4	7.041	7.44	18
8.733	10.969	5.6	278.0	7.886	8.57	19
9.259	9.140	8.5	291.8	5.838	8.83	20
5.217	5.769	7.7	311.1	4.989	8.43	21
5.844	6.452	7.1	336.4	5.265	8.02	22
7.309	7.658	6.1	364.2	7.221	8.73	23
8.477	11.259	5.8	390.5	10.041	9.63	24
8.964	13.523	7.1	415.6	11.506	11.94	25

12. *T* Time, measured as the number of years since 1955 (i.e., 1 in 1956, 2 in 1957, etc.); this is the same as the observation number.

The meanings of most of these variables are discussed in books on macroeconomics. Variables 1–6 are based on the National Income Accounts, 7 and 8 on U.S. Department of Labor concepts, and 9–11 on Federal Reserve and private business reports.

In 1985 the U.S. Department of Commerce made extensive statistical and definitional changes in the National Income Accounts data, as it does from time to time. These revisions change our information about how the economy has functioned. For example, real gross national product is now seen to have grown more slowly from 1972 to 1984 than was evident before the revisions. Economists working with time-series data must be especially careful that the data being used are based on consistent definitions and measurement techniques over time.

Constructed Variables

Macroeconomic time-series variables are not always simple reports of observed magnitudes. Instead, some original data are adjusted to conform to certain statistical or economic concepts. Our small data set reflects four types of these adjustments.

First, some variables are presented as indexes. An ***index*** is constructed to measure the average of an underlying set of price or quantity variables. It is a series of numbers whose information is carried in the relative size of different years' values, not in their individual magnitudes. In our data, *PGNP* is like an index of the price of products and services in the economy. By construction, the value of $PGNP = 100.0$ for 1972. The *PGNP* value for any other year measures the average level of prices in that year relative to the level in 1972. For example, in 1956 the average level of prices was 62.79 percent as high as the 1972 level; in 1975, the level was 125.56 percent as high as the 1972 level (i.e., 25.56 percent higher). Other well-known indexes are the Dow Jones Index of Industrial Stock Prices, the Fed's Index of Industrial Output, and the Consumer Price Index.

Second, some variables have been adjusted from their actually observed (***current*** or ***nominal***) values to an economic concept known as their ***real*** values. This adjustment aims to represent what the magnitude of the variable would have been in different years if the average level of prices had been constant at some particular level. In some cases, it is appropriate to work with aggregate variables measured in real terms in order to be consistent with theoretical considerations regarding economic behavior; in other cases nominal variables are appropriate. In the National Accounts data presented here, the benchmark year is 1972. For gross national product, the relation between its real value (*GNP*) and its nominal value (*NGNP*, say) is given by

$$GNP_i = \frac{NGNP_i}{PGNP_i} \cdot 100 \tag{2.1}$$

To get real GNP for any year (the ith), nominal GNP is deflated (i.e., divided) by the index of prices; the multiplication by 100 simply offsets the base value of $PGNP = 100$, so that $GNP = NGNP$ in 1972. Thus "real" magnitudes are sometimes described as "price-deflated."

Given a nominal time series, it is not always obvious what price index should be used for deflation. For example, the Federal Reserve releases data on the money supply in nominal terms (here, *M1*). It makes sense to deflate *M1* by *PGNP* to create real money supply data to reflect economy-wide behavior, but it also makes sense to deflate *M1* by the Consumer Price Index to reflect consumer behavior. Clearly, good economic judgment is needed in handling data even before statistical techniques are applied.

Third, some variables are expressed as the annual rate of change of some other variable. For example, in macroeconomics "inflation" means change in the level of prices. Usually, this is measured as the proportionate change from one period to another. The year-to-year inflation measured through the Consumer Price Index (*CPI*) is

$$RINF2_i = \frac{CPI_i - CPI_{i-1}}{CPI_{i-1}} \cdot 100 \tag{2.2}$$

By convention, the rate of inflation this year is the change in the level of the price index from last year to this year, divided by the level of the price index last year; multiplication by 100 converts the results to a percentage amount. Notice that to get *RINF2* as in Table 2.3 for the years 1956–1980, it is necessary to start with data on *CPI* from 1955; in computing the rate of change, the first observation is lost.

Fourth, some variables are presented as ratios of two other variables, when the ratio itself is an economic concept. For example, the Bureau of Labor Statistics calculates the unemployment rate as the ratio of the number of unemployed persons to the number of persons in the labor force (these two variables are determined from monthly surveys). The ratio is multiplied by 100 to express it as a percent.

2.4 Change and Growth in Time Series*

A common characteristic of time-series variables is that their values tend to grow over time. For example, in the data for *GNP* in Table 2.3 we see that there was positive growth from 1956 to 1980, although there were several years in which there was a decrease. Data for a time-series variable can be displayed

*The latter part of this section is relatively difficult and can be skipped without loss of continuity.

usefully in a ***time plot,*** in which the values of the variable are graphed on the vertical axis and a measure of time is graphed on the horizontal axis. The top panels of Figure 2.1 show some special cases, in which the time plots are quite smooth. By contrast, a time plot of *GNP* would look somewhat more jagged.

In the following discussion of change and growth we focus on a general time-series variable Y and we think about graphing it against T, which is the measure of time defined in our time-series data set.

There are two concepts of change, or growth, in common use, and this sometimes leads to confusion. ***Absolute change*** is the difference in the value of a variable from one period to another. One year's absolute growth (or change) in Y is given by the difference from the previous year:

$$\Delta Y_i = Y_i - Y_{i-1} \tag{2.3}$$

(Note that ΔY_i may be negative; we are not dealing with "absolute value" here.)

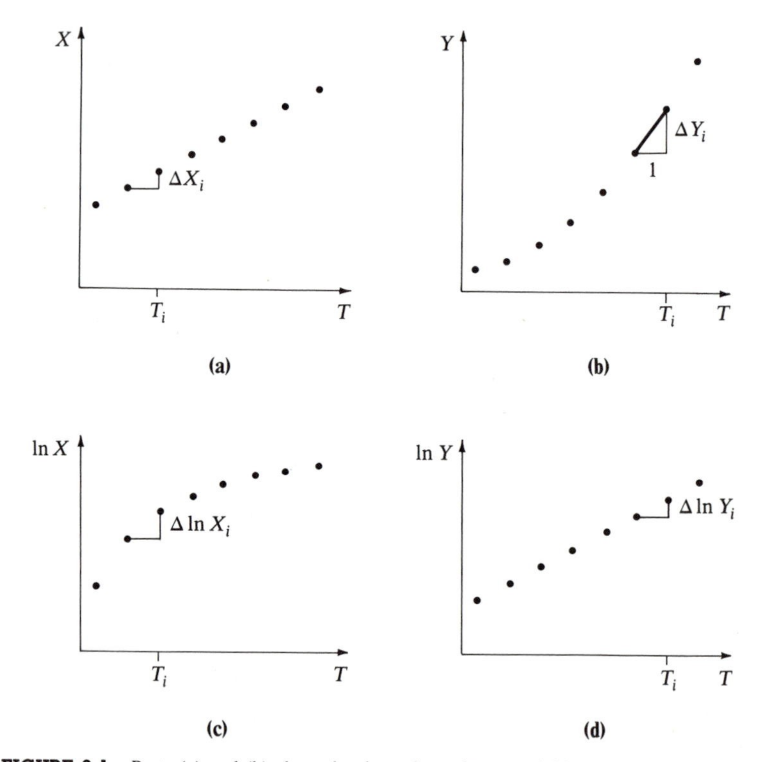

FIGURE 2.1 Parts (a) and (b) show the time plots of two variables: it is clear that X grows with a constant absolute change, whereas Y grows with an increasing absolute change. Parts (c) and (d) show the time plots of $\ln X$ and $\ln Y$. The slope of a logarithmic time plot is directly related to the rate of growth. Thus it is clear that X grows at a decreasing rate, whereas Y grows at a constant rate.

If we draw a line between the potted points for Y_{i-1} and Y_i, as in Figure 2.1b, its slope will be equal to ΔY_i because the corresponding ΔT_i equals 1.

Relative change is the proportionate change in a variable from one period to another. One year's relative growth (or change) in Y is specified by the ***rate of change*** or growth, which is denoted by r_i. By definition this is equal to the absolute change divided by the level of Y in the previous year:

$$r_i = \frac{\Delta Y_i}{Y_{i-1}} = \frac{Y_i - Y_{i-1}}{Y_{i-1}} = \frac{Y_i}{Y_{i-1}} - 1 \tag{2.4}$$

While the absolute change ΔY_i tells us how *much* Y is growing, the relative change $(\Delta Y_i)/Y_{i-1}$ tells us how *fast* Y is growing.

Suppose that we wish to make a graphical comparison of the growth of a variable at different periods. For visual clarity, it makes sense to draw lines between consecutive periods' plotted points. In a regular time plot, a comparison of slopes tells us something about growth: where the slope is steeper, the absolute growth is greater. This is because the slope is equal to ΔY_i, as noted above. It is usually difficult to compare the rates of growth in such a plot, because we must simultaneously compare terms like ΔY_i and Y_{i-1}. However, were we to plot the logarithm of Y against time, then a comparison of the slopes would show us immediately in which period the rate of growth is greater. (Logarithms are discussed in the appendix to this chapter.) Since relative growth is usually more interesting than absolute growth, graphs of time-series variables are often presented this way—even in newspapers.

To see why and how the slope of ln Y plotted against T gives us information about the rate of growth, we note that the slope of the line drawn between any two consecutive periods' ln Y values is:

$$\text{slope} = \frac{\Delta \ln Y_i}{\Delta T_i} = \frac{\ln Y_i - \ln Y_{i-1}}{1} = \ln \frac{Y_i}{Y_{i-1}} \tag{2.5}$$

[Reading right to left, we know from (2.17) that the logarithm of a ratio is equal to the logarithm of the numerator minus the logarithm of the denominator.] Equation (2.4) shows us that $r_i = (Y_i/Y_{i-1}) - 1$, so $Y_i/Y_{i-1} = 1 + r_i$. Thus

$$\text{slope} = \ln (1 + r_i) \tag{2.6}$$

Since $\ln (1 + r_i)$ increases with r_i, we see from (2.6) that the slope increases with r_i also. In other words, when ln Y is plotted against T, the slope of the line drawn between two consecutive periods' ln Y values is directly related to the rate of growth. Thus a time plot of ln *GNP* readily shows the rates of change of *GNP* if we look at the slopes along the plot: where the slope is steeper, the rate of growth is greater.

The impact of logarithmic plotting of time-series variables can be seen in some special cases. In Figure 2.1 the regular time plots of two variables, X and

Y, are given in the top panels. Beneath each is the time plot of the logarithm of that variable. Variable X displays a constant amount of absolute growth, since the year-to-year slope ΔX_i is constant throughout. As X increases, $\ln X$ increases also. But since X increases over time, the relative rate of growth of X decreases over time ($r_i = \Delta X_i/X_{i-1}$). This is reflected in the fact that the slope of $\ln X$ decreases as time passes. By contrast, variable Y is constructed to have a constant rate of growth, which is reflected in the fact that the slope of $\ln Y$ is constant. With Y increasing over time this means that ΔY_i is increasing too, which causes the plot of Y against T to have a slope that increases over time.

Problems

Section 2.1

2.1 Consider a data set containing information relating to the number of employees, amount of output, number of plants, and number of different product lines for each of 200 firms in a given industry. Thinking of these data as being in a rectangular data set:
(a) How many rows and columns are there?
(b) How many numbers constitute one observation?
(c) How many observations are there?

⋆ **2.2** In the data set of Problem 2.1, which variables are practically discrete, and which are practically continuous?

⋆ **2.3** Why is it important that the sample used to estimate the nation's unemployment rate be representative?

2.4 Explain why combining data from two decades governed by different marginal propensities to consume would be inappropriate for estimating a single consumption function. (A graph would help.)

Section 2.2

2.5 Look at the data on *ED* in Table 2.2:
(a) What are the minimum and maximum values of the variable?
(b) What value of the variable seems to be typical?
(c) If a person were chosen at random, what would be your guess of his *ED* value?

2.6 From among the variables in the cross-section data set, name one that is clearly discrete and one that is clearly continuous.

⋆ **2.7** Suppose that it is true that the size of savings accounts held by individuals increases as a linear function of their total wealth: $S = \beta_0 + \beta_1 W$. Based on the discussion of errors in the *SFC*, what impact would these errors have on the apparent (i.e., estimated) value of β_1—assuming that W is measured without error? (A graph would help.)

2.8 Looking down Table 2.2, describe the relation between *INCOME* and *WEALTH* in the data. What are the economic links between the two variables?

2.9 Looking down Table 2.2, describe the relation between *SAVING* and *WEALTH* in the data. What are the economic links between the two variables?

Section 2.3

2.10 Look at the data on *GNP* in Table 2.3:

(a) What are the minimum and maximum values of the variable?

(b) What value of the variable seems to be "typical"?

(c) What is your guess of the *GNP* value for 1981?

⋆ **2.11** Using the aggregate consumption function discussed in Section 1.1 as an example, explain why in making predictions about future consumption we must assume that the economic process represented in data from the past will continue unchanged. (A graph would help.)

2.12 Compute the real money supply for the years 1956–1958. For these years, is the real money supply greater or less than the nominal money supply? Why?

2.13 Compute the rate of inflation in *PGNP* for 1957 and for 1958. Compare your results with *RINF1*.

⋆ **2.14** Unlike the national accounts data, the Consumer Price Index is benchmarked in 1967 (i.e., *CPI* = 100 for 1967) in our data set. Using *RINF2*, determine the values of *CPI* for 1968 and 1969.

2.15 Based on the data for *PGNP*, how much higher were prices in 1980 than they were in 1970?

2.16 Determine the nominal values of *GNP* for the years 1971–1973.

Section 2.4

⋆ **2.17** For time periods 1 through 5, the five values for a time-series variable X are: 10, 15, 22, 31, and 42. Graphically or algebraically determine what is happening to the absolute growth in X as time passes. Now determine the five values of $\ln X$. What is happening to the rate of growth of X as time passes?

2.18 Let the value of X be 50 in time period 1. Determine the values of X in periods 2 through 5 if the absolute change in X is -20 each period. Now, determine the values of X in periods 2 through 5 if the rate of change is -20 percent per period.

⋆ **2.19** Extending Problem 2.18, sketch the graph of X and of $\ln X$ if the rate of change of X is -20 percent.

2.20 Compare the annual growth of Disposable Personal Income in 1961 with that in 1975, making both absolute and relative comparisons.

Appendix

2.21 Using a calculator, prepare a two-column table showing the values of x and $\ln x$ for $x = 1.00, 1.01, 1.02, \ldots, 1.20$ (giving $\ln x$ to three decimals). Now add another column to the table, showing the approximate value of $\ln x$ computed according to Equation (2.11). For what values of x is the difference between the actual and approximate values of $\ln x$ less than 0.01?

★ **2.22** If a number is doubled, how is its logarithm affected?

★ **2.23** If $Q = AP^b$ is the equation of a demand curve, reexpress this relation in terms of how $\ln Q$ is related to $\ln P$.

2.24 If $Y = Ab^X$, how is $\ln Y$ related to X?

2.25 If $Y = AX^bZ^{-c}$, reexpress this relation in terms of logarithms.

2.26 If $\ln Y = a + b \ln Z$, reexpress this relation showing Y as a function of Z.

★ **2.27** If $\ln Y = a + Z \ln b$, reexpress this relation showing Y as a function of Z.

★ **2.28** Suppose that $X = 10$ in an initial period and that four periods later $X = 42$. Determine the rate of growth of X per period.

2.29 Suppose that $X = 10$ initially, and that four years later $X = 42$. Determine the continuous rate of growth of X.

APPENDIX: Logarithms*

Many people find logarithms to be mysterious and difficult to incorporate into everyday thinking. This is unfortunate, because logarithms play a very useful role in econometric analysis: they simplify matters in certain cases.

To start with, we consider a relation between x and y of the form

$$x = b^y \tag{2.7}$$

In this relation, b is a ***base*** that is raised to a certain power, the ***exponent*** y. For example, for $b = 2$, the x values that correspond to the integer values for y from -3 to $+3$ are given in Table 2.4; these should be familiar computations. [Note that when an exponent is negative, the value of the expression is equal to the reciprocal of the base raised to the positive value of the exponent: $2^{-3} = 1/(2^3) = \frac{1}{8}$.] The information in this table is plotted as the points in Figure 2.2.

Applying a little mathematical finesse, we can view x as being a continuous function of y. This is suggested in Figure 2.2a by drawing a curve through the plotted points. For any possible value of y, the curve (function) determines a

*This section is relatively difficult and can be skipped without loss of continuity.

TABLE 2.4 Values of $x = 2^y$ and $y = \log_2 x$

y	x
−3	1/8 = 0.125
−2	1/4 = 0.25
−1	1/2 = 0.5
0	1
1	2
2	4
3	8

particular value of x. For example, if $y = 1.5$, $x = 2^{1.5} \approx 2.83$. Given that the base is positive, the x value of the function will always be positive. We will be interested in cases in which $b > 1$. In all these cases, the x value of the function will increase as y increases and the graph will be qualitatively similar to Figure 2.2a: $0 < x < 1$ for $y < 0$; $x = 1$ for $y = 0$; and $1 < x < \infty$ for $y > 0$.

This function, known as an ***exponential function,*** provides a unique one-to-one correspondence between x and y values. Hence it is possible to focus on the inverse relation: given a particular base, the exponent y is a function of x. This function is known as a ***logarithmic function.*** In other words, if x is the exponential of y, then y is the ***logarithm*** of x. Some simple notation to reexpress this is:

$$\text{if} \quad x = b^y, \quad \text{then} \quad y = \log_b x \tag{2.8}$$

where $y = \log_b x$ is known as the logarithmic function.

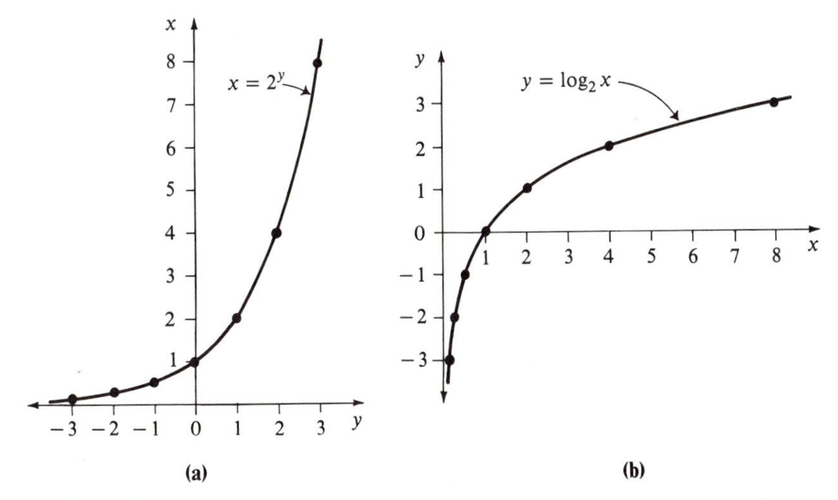

FIGURE 2.2 Part (a) shows an exponential function with base 2, and (b) shows its inverse—a logarithmic function. The plotted points are listed in Table 2.4.

Less formally, a logarithm is simply an exponent. The logarithm of x is the exponent to which a given base must be raised to yield the particular x value. The following tautology is useful to remember:

$$x = b^{\log_b x} \tag{2.9}$$

For base 2, Table 2.4 shows the y values that correspond to certain x values; these y values are the logarithms (to the base 2) of the x values. Figure 2.2b, which shows the logarithmic function, is the same as Figure 2.2a but with the axes transposed.

Exponential and logarithmic functions may be set up for any positive base. In scientific and mathematical work it is convenient to use a number approximately equal to 2.718 as the base. This number is known as e, and logarithms using the base e are called ***natural logarithms.*** They may not seem very natural at first, but that is their proper name. The symbol "ln" is often used as a shorthand symbol instead of "$\log_e$," and we adopt this convention:

$$\text{if} \quad x = e^y, \quad \text{then} \quad y = \ln x \tag{2.10}$$

where e is understood to be the base of natural logarithms.

Figure 2.3 shows the graph of the ***natural logarithmic function,*** $y = \ln x$. Note that x is always positive but that y, the value of $\ln x$, may be negative or positive. Also, $\ln 1 = 0$ because $e^0 = 1$, as is true of any nonzero number raised to the zero power. The slope of $y = \ln x$ is always positive but it decreases as x gets larger, meaning that as x increases $\ln x$ also increases but at a slower rate for larger values of x. Many electronic calculators can compute natural logarithms, and there also exist detailed tables to determine these values. Table 2.5 shows the natural logarithms of a few selected numbers. Note that the values

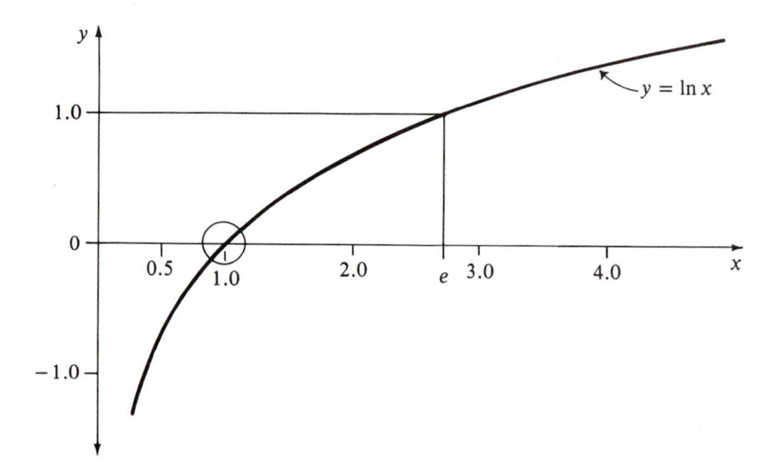

FIGURE 2.3 The natural logarithmic function, $y = \ln x$, is defined for positive values of x, but y ranges from $-\infty$ to ∞. The slope of the function is always positive, but it decreases as x increases. The circled portion of the function is examined in Figure 2.4.

TABLE 2.5 Selected Values of $y = \ln x$

x	$y = \ln x$
0.0	$-\infty$
0.0001	−9.21
0.001	−6.91
0.01	−4.61
0.1	−2.30
0.5	−0.69
0.90	−0.105
0.95	−0.051
1.00	0.000
1.05	0.049
1.10	0.095
1.5	0.41
2.0	0.69
2.5	0.92
(e) 2.718	1.00
3.0	1.10
5.0	1.61
10.0	2.30
50.0	3.91
100.0	4.61
1,000.0	6.91
10,000.0	9.21
1,000,000.0	13.82

of $\ln x$ do not get very large even when x does. Also, $\ln x$ gets very small (very negative) rapidly as x decreases below 1 toward zero.

Figure 2.4 enlarges the circled portion of Figure 2.3, which shows $y = \ln x$, and adds a reference line, $y = x - 1$. It turns out that this line is tangent to the curve $y = \ln x$ at $x = 1$, where $y = x - 1 = \ln x = 0$, and it lies above the curve everywhere else. For values of x quite close to 1, the graph of $y = \ln x$ lies just below that of $y = x - 1$, so that $\ln x$ is just slightly less than $x - 1$. In other words,

$$\ln x \approx x - 1, \quad \text{when} \quad x \text{ is close to } 1 \tag{2.11}$$

This relation serves as a simple and useful method of approximation for determining the value of $\ln x$ when x is close to 1, without the aid of a calculator or complicated formula.

It is useful to recast this finding in terms of differences. Let δ be the difference between x and 1: $\delta = x - 1$ or $x = 1 + \delta$. (Note that δ is lowercase "delta," the Greek "d.") Then the statement that the graph of $y = \ln x$ lies below that of $y = x - 1$, except at $x = 1$ where the graphs are tangent, can be restated as $\ln(1 + \delta) \leq \delta$. Thus the approximation (2.11) can be restated as

$$\ln(1 + \delta) \approx \delta \qquad \text{for } \delta \text{ close to zero} \tag{2.12}$$

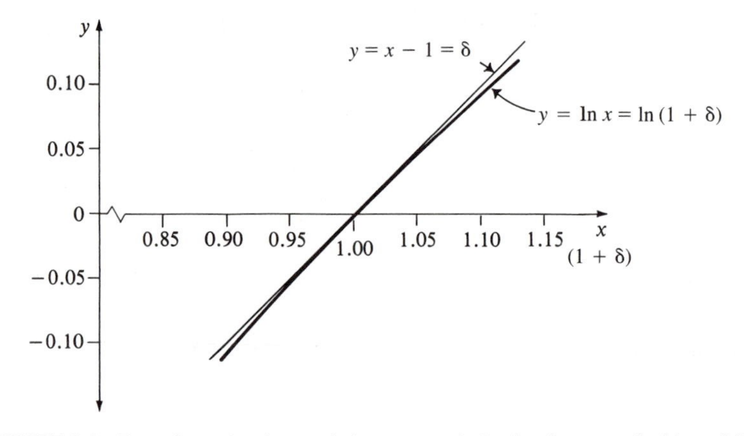

FIGURE 2.4 For values of x close to 1, $\ln x \approx x - 1$. Letting $\delta = x - 1$, this useful approximation can be restated as: $\ln (1 + \delta) \approx \delta$, for δ close to 0. In this figure we see that the height of $y = \delta$ is only slightly greater than that of $y = \ln (1 + \delta)$ for a range of values where δ is close to zero (i.e., where x is close to 1).

For example, the true value of ln (1.05) equals 0.049, rounded to three decimals. In this case $x = 1.05$ corresponds to $\delta = 0.05$, and this value is a good approximation of 0.049. Similarly, $x = 0.95$ corresponds to $\delta = -0.05$, and the true $\ln x = -0.051$ (rounded to three decimals) is well approximated by the δ value. Table 2.5 shows that the approximation is fairly good in the range of x values between 0.90 and 1.10, but that it is less accurate beyond there.

The ***antilog*** of a number is the value of the exponential function using that number as the exponent. In the natural logarithmic system,

$$\text{``}x \text{ is the antilog of } y\text{''} \quad \text{means} \quad x = e^y \tag{2.13}$$

In other words, x is the antilog of y when y is the log of x. For example, from Table 2.5 we see that the antilog of 1.1 is 3.0, and the antilog of 9.21 is 10,000.

Our main use of logarithms will be to transform certain nonlinear equations into new equations that are linear. The ***rules of logarithms,*** which are used in these transformations, are based on the familiar rules of exponents in algebra. In what follows, the symbols y and x play different roles than in the earlier part of this appendix.

The most frequently used rule is

$$\text{if} \quad y = xz, \quad \text{then} \quad \ln y = \ln x + \ln z \tag{2.14}$$

Put differently,

$$\ln xz = \ln x + \ln z \tag{2.15}$$

meaning that the logarithm of a product of two terms is equal to the sum of the logarithms of those terms. To see that (2.14) is true, we note that $y = e^{\ln y}$, $x = e^{\ln x}$, and $z = e^{\ln z}$ by definition. Thus $y = xz$ can be rewritten as $e^{\ln y} = (e^{\ln x})(e^{\ln z})$. By the rules of exponents, $(e^{\ln x})(e^{\ln z}) = e^{(\ln x + \ln z)}$, so $e^{\ln y} = e^{(\ln x + \ln z)}$. Thus $\ln y = \ln x + \ln z$.

A companion rule is based on division:

$$\text{if} \quad y = \frac{x}{z}, \quad \text{then} \quad \ln y = \ln x - \ln z \tag{2.16}$$

Put differently,

$$\ln \frac{x}{z} = \ln x - \ln z \tag{2.17}$$

meaning that the logarithm of a quotient of two terms is equal to the logarithm of the numerator minus the logarithm of the denominator.

Both (2.14) and (2.16) generalize to equations having more than two terms. Taking the logarithm of a complicated expression yields an expression involving the sums and differences of the logarithms of the terms. The logarithms of terms originally in the numerator are added together and the logarithms of terms originally in the denominator are subtracted. For example,

$$\text{if} \quad y = \frac{x}{wz} \cdot t \quad \text{then} \quad \ln y = \ln x - \ln w - \ln z + \ln t \tag{2.18}$$

Often the basic multiplicative terms themselves involve exponentiation, and a final rule is helpful:

$$\text{if} \quad y = x^z, \quad \text{then} \quad \ln y = z \ln x \tag{2.19}$$

Put differently,

$$\ln x^z = z \ln x \tag{2.20}$$

To see that (2.19) is true, we note that $y = x^z$ can be rewritten as $e^{\ln y} = (e^{\ln x})^z$. By the rules of exponents $(e^{\ln x})^z = e^{z \ln x}$, so $e^{\ln y} = e^{z \ln x}$. Thus $\ln y = z \ln x$. Two special cases arise in econometric work. First, if z is a constant, we may replace it with c, so that

$$\ln x^z = c \ln x \tag{2.21}$$

Second, if x is a constant, we may replace it with c, so that

$$\ln c^z = z \ln c = zc' \tag{2.22}$$

where $c' = \ln c$.

As an example of how these rules help us, we consider the specification of the Cobb–Douglas production function

$$Q = AK^aL^b \tag{2.23}$$

in which Q is output, K is capital services input, L is labor input, and A, a, and b are constants. Taking logarithms, we find that

$$\ln Q = \ln A + a \ln K + b \ln L \tag{2.24}$$

Thus we have transformed an equation that specifies Q as a particular nonlinear function of K and L into one that specifies $\ln Q$ as a linear function of $\ln K$ and $\ln L$. The importance of this will become apparent later.

As a second example, a constant-growth model in discrete terms is given in mathematical form by

$$P_i = P_0(1 + r)^i \tag{2.25}$$

where P_0 is an initial amount or level, r is the constant rate of growth, and P_i is the value of the variable i time periods later. In a discrete model, time is measured in whole periods, so that i takes on integer values only and P_i is the value of P at the end of period i. Taking logarithms of (2.25), we have

$$\ln P_i = \ln P_0 + i \ln (1 + r) \tag{2.26}$$

Since r is constant, $\ln P_i$ is a linear function of i.

In theoretical work, constant-growth models are often specified in continuous terms, such as

$$P_t = P_0 e^{gt} \tag{2.27}$$

where P_0 is the initial level, g is the constant rate of growth, t is a continuous measure of time, and P_t is the value of P at time t. In a continuous model every moment of time is numbered, so that t can take on fractional values (like 7.59) and P_t is the value of P at the exact time t. Taking logarithms, we have

$$\ln P_t = \ln P_0 + gt \tag{2.28}$$

Since g is constant, $\ln P_t$ is a linear function of t, just as $\ln P_i$ is a linear function of i in (2.26).

The growth rate g in the continuous model is analogous to the growth rate r in the discrete model, but they take on different values in comparable settings. For example, suppose that an amount of money P_0 is deposited in a bank and the interest earned is kept in the account. If the bank keeps the rate of interest constant, the amount of money on deposit can be analyzed with either growth model. We let our measures of time be consistent, so that $t = i$ at the end of each period (e.g., at the end of each year). After a certain number of years the amount of money in the account would be denoted by P_t in the continuous model and by P_i in the discrete model.

Since the amount of money is the same no matter how it is analyzed, and since $t = i$ at the end of a certain number of periods, inspection of (2.28) and (2.26) shows that

$$g = \ln (1 + r) \tag{2.29}$$

Our earlier analysis of the natural logarithmic function revealed that $\ln (1 + r) < r$ unless $r = 0$, so that (2.29) implies that $g < r$. Recall also that if r is small, $g \approx r$. Thus, to generalize, we see that in describing the same real growth process the continuous rate of growth will be a smaller value than the discrete rate of growth, but these values will be close to each other when the rates are small.

Descriptive Statistics

In this chapter we examine basic concepts relating to variables on which we have a set of data. To summarize such data, we develop statistics that measure the typical value of a single variable and how much variation there is among the values of that variable. Also, we develop statistics that measure how much relation there is between any two variables. In the sense used here, a ***statistic*** is a measure that is calculated from the values of variables in a given set of data. This technical definition contrasts with common usage, in which "statistics" is used to mean what we call data.

The appendix to this chapter explains the use of summation operator, Σ.

3.1 Univariate Statistics

Throughout this section we focus on a single variable for which we have n observations. Our examples use the variable measuring family income in the cross-section data set from Chapter 2; sometimes this variable is called X, and sometimes, *INCOME*. Recall that in the full data set $n = 100$.

Measures of Central Tendency

A ***measure of central tendency*** is a statistic that measures the typical value that a variable takes on in data. Since "typical" is not a precise concept, it should not be surprising that alternative measures have been developed.

The most common and useful measure is the ***mean.*** This is simply what most people call the "average," but that term is ambiguous to statisticians, who use it to refer to any measure of central tendency. The mean of a variable is conventionally symbolized by placing a bar over the variable name: the mean of X is $\bar{X}$, which is read as "X bar." Formally, the definition of the mean is

$$\bar{X} = \frac{1}{n} \sum_{i=1}^{n} X_i \tag{3.1}$$

which tells us that to calculate the mean of X we must add up all the values of X and multiply by $1/n$ (i.e., divide by n). It should be noted that the number produced by this calculation need not be a value that is actually taken on by any of the observations. In our data on family income, $\bar{X}^* = 9.941$, rounded to three decimals. (Throughout this book, when a symbol such as $\bar{X}$ defines a statistic, an asterisk is added to indicate its value in a specific set of data.)

An interesting property of the mean is based on the deviations, or differences, between the X values in the data and the mean value $\bar{X}$. For the ith observation, the deviation of X from its mean is

$$d_i = X_i - \bar{X} \tag{3.2}$$

We can easily prove that the sum of these deviations is always exactly equal to zero:

$$\begin{aligned} \sum d_i &= \sum (X_i - \bar{X}) \\ &= \sum X_i - \sum \bar{X} \\ &= n\bar{X} - n\bar{X} \\ &= 0 \end{aligned} \tag{3.3}$$

In going from the second line to the third we make two simple substitutions: first, $\Sigma X_i = n\bar{X}$, as may be seen from (3.1), and second, $\Sigma \bar{X} = n\bar{X}$, since the constant $\bar{X}$ is being summed n times.

Another useful measure of central tendency is the ***median.*** The median value of X is the X value of the middle observation, after the observations have been ordered. If there are an odd number of observations, the median is unambiguous. If there are an even number, there is no single middle observation, and by convention we take the two values of X that straddle the middle and average them to produce the median. In our data on *INCOME* the median is $(X_{50} + X_{51})/2 = (8.703 + 8.850)/2 = 8.777$ (rounded), where X_{50} and X_{51} are the two middle observations ranked by the value of X.

In econometric work, the mean plays an important role while the median is rather neglected. However, for simply describing the typical value of a variable the median is often a more useful statistic. For example, the mean income of families in a community is fairly sensitive to the presence or absence of a few families that are very well off, even though the situation of the typical family is not much affected. By contrast, the median is usually less sensitive to this kind of change. In our data on *INCOME*, if we delete the five families with the greatest income, the mean drops markedly from 9.941 to 8.950, while the median drops only from 8.777 to 8.573 thousand dollars.

A third measure of central tendency is the ***mode,*** which is the most frequently occurring value. The mode is not too useful for summarizing raw data on continuous variables, because every observation's value may be unique; this is the case for *INCOME*. For discrete variables, such as family size, the mode is sometimes a useful statistic.

Measures of Dispersion

After finding a measure of central tendency for the data on a variable, the next logical concern is to determine how spread out the values are. A statistic that conveys this information is called a ***measure of dispersion.***

One such measure is the ***range,*** which is simply the difference between the greatest and the smallest values of X among the n observations. In our example, the maximum X is 32.080 and the minimum is 0.750, so the range is 31.330. The range is a simple concept, but it does not tell us anything about the distribution of the values of X—whether they are concentrated near the middle, spread all over, or concentrated near the endpoints (the minimum and maximum).

The most useful approach to developing a measure of dispersion is to seek a statistic that represents the typical deviation of the values of X from their mean. The deviation for any observation, d_i, was defined in (3.2) as $d_i = X_i - \overline{X}$. For *INCOME,* these deviations take on values between -9 and 20, roughly. We might think that the mean deviation, $(1/n) \sum d_i$, would be a good candidate to represent the "typical deviation." However, as seen in (3.3), the positive and negative values of the d_i always cancel out in summation, so the mean deviation always equals zero. This suggests that we should seek a statistic that represents the "typical deviation, without regard to sign."

One possibility solves the sign problem by calculating the mean of the absolute deviations: $(1/n) \sum |d_i|$. In our data on *INCOME* the mean absolute deviation is about 3.771; since the absolute deivations range roughly from 0 to 20, with greater concentration at the smaller values, the value 3.771 seems reasonable as a measure of the typical deviation. Although this measure appeals to our common sense, direct mathematical treatment of absolute values is cumbersome and the mean absolute deviation is not frequently used.

A second possibility solves the sign problem by squaring all the deviations and calculating their mean: $(1/n) \sum d_i^2$. This ***mean squared deviation*** (*MSD*) is the basis for several measures of dispersion, but because of the squaring it yields a value that is not interpretable as the typical deviation. For our data on *INCOME* the *MSD* is 30.869.

To solve this difficulty, we take the square root of the *MSD* to arrive at a more useful way to define and calculate the typical value of d_i. This leads to the ***root mean squared deviation*** of X, which is denoted by $RMSD_X$:

$$RMSD_X = \sqrt{\frac{\sum_{i=1}^{n} d_i^2}{n}} = \sqrt{\frac{\sum_{i=1}^{n} (X_i - \overline{X})^2}{n}} \tag{3.4}$$

To understand this definition, one must be able to carry out the indicated operations. After the mean has been determined, the first step in calculating the *RMSD* is to compute the deviation (from the mean) for each observation. Each of these deviations is then squared. The n squared deviations are then added together, and the sum is divided by n to yield the mean squared deviation. Taking the square root yields the root mean squared deviation. In our data on *INCOME* the *RMSD* is 5.556, which can reasonably be interpreted as the typical deviation. The fact that the *RMSD* is substantially larger than the mean absolute deviation reflects the fact that the intermediate squaring step gives extra importance to the observations with large deviations.

For purposes of statistical inference, which later will be a major concern to us, it turns out that a minor variation on $RMSD_x$ is even more useful. Instead of dividing by n, we divide by $n - 1$. In this case, $n - 1$ is referred to as the ***number of degrees of freedom.*** This name arises from the fact that if we know the values of n and $\bar{X}$, then we can determine ΣX. When ΣX is known, then the value of the nth observation on X can be determined by subtraction once the other $n - 1$ observations are known. Hence the last observation is not "free." The reason for being concerned with this cannot be apparent now. Loosely speaking, however, when we use a sample of data to learn about the population from which it is drawn, dividing by $n - 1$ leads to a sample statistic that is "better" than the *RMSD* for estimating the dispersion of X values in the population.

The resulting statistic is called the ***standard deviation*** and its square is called the ***variance.*** The common symbol for the standard deviation of X is S_X, and for the variance it is S_X^2. The formal definitions are

$$\text{Variance:} \qquad S_X^2 = \frac{\sum_{i=1}^{n} (X_i - \bar{X})^2}{n - 1} \tag{3.5}$$

$$\text{Standard deviation:} \ S_X = \sqrt{\frac{\sum_{i=1}^{n} (X_i - \bar{X})^2}{n - 1}} = \sqrt{S_X^2} \tag{3.6}$$

It is useful to remember that the variance is simply the mean squared deviation except that n is replaced by $n - 1$, and that the standard deviation is simply the square root of that. In our data on incomes, $S_X^* = 5.584$ and $S_X^{2*} = 31.181$. When n is reasonably large, perhaps more than 20, then the computed value of S_X is very close to that of $RMSD_X$. S_X is simply $\sqrt{n/(n - 1)}$ times $RMSD_X$; when $n = 20$ this term is 1.026, and when $n = 100$ it is 1.005. Thus the standard deviation is as reasonable a measure of typical deviation as *RMSD* is.

An interesting property of the standard deviation is that when the data are rather symmetrically distributed about the mean and not concentrated near the extremes, about 68 percent of the observations have values within one standard deviation of the mean (i.e., in the interval $\overline{X} - S_X$ to $\overline{X} + S_X$). Similarly, about 95 percent of the observations have values within two standard deviations of the mean (i.e., in the interval $\overline{X} - 2S_X$ to $\overline{X} + 2S_X$). These general rules add considerable interpretative power to the standard deviation and *RMSD* as measures of typical deviation. It happens that *INCOME* is not symmetrically distributed, as shown in Chapter 4, and the first approximation does not work too well. The one-standard-deviation interval is (9.941 − 5.584) to (9.941 + 5.584), or 4.357 to 15.525, and by actual count 85 percent of the observations lie in this interval. The two-standard-deviation interval is −1.227 to 21.109, and 95 percent of the observations lie therein.

Computation Considerations

In practice, if the mean and standard deviation are to be calculated by hand it makes sense to do it in a tabular format rather than by just substituting a long string of numbers into the appropriate formulas. Suppose that we wish to compute the mean and standard deviation for a sample consisting of the first 10 observations in our data set. Table 3.1 shows the X values listed in column (2). Summing these, we find that $\Sigma X = 106.768$, which yields a mean of $\overline{X} = 10.6768$. For each observation, separately, the deviation from the mean and its square are computed, as indicated in columns (3) and (4). The squared deviations are summed, $\Sigma d_i^2 = 420.388$, which then yields a variance of $S_X^{2*} = 46.710$

TABLE 3.1 First 10 Observations on Family Income and Computation of $\overline{X}$ and S_X

(1) i	(2) X_i	(3) $d_i = (X_i - \overline{X})$	(4) $d_i^2 = (X_i - \overline{X})^2$
1	1.920	−8.7568	76.682
2	12.403	1.7262	2.980
3	6.396	−4.2808	18.325
4	7.005	−3.6718	13.482
5	6.990	−3.6868	13.592
6	6.500	−4.1768	17.446
7	26.007	15.3302	235.015
8	15.363	4.6862	21.960
9	14.999	4.3222	18.681
10	9.185	−1.4918	2.225
	106.768	0.0	420.388

Mean: $\overline{X} = \Sigma X/n = 106.768/10 = 10.6768$

Variance: $S_X^2 = \Sigma d_i^2/(n - 1) = 420.388/9 = 46.710$

Standard Deviation: $S_X = \sqrt{S_X^2} = 6.834$

Median = 8.095 Range = 24.087

Note: d_i is calculated to four decimals to illustrate that $\Sigma d_i = 0$; d_i^2 is rounded to three decimals for convenience.

and a standard deviation of $S_X^* = 6.834$. Finally, a quick sorting yields a median of 8.905 and a range of 24.087.

This procedure, which is based on the direct application of (3.1) and (3.5), involves passing through the X values twice: first $\bar{X}$ is computed, and then we go back to each X_i to calculate d_i and d_i^2 in order to get the numerator of (3.5), which defines the variance. Is it possible to calculate these statistics in just one pass through the data? If so, this would provide a more efficient calculating procedure, which would be important if we were handling many observations (imagine having 25,000!). The answer is yes, as demonstrated by the following derivation:

$$\begin{aligned}
\sum (X_i - \bar{X})^2 &= \sum (X_i^2 - 2\bar{X}X_i + \bar{X}^2) \\
&= \sum X_i^2 - 2\bar{X} \sum X_i + n\bar{X}^2 \\
&= \sum X_i^2 - 2\left(\frac{\sum X_i}{n}\right) \sum X_i + n\left(\frac{\sum X_i}{n}\right)^2 \qquad (3.7) \\
&= \sum X_i^2 - 2\frac{\left(\sum X_i\right)^2}{n} + \frac{\left(\sum X_i\right)^2}{n} \\
&= \sum X_i^2 - \frac{1}{n}\left(\sum X_i\right)^2
\end{aligned}$$

In this derivation, we move from line to line by applying the laws of algebra and summation, and substituting $\Sigma X_i/n$ for $\bar{X}$ in the third line. Focusing on

$$\sum d_i^2 = \sum (X_i - \bar{X})^2 = \sum X_i^2 - \frac{1}{n}\left(\sum X_i\right)^2 \qquad (3.8)$$

we note that ΣX_i^2 is a sum that can be accumulated in one pass through the X values; at each step we simply square the X value and accumulate it. Similarly, ΣX_i can be accumulated in the same pass, simultaneously with but independently of ΣX_i^2. Hence in one pass we can calculate the mean (from ΣX_i) and the standard deviation [from (3.8)]. This computational procedure is appropriate for writing an efficient computer program, but it does not lead us through the basic concepts in the way that the tabular approach does. For our purposes, being able to work through the derivation (3.7) is more important than applying result.

Finally, Figure 3.1a presents a graphical review of the basic descriptive statistics calculated in Table 3.1. The graph is one-dimensional, plotting values of X along the horizontal axis. (Note that four observations clump together near $X = 7$.) The range of 24.087 extends from $X_{\text{min}} = 1.920$ to $X_{\text{max}} = 26.007$. The median, $X_{\text{med}} = 8.095$, is halfway between the fifth and sixth ordered

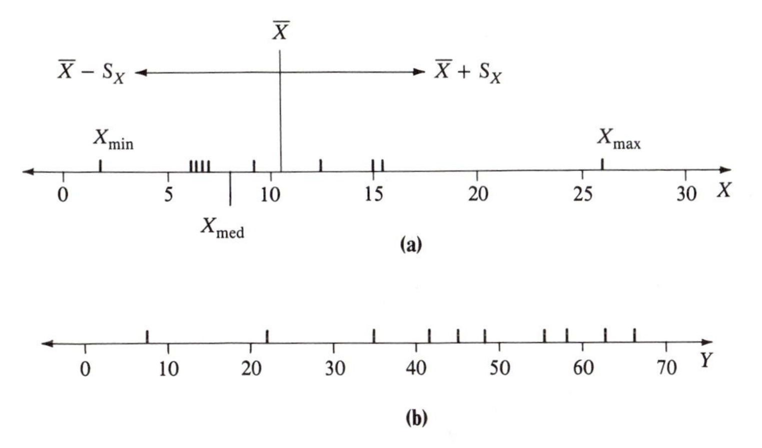

FIGURE 3.1 The one-dimensional graph in (a) plots the values of family income (X) from Table 3.1, and the values or positions of various statistics are indicated. The analysis of the data values plotted in (b) is left as a problem.

observations. The mean, $\overline{X} = 10.6768$, is larger. The span of the one-standard-deviation interval about the mean includes 8 of the 10 observations.

A good intuitive understanding of the mean and standard deviation allows one to guess their values simply by looking at a plot of data. For example, temporarily ignore the calculated statistics shown in Figure 3.1a, and look just at the plotted X values. It appears that the mean must be about 10—this is a middle-ish value, somewhat affected by one extremely high value of X. The standard deviation appears to be about 5, because the one-standard-deviation interval about the mean must contain roughly two-thirds of the observations. As we know in fact, these guesses are quite good. We leave as a problem the task of guessing $\overline{Y}$ and S_Y in Figure 3.1b.

3.2 Linear Transformations

Sometimes it is convenient to define and create a new variable as a transformation of an existing one. For example, consider a set of data for n earthquakes. Seismologists measure the energy dissipated in earthquakes and then transform these energy values into a measure known as the Richter scale. If X is the energy of an earthquake and Y is the Richter measure, there is a defining rule that relates each Y value to the corresponding X value:

$$Y_i = g(X_i) \qquad \text{for all } i \tag{3.9}$$

For the Richter measure the transformation is logarithmic.

Now, if we start with n values of X, applying the transformation creates n values of Y. Just as the data for X can be summarized by the descriptive statistics $\overline{X}$ and S_X, the n values of Y can be summarized by computing $\overline{Y}$ and S_Y using the basic defining formulas, (3.1) and (3.6). Since each Y_i is related to the corresponding X_i in a known way, we might consider whether there is a simple way to determine the values $\overline{Y}$ and S_Y directly from their counterparts $\overline{X}$ and S_X. The answer is no, except in special cases. For example, the mean Richter value for a set of earthquakes is not simply the Richter transformation of the mean energy dissipated in those earthquakes:

$$\overline{Y} \neq g(\overline{X}) \qquad \text{except in special cases} \tag{3.10}$$

In econometric and statistical work, a common and useful type of transformation is the ***linear transformation.*** If Y is related to X according to

$$Y_i = a + bX_i \qquad \text{for all } i \tag{3.11}$$

then Y is a linear transformation of X, and vice versa. In (3.11), a and b are constants, or parameters, while Y and X are variables. For each and every observation, the Y value is related to the X value in this way. For example, let X be family income measured in thousands of dollars. If b is the marginal tax rate and a is the (negative) tax credit in a simple linear tax system, then Y as defined by (3.11) is the net tax payment due. Similarly, the Celsius measure of temperature is a linear transformation of the Fahrenheit measure.

As in the general case, the n values of Y can be summarized by $\overline{Y}$ and S_Y calculated according to basic procedures. However, in contrast to the general case treated above, when Y is a linear transformation of X there are simple relations between their means, their variances, and their standard deviations. These are:

$$\text{if } Y_i = a + bX_i, \text{ then } \overline{Y} = a + b\overline{X} \tag{3.12}$$

$$S_Y^2 = b^2 S_X^2 \tag{3.13}$$

$$S_Y = |b| S_X \tag{3.14}$$

Thus in the income tax example with $a = -2$ and $b = 0.25$,

$$\overline{Y} = -2 + 0.25\overline{X} \quad \text{and} \quad S_Y = 0.25 S_X \tag{3.15}$$

In other applications, either a or b may take on the special values of zero or 1. For example, if every family is given 1.5 thousand dollars as a national dividend, the new income might be called Y, with $Y_i = 1.5 + X_i$. Implicitly, $b = 1$. The relations above enable us to see quickly that $\overline{Y} = 1.5 + \overline{X}$ and $S_Y = S_X$. Adding \$1500 increases every family's income by the same amount, so the dispersion among them remains the same. By contrast, suppose that every family's income is increased by 15 percent, leading to $Y_i = 1.15X_i$ (note that to increase by 15 percent is to multiply by 1.15). Here, $a = 0$. Then $\overline{Y} = 1.15\overline{X}$ and $S_Y = 1.15S_X$. Increasing all incomes proportionately leads to an increase in dispersion (i.e., differences) in incomes as well as an increase in the mean.

As another example we might define Y as income expressed in dollars, while X is still income in thousands of dollars. Clearly, $Y_i = 1000X_i$. Thus, extending the results from earlier in the chapter, $\overline{Y}^* = 9941$ and $S_Y^* = 5584$, compared with $\overline{X}^* = 9.941$ and $S_X^* = 5.584$. Hence, as promised, measuring financial variables in thousands of dollars or other convenient units does not distort the statistical findings; it merely changes the units of measurement.

The nature of linear transformations is illustrated in Figure 3.2. These are graphs of relations, not scatter diagrams of observations. The Y values corresponding to any given X values are given by heights up to the graph of the transformation, and these values are traced over to the Y axis. Equation (3.12) states that $\overline{Y}$ is the transformation of $\overline{X}$, and this is illustrated in the figure. To get some insight into (3.14), look at Figure 3.2a, in which $b > 0$. For a given set of X values, consider what would happen if the slope b were steeper: the range between Y_{max} and Y_{min} would increase, as would the difference between all Y values. Hence (for $b > 0$) for a given S_X, the greater b is, the greater S_Y is. Figure 3.2b shows another transformation, $Y_i = a' + b'X_i$, applied to the same X values. The slope b' is negative, but it has the same magnitude as b in Figure 3.2a. Graphically, we see that the range between the new Y_{max} and Y_{min} is the same as in the first case, and this holds true for the standard deviation also. As (3.14) states, it is the magnitude of the slope parameter—not its sign—that determines how S_Y is related to S_X.

The proofs of (3.12) and (3.13) are straightforward. For the mean,

$$\begin{aligned}\overline{Y} = \frac{1}{n}\sum Y_i = \frac{1}{n}\sum (a + bX_i) &= \frac{1}{n}\sum a + \frac{1}{n}\sum bX_i \\ &= \frac{1}{n}\, na + \frac{1}{n}\, b \sum X_i \qquad (3.16) \\ &= a + b\overline{X}\end{aligned}$$

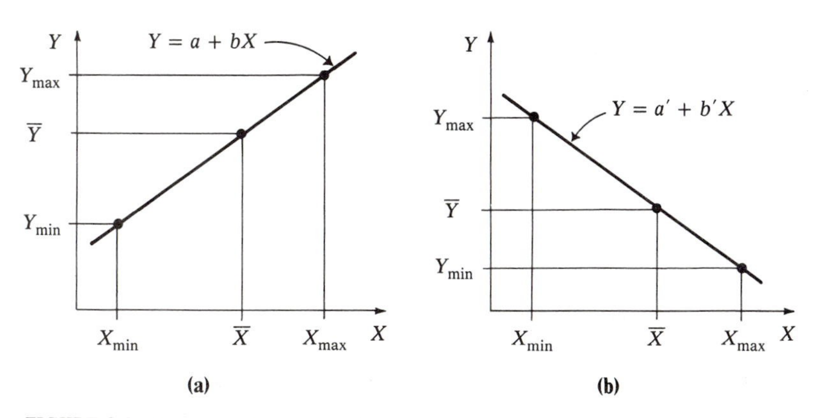

FIGURE 3.2 The heavy lines graph the equations of two linear transformations applied to the same X values. The constants a and a' bear no particular relation to each other, but b' has the same absolute value as b. Consequently, S_Y is the same in both cases. For a given set of X values, we see that if $|b|$ or $|b'|$ were greater, then the range between Y_{min} and Y_{max} would be greater. Similarly, S_Y would be greater also.

For the variance,

$$\begin{aligned} S_Y^2 &= \frac{1}{n-1} \sum (Y_i - \overline{Y})^2 \\ &= \frac{1}{n-1} \sum (a + bX_i - a - b\overline{X})^2 \\ &= \frac{1}{n-1} \sum (bX_i - b\overline{X})^2 \\ &= \frac{1}{n-1} b^2 \sum (X_i - \overline{X})^2 \\ &= b^2 S_X^2 \end{aligned} \tag{3.17}$$

The proof for the standard deviation follows from taking the square root of the variance.

3.3 Bivariate Statistics

Our ultimate objective in econometrics is to determine how some variables are related to others, and in this section we take a first step in that direction. We examine how much, or to what degree, two variables are related in a certain way.

We start with a data set that has n observations on two variables, X and Y. This means that there are n pairs of numbers, each of the form (X_i, Y_i). Each pair of numbers can be plotted, as in Figure 3.3, in a ***scatter diagram*** with X measured on the horizontal axis and Y on the vertical axis. In Figure 3.3a it looks as through X and Y are positively related: there is a tendency for high values of X to be associated with high values of Y. For a different set of data plotted in Figure 3.3b, it looks as though X and Y are negatively related.

One measure that quantifies how much the values of X and Y vary together is the ***covariance***. The symbol used for the covariance of X and Y is S_{XY}, and its definition is

$$S_{XY} = \frac{1}{n-1} \sum_{i=1}^{n} (X_i - \overline{X})(Y_i - \overline{Y}) \tag{3.18}$$

Notice that X and Y enter this relation in basically the same way, so that there is no difference between the covariance of X and Y and that of Y and X. Neither variable is special in its relation to the other.

The values of X and Y enter the covariance formula as deviations from their means [i.e., as $(X_i - \overline{X})$ and $(Y_i - \overline{Y})$]. The impact of this may be seen by graphing a vertical line through $\overline{X}$ and a horizontal line through $\overline{Y}$, as in Figure 3.3a. Each point, or observation, can be viewed in relation to these auxiliary

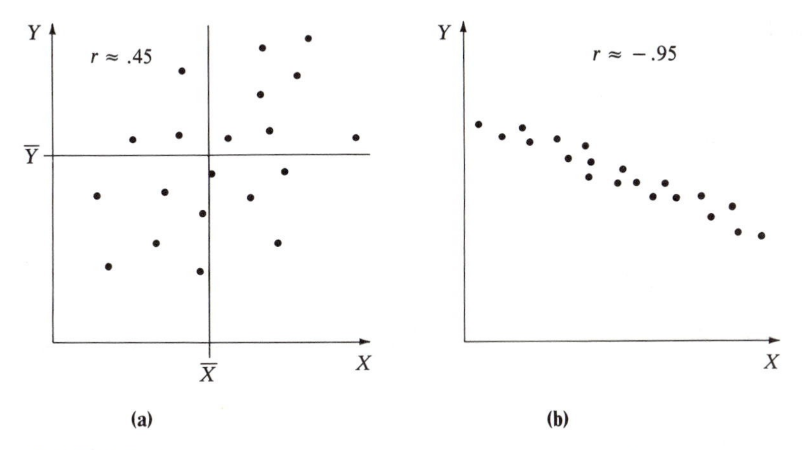

FIGURE 3.3 Each point in a scatter diagram represents a pair of numbers of the form X_i, Y_i. In (a), Y and X are positively but loosely correlated; the auxiliary axes show the positions of the data points relative to the means $\overline{X}$ and $\overline{Y}$. In (b), Y and X are negatively and strongly correlated.

axes. A point in the upper right quadrant has a positive X deviation $(X_i - \overline{X})$ and a positive Y deviation $(Y_i - \overline{Y})$. Thus the product $(X_i - \overline{X})(Y_i - \overline{Y})$ is positive, and its contribution to the summation in (3.18) tends to make the covariance positive. By contrast, a point in the upper left quadrant has a negative $(X_i - \overline{X})(Y_i - \overline{Y})$ product, because $(X_i - \overline{X})$ is negative and $(Y_i - \overline{Y})$ is positive, and its contribution tends to make the covariance negative. The signs for the products of points in the other quadrants are determined similarly. We come to the conclusion that if most of the points in the scatter diagram are in the upper right and lower left quadrants formed by the auxiliary axes through $\overline{X}$ and $\overline{Y}$, then the covariance probably will be positive; this is because most of the terms being summed will be positive, while a few will be negative. Similarly, if most of the points are concentrated in the upper left and lower right quadrants, the covariance probably will be negative. And if the points are scattered all over, it is hard to guess the sign of the covariance before it is actually calculated.

A difficulty with the covariance statistic is that the magnitude of the number is not easily interpreted. To remedy this, it is common to use an associated statistic, the ***correlation coefficient,*** which is defined as

$$r = \frac{S_{XY}}{(S_X)(S_Y)} \tag{3.19}$$

If we substitute the defining formulas for S_{XY}, S_X, and S_Y into this equation we end up with a formidable display of X's, Y's, Σ's, and $\sqrt{}$'s. We prove in the appendix to Chapter 5 that the effect of dividing the covariance of X and Y by the product of their standard deviations is to limit the possible values of r to be in the range from -1 to 1.

TABLE 3.2 Descriptive Statistics for *SAVING* (*Y*) and *INCOME* (*X*)

(1) Y_i	(2) X_i	(3) $(X_i - \bar{X})$	(4) $(X_i - \bar{X})^2$	(5) $(Y_i - \bar{Y})$	(6) $(Y_i - \bar{Y})^2$	(7) $(X_i - \bar{X})(Y_i - \bar{Y})$
0.0	1.9	−5.04	25.4016	−0.56	0.3136	2.8224
0.9	12.4	5.46	29.8116	0.34	0.1156	1.8564
0.4	6.4	−0.54	0.2916	−0.16	0.0256	0.0864
1.2	7.0	0.06	0.0036	0.64	0.4096	0.0384
0.3	7.0	0.06	0.0036	−0.26	0.0676	−0.0156
2.8	34.7	0.0	55.5120	0.0	0.9320	4.7880

$$\bar{X} = \Sigma X_i/n = 34.7/5 = 6.94$$

$$\bar{Y} = \Sigma Y_i/n = 2.8/5 = 0.56$$

$$S_X = \sqrt{\Sigma (X_i - \bar{X})^2/(n - 1)} = \sqrt{55.512/4} = \sqrt{13.878} = 3.725$$

$$S_Y = \sqrt{\Sigma (Y_i - \bar{Y})^2/(n - 1)} = \sqrt{0.932/4} = \sqrt{0.233} = 0.483$$

$$S_{XY} = \Sigma (X_i - \bar{X})(Y_i - \bar{Y})/(n - 1) = 4.788/4 = 1.197$$

$$r = S_{XY}/(S_X S_Y) = 1.197/[(3.725)(0.483)] = 0.67$$

The interpretation of the correlation coefficient, r, relates to an imaginary line that can be fit through the data points. The correlation carries two pieces of information about the data and the line: the sign of r is the same as the sign of the slope of that line, and the magnitude (absolute value) of r measures the degree to which the points lie close to the line. The magnitude of r does *not* tell us the slope of the line.

If the points in the scatter diagram lie exactly along some straight line with a positive slope, then $r = 1$; if they lie exactly along a line with a negative slope, then $r = -1$. (If they lie along a vertical or horizontal line, then r will not be defined because S_X or S_Y is zero.) At the other extreme, if the points are

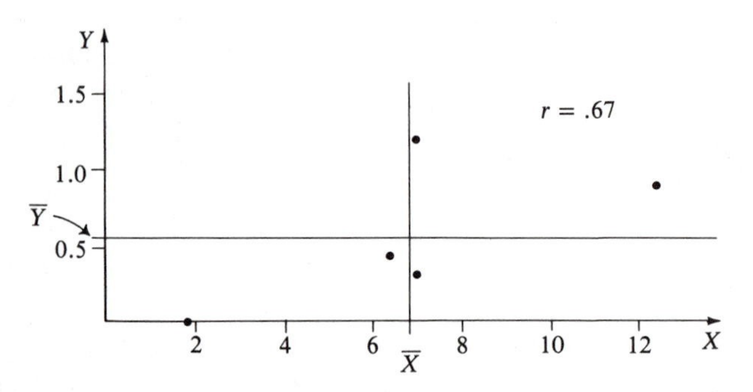

FIGURE 3.4 The scatter diagram displays the five observations on *SAVING* (Y) and *INCOME* (X) from Table 3.2, and the auxiliary axes showing the means, $\bar{X}$ and $\bar{Y}$, are drawn in. There is a moderately strong, positive correlation between Y and X in these data.

TABLE 3.3 Correlation Matrix from the Cross-Section Data Set (n = 100)

	EARNS	*ED*	*AGE*	*INCOME*	*SIZE*	*WEALTH*
EARNS	1.000					
ED	.534	1.000				
AGE	.057	−.214	1.000			
INCOME	.886	.457	.252	1.000		
SIZE	.002	.003	−.216	−.102	1.000	
WEALTH	.625	.251	.359	.673	−.132	1.000

scattered all over rather randomly, the correlation will be close to zero. In Figure 3.3a $r \approx .45$ and we would say that Y and X are positively but loosely correlated; in Figure 3.3b $r \approx -.95$ and we would say that Y and X are negatively and strongly correlated. It should be noted again that the correlation does not tell us whether a line drawn through the data would be steep or flat; it tells us only the sign of the slope.

Although the correlation coefficient is very useful, its weakness is that it measures only the degree of *linear* relation between the variables. Often this is interesting enough. However, if the data points in the diagram make a perfect V shape, or U shape, the correlation will be zero. If they lie neatly along an upward sloping snakelike shape, the correlation will be positive but not unity. Hence r is not a measure of general relation.

As an example of computation, consider the five observations on Y and X in Table 3.2, which are plotted in Figure 3.4. These are the first five observations on *SAVING* (Y) and *INCOME* (X) from the cross-section data set, with the values rounded to one decimal position for simplicity. The correlation $r = .67$ indicates a moderately strong relation between *SAVING* and *INCOME*, which we investigate further in Chapter 5.

It is usually interesting, but never conclusive, to look at the correlations among possibly related variables in any data set. The correlation matrix for a selected set of the variables in the cross-section data set is shown in Table 3.3, in which each entry in the table gives the correlation between the variables indicated on the row and column headings. Only half the matrix is filled in, to avoid repetition. The correlation matrix for a selected set of our time-series variables is shown in Table 3.4.

TABLE 3.4 Correlation Matrix from the Time-Series Data Set (n = 25)

	GNP	*CON*	*DPI*	*RAAA*	*M1*	*RINF1*	*UPCT*
GNP	1.000						
CON	.998	1.000					
DPI	.998	.999	1.000				
RAAA	.945	.952	.954	1.000			
M1	.963	.974	.966	.948	1.000		
RINF1	.860	.866	.874	.915	.875	1.000	
UPCT	.351	.396	.390	.434	.488	.354	1.000

It should be emphasized that finding a high correlation between two variables does not prove that there is a cause-and-effect relation between them. Why not? It could be, and often is, that two variables are both strongly affected by a third variable. They may be different effects of the same cause and may have no direct link between them. In time-series data, especially, there often are high correlations between variables even when a true relation is lacking. This is because most variables tend to go steadily up, or steadily down, over time.

Problems

Section 3.1

★ **3.1** Calculate the mean and median of *UPCT* for a sample consisting of the first 10 observations in the time-series data set.

3.2 Calculate the mean and median of *UPCT* for a sample consisting of the last 15 observations in the time-series data set.

3.3 Can the mean *UPCT* for all 25 observations be determined from your answers to Problems 3.1 and 3.2? If so, show how.

★ **3.4** Can the median *UPCT* for all 25 observations be determined from your answers to Problems 3.1 and 3.2? If so, show how.

3.5 Suppose that among a group of 10 families the mean number of persons in a family is 3.8. If all the families came together, how many people would be present?

3.6 Determine the mode of *SIZE* in the cross-section data set.

3.7 Using the tabular method presented in the text, calculate the mean, variance, and standard deviation of the first 10 observations on *UPCT*. Also calculate the range.

★ **3.8** Extending Problem 3.7, exactly what proportion of these 10 observations lie within one standard deviation of the mean? What proportion lie within two?

3.9 Suppose that data on a variable X consist of five observations for which $X = -1$ and five for which $X = 1$. For the whole data set of 10 observations, calculate the mean, root mean squared deviation, and standard deviation of X. In this case, is $RMSD_X$ or S_X a better measure of the typical deviation of X?

3.10 Prove that $S_X = \sqrt{n/(n-1)}RMSD_X$.

★ **3.11** Use the procedure associated with Equation (3.8) to help calculate the mean, variance, and standard deviation of the first 10 observations on *UPCT*. Compare the results with those from Problem 3.7.

★ **3.12** Guess the values of the mean and standard deviation of Y in Figure 3.1b.

3.13 Using the tabular method presented in the text, calculate the mean and standard deviation of the last 15 observations on *UPCT* in the time-series data set. What proportion of the observations lie within one standard deviation of the mean?

3.14 Using the data from Problem 3.13, plot the data on *UPCT* in a one-dimensional graph and indicate the values of the median, mean, and standard deviation.

Section 3.2

3.15 Suppose that X in Table 3.5 were survey data on the number of years of work experience of the head of each family and that each person had worked exactly X years (i.e., not X plus or minus a few months). Assume that $\bar{X}^* = 18$ and $S_X^* = 8.9$. Now, let Y_i stand for the number of months that the ith head of family has worked.

(a) How is each Y_i related to the corresponding X_i?

(b) Determine the mean and standard deviation of Y.

3.16 Let Z_i be the work experience, in years, of these same persons two years from the original survey.

(a) How is each Z_i related to the corresponding X_i?

(b) Determine the mean and standard deviation of Z.

⋆ **3.17** Let W_i be the work experience, in months, of these same persons two years from the original survey.

(a) How is each W_i related to the corresponding X_i?

(b) Determine the mean and standard deviation of W.

⋆ **3.18** Two common measures of cross-section income inequality are the standard deviation of income and the standard deviation of the logarithm of income. If all persons' incomes are increased by 20 percent, what happens to these measures of inequality?

3.19 Solve Equation (3.11) for X in terms of Y and write the equation in a form that makes it clear that X is a linear transformation of Y.

Section 3.3

3.20 Make a rectangular data set consisting of the first five observations on *UPCT* and *RINF1* from the time-series data set (for convenience, round *RINF1* to one decimal position). Make a scatter plot of *RINF1* against *UPCT* and guess the sign of the correlation coefficient.

⋆ **3.21** Calculate the covariance and the correlation between the data variables of Problem 3.20.

3.22 Suppose that the data on Y and X are identical (i.e., that $Y_i = X_i$ for all i). How does the covariance of Y and X compare with the variance of X?

★ **3.23** Suppose that we have data on Y and X, and then we create a new variable $Z = a + bX$ (with $b > 0$). How does the covariance of Y and Z compare with the covariance of Y and X?

3.24 In the context of Problem 3.23, prove that the correlation between Y and Z is equal to the correlation between Y and X.

3.25 Table 3.3 shows a negative correlation between *SIZE* and *INCOME*. What economic forces or patterns of behavior might lead to this? What might lead to a positive correlation?

APPENDIX: Summation

Statistical calculations often involve adding together many numbers. When the addition is complicated by other algebraic operations, the use of a special summation notation simplifies matters considerably.

Suppose that we have data on the annual incomes, measured in thousands of dollars, of five families. We let the families be identified by a number from 1 to 5; this number is called the ***index,*** and its symbol for the typical case is i. We let X be the symbol denoting annual income, so X_i stands for the income of the ith family. The second column of Table 3.5 shows the hypothetical data: the income of the first family (X_1) is 21 thousand dollars, that of the second is 10, $X_3 = 32$, and so on.

The total income of this group of five families is 90 thousand dollars, this being the sum of the five X_i values. To symbolize this summation operation, we write

$$\begin{aligned}\sum_{i=1}^{5} X_i &= X_1 + X_2 + X_3 + X_4 + X_5 \\ &= 21 + 10 + 32 + 12 + 15 = 90\end{aligned} \tag{3.20}$$

(Σ is uppercase "sigma," the Greek "S.") The left-hand side of (3.20) is read as "the sum of X_i, for i running from 1 to 5," and the first line of the right-hand side shows in simple algebraic terms what this operation means. The second line shows how the operation is performed. When no confusion is apt

TABLE 3.5 Hypothetical Income Data

(1) i	(2) X_i	(3) $0.25X_i$	(4) $-2 + 0.25X_i$
1	21	5.25	3.25
2	10	2.5	0.5
3	32	8.0	6.0
4	12	3.0	1.0
5	15	3.75	1.75

to occur, the indexing information below and above Σ may be partly or completely eliminated. For example, when it is clear that the operation involves all five families, the symbols

$$\sum_{i=1}^{5} X_i, \sum_{i} X_i, \sum X_i, \text{ and } \sum X \tag{3.21}$$

are understood to be equivalent. The first form gives a complete statement of the summation, while the third strikes a nice balance between complexity of notation and possibility of confusion.

When the laws of algebra are brought into play in the summation of complicated expressions involving the X_i values, various properties of the summation operator become apparent and are worth remembering. To be general, let X be the name of a variable for which we have n different values, numbered 1 through n: $X_1, X_2, \ldots, X_n$. Let a and b be two constants—numbers that are the same for all i. Then, starting with an original X_i value we can think of a new value being defined as $a + bX_i$ and we can consider taking the sum of these n values: $\Sigma (a + bX_i)$. We can prove without much ado that

$$\sum_{i=1}^{n} (a + bX_i) = na + b \sum_{i=1}^{n} X_i \tag{3.22}$$

To do any simple proof, the basic method is to start with the expression on the left-hand side of the equation and use definitions and the laws of algebra to get to the right-hand side. To wit:

$$\begin{aligned} \sum_{i=1}^{n} (a + bX_i) &= (a + bX_1) + (a + bX_2) + \cdots + (a + bX_n) \\ &= \underbrace{(a + a + \cdots + a)}_{n \text{ terms}} + (bX_1 + bX_2 + \cdots + bX_n) \\ &= na + b(X_1 + X_2 + \cdots + X_n) \\ &= na + b \sum_{i=1}^{n} X_i \end{aligned} \tag{3.23}$$

Two important special cases arise. If $a = 0$, then $a + bX_i = bX_i$, and it should be clear from (3.22) that

$$\sum_{i=1}^{n} bX_i = b \sum_{i=1}^{n} X_i \tag{3.24}$$

If $b = 0$, then $a + bX_i = a$, and substituting in (3.22) we find that

$$\sum_{i=1}^{n} a = na \tag{3.25}$$

Although this is an odd-looking use of summation notation, it clearly states that the sum of a constant added up n times is simply equal to that constant multiplied by n.

To return to our hypothetical income data, suppose that the government imposes a 25 percent tax on all incomes and simultaneously declares a 2 thousand dollar tax credit (i.e., negative tax levy) for every family. The net tax due for a family with income X_i is $-2 + 0.25X_i$; note that this is of the form $a + bX_i$, with $a = -2$ and $b = 0.25$. The total net tax collection will be

$$\begin{aligned}\sum_{i=1}^{5} (-2 + 0.25X_i) &= (5)(-2) + 0.25 \sum_{i=1}^{5} X_i \\ &= -10 + (0.25)(90) = 12.5\end{aligned} \tag{3.26}$$

thousand dollars, by (3.22). The gross income tax collection is of the form $\sum bX_i = b \sum X_i$, which is 22.5 here, and the total tax credit is of the form $\sum a = na$, which is 10 here. Table 3.5 shows the gross and net tax payments for each family, in columns (3) and (4). The totals may be verified.

For a slight variation, let us reconsider (3.24). If b happens to be the inverse of another constant ($b = 1/c$), then the summation property still holds, so that

$$\sum_{i=1}^{n} \frac{X_i}{c} = \frac{\sum_{i=1}^{n} X_i}{c} \tag{3.27}$$

Summations involving more complicated expressions than those considered here can be simplified or factored following the basic rules of algebra, along lines similar to those in (3.23).

Frequency Distributions

In this chapter we examine grouping and classification techniques for organizing and summarizing data on a single variable. The techniques are similar in the cases of discrete and continuous variables, but there are some important differences. Recall from Chapter 2 that a discrete variable usually results from counting and often it has relatively few distinct values actually occurring in the data. By contrast, a continuous variable usually results from measuring and it usually has many different values occurring in the data.

Most of the statistical methods of econometrics are based on calculations from raw data, as were those in Chapter 3. The techniques presented here are of interest to us because they are useful tools of economic statistics and because they provide a background for understanding the more abstract subject of probability theory that is presented in Chapter 9.

4.1 Discrete Data Variables

Suppose that we have data for n observations on a discrete variable X that takes on only m different values. In this case, it is possible to arrange the n observations into m different groups within which all the observations have exactly the same X value. The different groups ordered by their X values can be indexed by k, with $k = 1, 2, \ldots, m$, so that the typical value can be denoted by X_k.

For example, we examine the 100 observations on *SIZE*, the size of families in our cross-section data set, and denote it by X. Inspection of the raw data reveals that X ranges from 2 to 10, so that $m = 9$. Table 4.1 displays the result

TABLE 4.1 Frequency Table for *SIZE* ($n = 100$)

(1) k	(2) Value of X X_k	(3) Absolute Frequency n_k	(4) Relative Frequency f_k	(5) $X_k n_k$	(6) $(X_k - \bar{X})^2 n_k$
1	2	10	0.10	20	55.2250
2	3	16	0.16	48	29.1600
3	4	34	0.34	136	4.1650
4	5	21	0.21	105	8.8725
5	6	13	0.13	78	35.3925
6	7	3	0.03	21	21.0675
7	8	1	0.01	8	13.3225
8	9	1	0.01	9	21.6225
9	10	1	0.01	10	31.9225
		100	1.00	435	220.7500

$$\bar{X} = \Sigma X_k n_k / n = 4.35$$

$$S_X = \sqrt{\Sigma (X_k - \bar{X})^2 n_k / (n - 1)} = \sqrt{220.75/99} = 1.493$$

of grouping. Each row corresponds to a different value of X, and the rows are ordered with $X_1 = 2$, $X_2 = 3$, . . . , $X_9 = 10$. The first column shows the value of the group index k for each row and the second gives the value of X_k for that group. The third column gives the count of the number of observations among the original n that take on this value (here $n = 100$). This count is called the ***absolute frequency,*** and it is denoted by n_k. As a check on our counting we might add up the absolute frequencies, because in any frequency table

$$\sum_{k=1}^{m} n_k = n \qquad (4.1)$$

That is, the sum of the number of observations having each particular value must equal the total number of observations.

The fourth column in the table contains the ***relative frequency,*** denoted by f_k, with which each X_k occurs. The relative frequency of X_k is the proportion of the n observations that have this value, and it is defined by

$$f_k = \frac{n_k}{n} \qquad (4.2)$$

It is always true that the sum of the relative frequencies equals 1:

$$\sum_{k=1}^{m} f_k = 1 \qquad (4.3)$$

except that rounding error, if allowed, may cause a slight difference. Very simple algebra proves this:

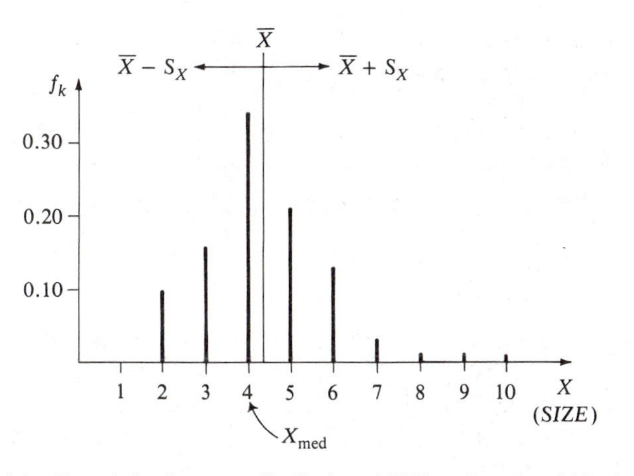

FIGURE 4.1 The relative frequency distribution of *SIZE*, a discrete variable, is graphed as a line chart. The height gives the relative frequency, f_k, corresponding to each group's X_k value. The mean and standard deviation calculated in Table 4.1 are displayed for reference.

$$\sum_{k=1}^{m} f_k = \sum_{k=1}^{m} \left(\frac{n_k}{n}\right) = \frac{1}{n} \sum_{k=1}^{m} n_k = \frac{1}{n} n = 1 \qquad (4.4)$$

The values of X_k coupled with the corresponding f_k make up the ***relative frequency distribution*** of X, which is represented graphically in a line chart, as in Figure 4.1. The value of f_k for each class is measured vertically, and it is graphed as a vertical line rising from the horizontal X axis at the value X_k. This distribution is a concise restatement of the original data, and the only loss of information is the value of n.

Since the relative frequency distribution implicitly orders the X values, it is easy to find the median. We see that 26 percent of the observations are in groups 1 or 2, while another 34 percent are in group 3; hence the middle observations are in group 3, with $X_{med} = 4$ being the median value of X. Also, we see that 4 is the most frequently occurring value, making it the mode. Finally, we see that very few families have seven or more members—only 6 percent do.

The mean and standard deviation of X can be calculated from the frequency table, instead of the raw data, according to

$$\overline{X} = \frac{1}{n} \sum_{k=1}^{m} X_k n_k \qquad (4.5)$$

$$S_X = \sqrt{\frac{\sum_{k=1}^{m} (X_k - \overline{X})^2 n_k}{n - 1}} \qquad (4.6)$$

Since grouping the data on a discrete variable involves no loss of information about the exact values of the variable, the statistics calculated from the frequency table are identical to those calculated from the underlying raw data.

To see that (4.5) is valid, first think of rearranging the n observations so that they are ordered by the value of X, with the smallest value being first and the largest being last. The mean defined in terms of the raw data as

$$\overline{X} = \frac{1}{n}\sum_{i=1}^{n} X_i = \frac{1}{n}(X_1 + X_2 + \cdots + X_n) \tag{4.7}$$

is $1/n$ times a summation of n terms, each of which is an observed X value. Now, the first n_1 of the ordered observations are equal to the same value (X_k with $k = 1$), the second n_2 are equal to another value (X_k with $k = 2$), and so on. Thus the summation is equal to $X_k n_k$ with $k = 1$, plus $X_k n_k$ with $k = 2$, and so on, and

$$\overline{X} = \frac{1}{n}\sum_{i=1}^{n} X_i = \frac{1}{n}\sum_{k=1}^{m} X_k n_k \tag{4.8}$$

The validity of (4.6) is similarly demonstrated.

The basic univariate descriptive statistics can also be stated in terms of relative frequencies. The mean and the root mean squared deviation of X are

$$\overline{X} = \sum_{k=1}^{m} X_k f_k \tag{4.9}$$

$$RMSD_X = \sqrt{\sum_{k=1}^{m} (X_k - \overline{X})^2 f_k} \tag{4.10}$$

and the standard deviation can be obtained from the $RMSD_X$:

$$S_X = \sqrt{n/(n-1)}RMSD_X \tag{4.11}$$

Although the definitions in terms of absolute frequencies provide a slightly easier method for computation, the definitions in terms of relative frequencies provide us with a better background for understanding probability distributions in Chapter 9.

In Table 4.1, columns (5) and (6) show the intermediate steps involved in calculating the mean and standard deviation for family size, according to (4.5) and (4.6). The calculated values are $\overline{X}^* = 4.35$ and $S_X^* = 1.49$, respectively. (Recall that an asterisk is used to denote the actual value of a statistic calculated in a data set.) The one-standard-deviation interval around the mean extends from $4.35 - 1.49$ to $4.35 + 1.49$; that is, it extends from 2.86 to 5.84. Since *SIZE* is discrete, only those observations with $X = 3$, 4, or 5 lie within this interval. By adding $f_2 + f_3 + f_4$ in Table 4.1 we see that 71 percent of the observations lie within one standard deviation of the mean. All the calculations in this paragraph yield exactly the same results as would be obtained by using the definitions in Chapter 3 with the raw data on *SIZE*. As noted above, grouping the data on a discrete variable involves no loss of information about the exact values of the variable; no approximations are introduced.

What is achieved by constructing the frequency table? First, we have transferred all the information in the original 100 observations into a nine-line table. If we had started with 25,000 observations, the economy resulting from this transfer would be even more striking. Second, the arrangement of the data in a frequency table displays the salient characteristics of the data on a single variable more readily than does presentation of the raw data: it is difficult to see any pattern in family size just by looking down the long column on *SIZE* in Table 2.2, but Table 4.1 and Figure 4.1 convey the pattern very clearly.

4.2 Continuous Data Variables

With a continuous variable (or with a discrete variable that is considered practically continuous), it would be impractical to set up groups within which all the observations have exactly the same value, because the number of groups would be very large. Instead, m classes are determined in such a way that every observation can be placed into one and only one class, which is defined to include an interval of values for the variable. For our purposes, it is convenient to follow three principles in setting up the class intervals: (1) the number of classes should be between 5 and 15, (2) the width of each interval should be the same, and (3) the midpoint of each interval should be a convenient number.

For example, letting X now denote the income of families in our cross-section data set (*INCOME*), we recall that X ranges from 0.75 to 32.080 thousand dollars. One convenient classification is to set up nine intervals, each with a range of 4 thousand dollars: the first runs from 0 to 4, the second from 4 to 8, and so on. A difficulty with this scheme is that some observations may lie right on the boundary between two classes. Rather than adopt an arbitrary rule such as putting all such boundary cases into the lower class, it is better to randomize—such as by putting the first boundary case into the lower class, the second into the higher class, the third into the lower class, and so on. In many cases clever choice of the class intervals avoids the boundary problem. For example, if we were to treat *AGE* from Table 2.2 as practically continuous, then we might choose class boundaries such as 22.5 to 27.5, 27.5 to 32.5, and so on. The midpoints would be 25, 30, and so on, and no observations would lie on a boundary.

The ***class mark***, X_k, is defined for each class as the midpoint of the class interval. The class marks here are $X_1 = 2$, $X_2 = 6$, and so on. The class mark is used to represent the value of each observation that is placed into a class, and as a result of this classification process the true value of each observation is lost. This loss of information contrasts sharply with the discrete case, in which there is no such loss.

TABLE 4.2 Frequency Table for *INCOME* ($n = 100$)

(1)		(2) Class Mark	(3) Absolute Frequency	(4) Relative Frequency	(5)	(6)
k	Bounds	X_k	n_k	f_k	$X_k n_k$	$(X_k - \bar{X})^2 n_k$
1	(0– 4)	2	5	0.05	10	329.672
2	(4– 8)	6	35	0.35	210	594.104
3	(8–12)	10	35	0.35	350	0.504
4	(12–16)	14	16	0.16	224	240.8704
5	(16–20)	18	4	0.04	72	248.3776
6	(20–24)	22	0	0.00	0	0.0
7	(24–28)	26	2	0.02	52	504.3488
8	(28–32)	30	2	0.02	60	790.4288
9	(32–36)	34	1	0.01	34	570.2544
			100	1.00	1012	3278.56

$$\bar{X} = \Sigma\, X_k n_k / n = 10.12$$

$$S_X = \sqrt{\Sigma\,(X_k - \bar{X})^2 n_k/(n-1)} = \sqrt{3278.56/99} = 5.755$$

The frequency table for family income (X) in our data is given in Table 4.2. The first column gives both the row index k and the boundary values for X that define the class. The second column gives the class mark, X_k. The third column gives the absolute frequency, n_k, which is the number of observations in our data whose X value falls in each class. To construct a frequency table we must look at each of the observations, decide which class it corresponds to, and keep a running tally. After we have looked at all the observations, our tallies give us the absolute frequencies.

The relative frequency for a particular class, f_k, is the proportion of all the n observations that are in that class. The values of X_k coupled with the corresponding f_k make up the ***relative frequency distribution*** of X, which is represented graphically in a ***histogram,*** or bar chart, as in Figure 4.2. The value of f_k for each class is measured vertically and is graphed as a horizontal line directly over the interval of X values that defines the class; then the area under each line is enclosed to make a bar.

Figure 4.3 shows some shapes that histograms commonly take on. When there is only one peak the resulting distribution is said to be ***unimodal,*** and when there are two it is ***bimodal.*** Among unimodal distributions, some are fairly symmetrical (Figure 4.3a) but others are not. The distribution in Figure 4.3c is said to be ***skewed*** to the right because it has a longer tail in that direction, and that in Figure 4.3d is skewed to the left.

Since all the class intervals are made to have the same width, denoted by w, the widths of all the bars in the histogram are equal. Hence the area of each bar, wf_k, is proportional to the relative frequency of that class. And the area of any one bar relative to the total area of the histogram is simply the relative

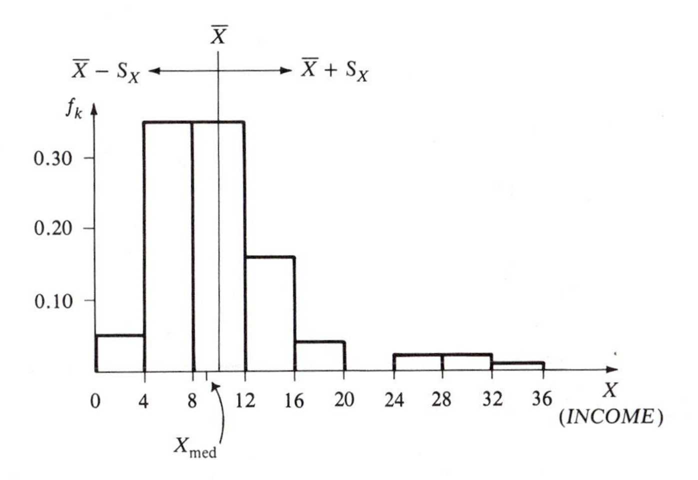

FIGURE 4.2 The relative frequency distribution of *INCOME*, a continuous variable, is graphed as a histogram or bar chart. The height of each bar gives the relative frequency, f_k, for each class. The mean and standard deviation calculated in Table 4.2 are displayed for reference.

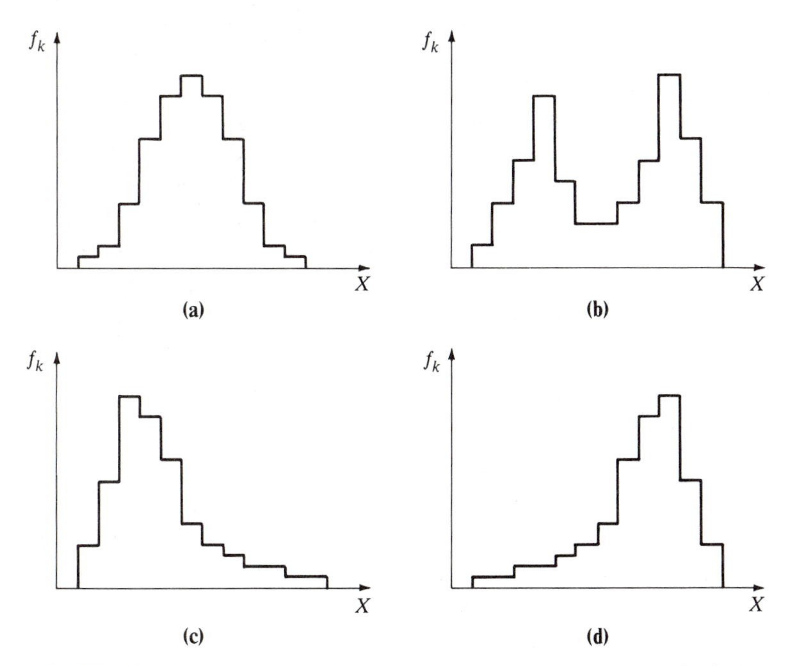

FIGURE 4.3 Histograms may take on a variety of shapes. The histogram in (b) is bimodal, whereas the others are unimodal. The histogram in (a) is symmetrical. The histogram in (c) is skewed to the right, and that in (d) is skewed to the left.

frequency for that class. To see this, note that the area of the bar for the kth class is wf_k, and the total area is $\Sigma\, wf_k$. Thus the ratio of the area of the kth bar to the total area is

$$\frac{wf_k}{\sum wf_k} = \frac{wf_k}{w \sum f_k} = \frac{wf_k}{w} = f_k \tag{4.12}$$

The notion that areas under part of a histogram represent relative frequencies may seem odd, but it is an important feature of the classification technique.

The mean and standard deviation of a continuous variable whose data are classified in a frequency table are defined in exactly the same way as for a discrete variable, by (4.5) and (4.6). In Table 4.2 we find that $\overline{X}^* = 10.120$ and $S_X^* = 5.755$. These values are somewhat different from the statistics calculated in Chapter 3 from the raw data on *INCOME*, which were $\overline{X}^* = 9.941$ and $S_X^* = 5.584$. The differences are due to our now using the class marks to stand for the actual X values of the observations, and the error thereby introduced is part of the cost of achieving summarization through classification.

Finally, it was noted in Chapter 3 that the interval $\overline{X} - S_X$ to $\overline{X} + S_X$ contains 85 percent of the observations in the raw data for family income, and the discrepancy between this amount and the 68 percent approximation was attributed to the lack of symmetry. Table 4.2 indicates more clearly what is occurring. The bulk of the observations lie in the first five classes, having *INCOME* less than 20 thousand dollars. However, in column (6) we see that more than half of the dispersion in X is attributable to the five observations with the greatest amounts of income. These extreme observations boost up the standard deviation and therefore they increase the range of the one-standard-deviation interval, leading such a great proportion of the observations to lie therein. To explore this further, we consider the frequency distribution for the sample with the five greatest X values deleted, leaving $n = 95$. The histogram for this reduced sample is the same as that for the whole sample, Figure 4.2, except that the three small bars lying at the right-hand side of the graph would not be present. The shape of the distribution for the reduced sample is clearly more symmetrical than that for the whole sample, although it is not perfectly so. Going back to the raw data ($n = 95$), we find $\overline{X}^* = 8.950$ and $S_X^* = 3.529$, and we find that 73 percent of the observations in the reduced sample lie within one S_X of $\overline{X}$. Now the 68 percent approximation looks much better.

Proportions

An interesting use of relative frequency distributions and histograms is making calculations that answer such questions as "What proportion of the observations have X values between X_a and X_b?" We may rephrase this question notationally as

$$\text{Prop}\,(X_a \leq X \leq X_b) = ? \tag{4.13}$$

where $X_a < X_b$. For example, from the classified data on *INCOME* we might want to determine the proportion of observations that lie in the one-standard-deviation interval 10.120 ± 5.755; that is, in (4.13), $X_a = 4.365$ and $X_b = 15.875$, with X being *INCOME*. Or, we might want to determine the proportion of observations having incomes less than or equal to 10 thousand dollars; in other words, Prop $(0 \leq X \leq 10)$.

If we were working with the raw data on X (or with the frequency distribution of a discrete variable), our method of answering this question would be to count up the number of observations that lie in the interval and reexpress this count as a proportion. The corresponding exact count generally cannot be made from the frequency distribution of a continuous variable, however, because we have thrown away information about the observations' exact X values and have let the class marks stand in their place. Thus whatever answer we come up with will be only an approximation to the answer obtained from the raw data.

The general approach to determining the proportion in question is to calculate the sum of the relative frequencies of the class intervals and parts of class intervals that make up the interval from X_a to X_b. The relative frequency of any whole class interval contained in the (X_a, X_b) interval is already known in the frequency table. If part of a class interval is involved, we must use some approximation technique to determine the relative frequency assignable to it.

For our purposes, we adopt the ***uniform distribution assumption*** that the X values of the observations in any interval are spread smoothly, or uniformly, throughout it. (It helps to think that the number of observations is very large, so that there are no questions of spacing within the interval.) This assumption implies that the number of observations in any part interval is proportional to the range of that part interval. For example, if the range of a part interval constitutes one-third of the range (width) of the interval, we say that one-third of the observations in the interval lie in the part interval. Since the number of observations can be reexpressed as a proportion, the uniform distribution assumption implies that the relative frequency assigned to any part interval is proportional to the range of that part interval. If the relative frequency for the whole interval is .30, a relative frequency value of .10 will be assigned to a part interval whose range constitutes one-third of the range of the whole interval.

As an example of this approach, we determine the proportion of observations lying in the one-standard-deviation interval for *INCOME*: Prop $(4.365 \leq X \leq 15.875)$. Referring to Table 4.2, we see that the interval (4.365, 15.875) corresponds to part of the second interval, all of the third, and part of the fourth interval defined in the frequency table. First focusing on the second interval in the table, we see that the whole-width range of 4.0 (i.e., from 4 to 8) contains .35 of all the observations. The part interval of interest, from 4.365 to 8, has a range of 3.635; this range amounts to a fraction, 3.635/4.0, of the range of the whole interval. By the uniform distribution assumption we assign to the part interval a relative frequency equal to that fraction times the relative frequency that is contained in the whole interval. Since the second interval contains

$f_2 = .35$ of all the observations, the part interval of interest is considered to have $(3.635/4.0)(.35) = .318$ of all the observations in the data set. Now focusing on the whole third interval, the table shows us that it contains $f_3 = .35$ of all the observations. Finally focusing on the fourth interval, we see that we are interested only in the part interval from 12 to 15.875. This has a range of 3.875, and by our assumptions we consider that it has $(3.875/4.0)(.16) = .155$ of all the observations in the data. Adding the three elements together we find that a total of $.318 + .35 + .155 = .823$ of all the observations on *INCOME* lie within one standard deviation of the mean in the frequency table. We recall from Chapter 3 that $\overline{X} - S_X$ to $\overline{X} + S_X$ contained 85 percent of the observations in the raw data. The discrepancy between .823 and .85 is attributable to our having lost information when we organized the original raw data into classes.

The proportion problem stated in (4.13) can be answered graphically by comparing areas in a histogram. We saw above, in (4.12), that in a histogram the relative frequency for an interval defined in the frequency table is equal to the area of the bar graphed above that interval, divided by the total area of the histogram. Similarly, the relative frequency assigned to a part interval is given by the area of the bar (part bar) standing above it, divided by the total area of the histogram. Since the proportion of observations in the (X_a, X_b) interval is given by the sum of the relative frequencies assigned to the intervals and part intervals that make it up, this same proportion is also equal to the sum of the areas corresponding to these intervals and part intervals, divided by the total area of the histogram. This is illustrated in Figure 4.4a, in which Prop $(X_a \leq X \leq X_b)$ is given by the ratio of the shaded area to the total area in the histogram. The graphical analogy can help us make some quick and approximate

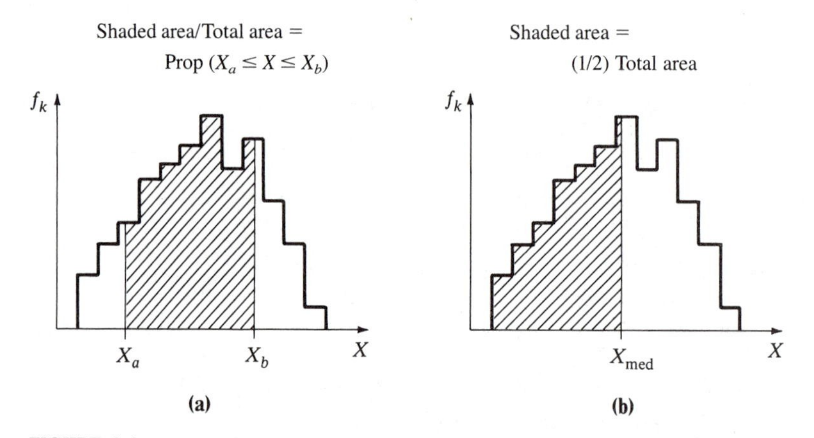

FIGURE 4.4 A portion of the area under a histogram, relative to the total area, gives the proportion of observations having values within the particular interval of X values. This idea is based on the uniform distribution assumption; a tally of the raw data would give a slightly different answer. Part (a) illustrates the general form of the proportion problem, given as Equation (4.13). Part (b) illustrates how this idea is used to determine the median, according to (4.14).

assessments of proportions, and it is helpful for understanding the probability theory in Chapter 9.

Another application of proportion calculations is determining the median. For a classified continuous variable, the median is defined as the value X_{med} such that

$$\text{Prop}\ (X \leq X_{med}) = .5 \tag{4.14}$$

Figure 4.4b illustrates this definition for a general case. In our data for family income we see that the median is somewhere in the third class, since Prop $(X \leq 8) = .40$ and Prop $(X \leq 12) = .75$. The median value X_{med} can be visualized as lying at the end of a part interval that begins at the value $X = 8$ and stretches into the third interval in the frequency table. This part interval is of such a size that exactly .10 relative frequency is assigned to it, thus assuring that X_{med} satisfies (4.14). The median is calculated to be

$$X_{med} = 8 + \left(\frac{.10}{.35}\right)(4) = 9.14 \tag{4.15}$$

In this calculation, the fraction (.10/.35) is a ratio of relative frequencies, and by the uniform distribution assumption the ratio of the range of the part interval to the range of the interval is also equal to this. Multiplying this fraction by the range of the interval (4) gives us the range of the part interval. By construction, the median lies at the end of a part interval beginning at 8.

Problems

Section 4.1

4.1 Treating *REG* (from the cross-section data set, $n = 100$) as though it were a meaningful discrete variable, construct a frequency table and graph the relative frequency distribution.

4.2 Based on your answer to Problem 4.1, determine the mean, median, and standard deviation of *REG*.

* **4.3** Based on your answers to Problems 4.1 and 4.2, what proportion of the observations have a value for *REG* that is within one standard deviation of the mean? What proportion are within two?

4.4 Starting from Equation (3.4) as the definition of *RMSD*, prove that Equation (4.10) is true for grouped discrete data.

4.5 Suppose that the frequency distribution of X is symmetric. Let $Y = 2X$. Explain why the distribution of Y is also symmetric.

* **4.6** Suppose that the frequency distribution of X is symmetric. Let $Y_i = \ln X_i$, that is, let each Y be equal to the natural logarithm of X_i. Explain why the distribution of Y is skewed to the left.

4.7 Considering only those 60 families with *SIZE* 2, 3, or 4 in Table 4.1, construct a frequency table and graph the relative frequency distribution for *SIZE*. From this new table, determine the mean, median, mode, and standard deviation of *SIZE* for these observations.

4.8 Starting from Equation (4.5), show that Equation (4.9) is true.

Section 4.2

⋆ **4.9** Treating *ED* ($n = 100$) as a continuous variable, construct a frequency table and carefully graph the relative frequency histogram. Choose as class boundaries 1.5–4.5, 4.5–7.5, and so on. Why is this a good choice?

⋆ **4.10** Based on your answers to Problem 4.9, determine the mean and standard deviation of *ED*. Should we expect that these calculated values will be exactly equal to the statistics calculated from the raw data on *ED*? Why?

4.11 On the histogram from Problem 4.9, highlight the area under the histogram that is bounded by vertical lines drawn through the value for *ED* equal to one standard deviation below the mean and the value for *ED* equal to one standard deviation above the mean.

4.12 Based on your answers to Problems 4.9 and 4.10, what proportion of the observations have a value for *ED* that is within one standard deviation of the mean?

⋆ **4.13** Based on your answer to Problem 4.9, determine the median value of *ED*.

4.14 Suppose that within each class interval the actual X values are concentrated toward the left side. How would the mean calculated from the frequency distribution compare with the mean calculated from the raw data?

4.15 Treating the frequency distribution in Table 4.1 as though *SIZE* were a continuous data variable, determine the proportion of observations that could be considered to lie within one standard deviation of the mean. Compare this with the proportion of actual observations that lie in this range.

4.16 Treating the frequency distribution in Table 4.1 as though *SIZE* were a continuous data variable, determine the median of size. Compare this with the true median of *SIZE*.

4.17 The ninetieth percentile of any variable is that value of the variable below which 90 percent of the observations lie. Based on Table 4.2, determine the ninetieth percentile of *INCOME*.

Specification and Estimation of Regression Models

Simple Regression: Theory

As previewed in Chapter 1, regression analysis is a technique for estimating the values of the structural coefficients in a model of an economic process. In this chapter and the next we discuss simple regression, which applies to models involving just two variables. In Chapter 7 we discuss multiple regression, which applies to models that are more complex.

The discussion in this chapter is limited to the theory and mechanics of estimating a simple regression model of regular form. In the next chapter we apply this theory in a variety of cases, and we see how the estimates and measures developed here are put into practice.

5.1 Specification of the Model

The econometric use of simple regression starts from the theory or presumption that there is some true relation between two variables. In economics a relation is usually based on behavior, as in the case of a consumption function, but sometimes the relation is based on technology, as in the case of a production function. For simplicity we refer to either type as a behavioral relation.

In technical discussion we think of such a relation as being an ***economic process*** that can be described very concretely in mathematical terms. As an analogy, it helps to think of some engineering process, with an input and an output. The input and output are observable, but the operations of the process itself are not.

Suppose that we have data for n observations on two variables, Y and X, that we believe are related in such a way that we would say that Y is determined by X. Our thinking is that there is some process occurring in which a value of X is fed in and a value for Y is produced. If n values of X are fed in one at a time, n corresponding values for Y are determined one at a time. The n values of X and Y are all observable, and they can be collected as a set of data.

Two aspects of this process should be noted. First, X is the only identifiable and observable variable that affects Y; it is the only variable that feeds into the process. By explicit exclusion from the discussion, other variables that might be measured for each observation are understood to play no direct role in the determination of Y. Second, the values of X are taken as given. The economic process under consideration does not affect those values and it plays no role in their determination. In addition, we make no effort to understand why X takes on whatever values it does. Our interest is in how the value of Y is related to the value of X for any observation.

We move toward statistical analysis by making a more concrete specification of the process by which Y is determined; this specification is known as the ***simple regression model.*** We theorize that the value of variable Y for each observation is determined by the equation.

$$Y_i = \beta_0 + \beta_1 X_i + u_i \tag{5.1}$$

In this regression model, Y is called the ***dependent variable,*** and X is called the ***explanatory variable*** (sometimes X is called the ***independent variable***). In the model, β_0 and β_1 are parameters that have fixed values throughout (β is lowercase "beta," the Greek "b"); they are called the ***coefficients*** of the regression model. The term u_i is called the ***disturbance.***

The disturbance u_i is considered to be an unobservable random term that does not depend on the value of X_i. The disturbances are meant to represent pure chance factors in the determination of Y. Among these factors there may be one that is best described as luck. Also, we might believe that Y is affected by a host of minor factors that we cannot identify and whose combined impact is indistinguishable from pure chance. Finally, we recognize possible measurement error in Y (but we presume that there is no measurement error in X). For any particular observation, u_i can be positive or negative, small or large. Overall, the average disturbance is anticipated to be close to zero, but only by coincidence would it be exactly zero in any given set of data.

For example, consider the process by which families determine their annual saving. Economic theory suggests that saving depends mostly on income. Letting Y be family saving and X be family income, (5.1) is a simple model of family saving behavior in which different families are the observations. The model does not explain the level of income of each family; it takes this as given. The parameter β_0 is the saving (probably negative) of families with zero income, and β_1 is the increase in saving resulting from a unit increase in family income (i.e., β_1 is the marginal propensity to save).

Based on the simple regression specification (5.1), we can decompose each Y_i value into a systematic component $\beta_0 + \beta_1 X_i$ and a random component u_i. The value of the systematic component is called the ***expected value*** of Y for each observation:

$$E[Y_i] = \beta_0 + \beta_1 X_i \tag{5.2}$$

This concept will be given a precise statistical meaning in Chapter 11, but for now it seems fairly intuitive: given (5.1), it is the value that Y_i would take on if the disturbance were equal to zero. This decomposition of each Y_i value into its systematic and random components can be compactly rewritten as

$$Y_i = E[Y_i] + u_i \tag{5.3}$$

The model specified in (5.1) is represented graphically in Figure 5.1. The systematic part of relation between Y and X is graphed as the line

$$E[Y] = \beta_0 + \beta_1 X \tag{5.4}$$

which is called the ***true regression line.*** We consider only a specific set of n observations whose X values are somehow given and whose Y values are determined according to (5.1). The resulting data points are plotted.

Following (5.3), the value Y_i for a single observation can be decomposed vertically into the distance up to a point on the true regression line and the distance from that point to the observation. The decomposition is shown explicitly for two observations. For the first, which has the values X_1 and Y_1 for

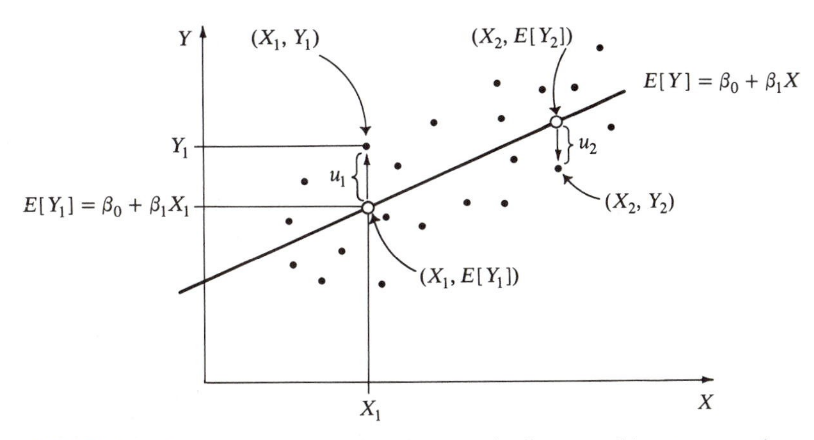

FIGURE 5.1 For each observation, the value Y_i can be decomposed into two parts: the systematic part $E[Y_i]$ (which equals $\beta_0 + \beta_1 X_i$), and the disturbance u_i. The systematic part is given by the height of the true regression line above the observation's X value, and the disturbance is given by the distance from the point on the regression line to the data point. The disturbance is positive for the first observation and negative for the second.

the two variables, the plotted point is shown and the X and Y values are traced to the axes. Directly above the value X_1 on the X axis we find a point on the true regression line; its vertical value is $E[Y_1]$, and this is traced over to the Y axis. The vertical distance from $E[Y_1]$ to Y_1 is u_1, which is shown also. The situation for the second observation (X_2, Y_2) is the same except that the data point lies below the true regression line, so the disturbance u_2 is negative.

In summary, the simple regression model illustrated in Figure 5.1 is a specification of the theoretical process that we use to describe the relation between two observable variables. For each observation the value X_i is determined outside the process. Given this X_i, the value Y_i is determined by (5.1). It will be convenient to say that the process produces the data on Y and X, even though the X values are determined outside.

Our theory is that the simple regression model actually reflects the way some economic behavior works. Surely this is a very simple model of how the values of a variable are determined: most economic variables systematically depend on more than one other variable, and that is why multiple regression is usually more appropriate. However, there are many instances in which simple regression is quite reasonable, and it is a useful tool of econometrics. Also, the presumption that the relation is strictly linear is not so constraining as it might seem. In some cases, we might well accept a linear model if we believe that the true relation is approximately linear. More important, we see in Chapter 6 that some truly nonlinear relations can be transformed into linear ones. In these cases we treat the transformed relation just as we treat the model specified in (5.1).

5.2 Estimation of the Model

When we have a set of data on Y and X that we presume was generated by a process described by (5.1), the values of β_0 and β_1 are unknown. Our interest focuses on estimating the values of these parameters, based on the data we have.

Our job now is pictured in Figure 5.2. Temporarily ignore the line drawn there. The data we have are plotted, and these points represent all that we can observe. To say, as above, that we presume or theorize that the data were generated by the process (5.1) means that we presume that there is some unobservable true regression line underlying the data. Figure 5.1, which illustrates the theory, explicitly shows a true regression line. Careful comparison of Figures 5.1 and 5.2 shows that the data points are exactly the same in both. In other words, Figure 5.1 presents our theory of how the data in Figure 5.2 were generated. Hence the true regression line in Figure 5.1 underlies the data in Figure 5.2. Of course, we do not see it there, because it is not observable. But if we accept the theory that the true regression line underlies the data, we can use the data to estimate the parameters of the line.

The task of estimating the parameters β_0 and β_1 of the unobservable true regression line is carried out by drawing or fitting an actual line through the

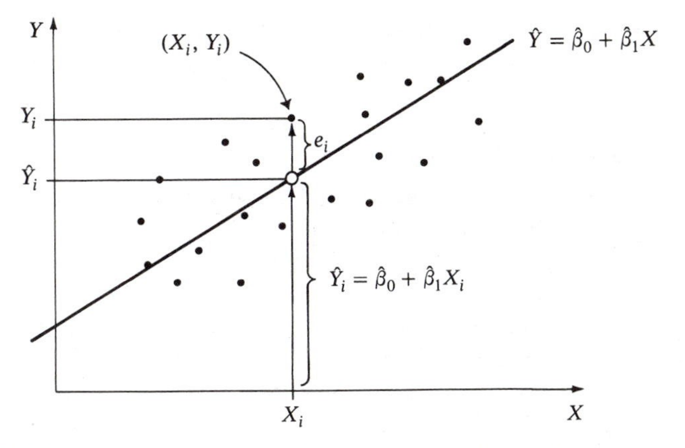

FIGURE 5.2 The estimated regression line is based on the data on Y and X, as plotted. It serves as an estimate of the true regression line in Figure 5.1, but it is generally different from it. For each observation, the value Y_i can be decomposed into two parts: the fitted value $\hat{Y}_i$ (which equals $\hat{\beta}_0 + \hat{\beta}_1 X_i$), and the residual e_i. The ordinary least squares criterion is to minimize the sum of the squares of these residuals.

data. The intercept of this actual line is denoted by $\hat{\beta}_0$ and it serves as our estimate of β_0, the intercept of the true regression line. Similarly, the slope of the actual line is denoted by $\hat{\beta}_1$ and it serves as our estimate of β_1. (The circumflex looks like a hat, so $\hat{\beta}_1$ is conventionally read as "beta-one-hat.")

Suppose that the $\hat{\beta}_0$ and $\hat{\beta}_1$ we end up with correspond to the line

$$\hat{Y} = \hat{\beta}_0 + \hat{\beta}_1 X \tag{5.5}$$

drawn through the data in Figure 5.2. This is called the ***estimated regression line*** or the ***fitted regression line.*** For any data point such as (X_i, Y_i), this line decomposes the total value of Y_i into two parts: $\hat{\beta}_0 + \hat{\beta}_1 X_i$, which is the height of the line above X_i, and e_i, which is the vertical distance from the line to the data point. The first part is the ***fitted*** or ***predicted value*** for Y:

$$\hat{Y}_i = \hat{\beta}_0 + \hat{\beta}_1 X_i \tag{5.6}$$

The second part, e_i, is called the ***residual,*** or the ***error*** of fit or prediction:

$$e_i = Y_i - \hat{Y}_i \tag{5.7}$$

Clearly,

$$Y_i = \hat{Y}_i + e_i \tag{5.8}$$

How can we determine the parameters $\hat{\beta}_0$ and $\hat{\beta}_1$ of the estimated regression line? It seems reasonable to try to find the line that best fits the scatter of data. There are a variety of alternative techniques that might be used, and we consider several before focusing on the one we adopt.

One possibility is just to draw a line that seems to fit pretty well and accept that. In some cases this might be satisfactory, but it is not very precise. Two different persons analyzing the same data would undoubtedly come up with different estimates of the parameters, and we would have no basis for deciding which were better.

Another possibility is to measure the perpendicular distance from each point to the fitted line and then develop a method that calculates values for the parameters so as to minimize some overall measure of these distances. The results might be called "perpendicular estimators."

Instead of dealing with the perpendicular distances, another possibility is to measure the vertical distances from each point to the fitted line. These distances are the residuals, defined by (5.7). We might try to find a line for which the sum of the residuals is zero. It turns out that many lines satisfy this criterion, and some of them definitely do not fit very well. Alternatively, we might try to minimize the sum of the absolute values of the residuals. This is reasonable but somewhat awkward.

The standard approach in much practical work, which is the approach we adopt, is called the method of ***ordinary least squares*** (OLS). With this method, the criterion for being the best fit is that the line must make the ***sum of the squared residuals*** (*SSR*) as small as possible:

$$\text{OLS criterion:} \quad \text{minimize } SSR = \sum_{i=1}^{n} e_i^2 \tag{5.9}$$

Based on this criterion, we can develop mathematical rules or formulas for calculating $\hat{\beta}_0$ and $\hat{\beta}_1$.

Putting this more formally, suppose that we have n observations on Y and X, as illustrated in Figure 5.2. Any line drawn through the data will be of the form (5.5), and associated with particular $\hat{\beta}_0$ and $\hat{\beta}_1$ values are a set of residuals, e_i, that are determined by

$$e_i = Y_i - \hat{\beta}_0 - \hat{\beta}_1 X_i \tag{5.10}$$

Consider the sum of the squared residuals around a fitted line:

$$\sum_{i=1}^{n} e_i^2 = \sum_{i=1}^{n} (Y_i - \hat{\beta}_0 - \hat{\beta}_1 X_i)^2 \tag{5.11}$$

For a given set of data the X_i and Y_i are specific numbers, and our interest is in finding the $\hat{\beta}_0$ and $\hat{\beta}_1$ values that minimize this expression.

A calculus derivation, given in the appendix to this chapter, leads us to the following:

$$\hat{\beta}_1 = \frac{\sum_{i=1}^{n} (X_i - \overline{X})Y_i}{\sum_{i=1}^{n} (X_i - \overline{X})^2} \tag{5.12}$$

and

$$\hat{\beta}_0 = \overline{Y} - \hat{\beta}_1\overline{X} \tag{5.13}$$

These are known as the OLS ***estimators*** of β_1 and β_0: they are the formulas for calculating the $\hat{\beta}_1$ and $\hat{\beta}_0$ of the estimated regression line. In practice we calculate $\hat{\beta}_1$ first, and then using that value we calculate $\hat{\beta}_0$. Whenever we use these estimators we can be confident of getting the best-fitting line.

In the derivation of the OLS estimators, the only special assumption made is that not all the X values are identical. If this were the case, the denominator in (5.12) would be zero and $\hat{\beta}_1$ would be undefined. Graphically, this would be a case in which all the data points lie along a vertical line.

Three properties of the least squares fit always hold true. First, the sum of the residuals is exactly zero: $\Sigma\, e_i = 0$. Thus the average error of fit is also zero: $\overline{e} = 0$. This property holds true even though the sum of the squared residuals, which has been minimized, is some positive amount. Interestingly, the least squares line is not the only line for which the associated sum of residuals is zero; one could construct a very poor fitting line for which this property also holds.

Second, the fitted regression line goes through the ***point of means,*** which is the point $(\overline{X}, \overline{Y})$. This property makes it easy to graph the line: one point on the estimated line is its intercept with the vertical axis $(0, \hat{\beta}_0)$ and now we know that a second point is $(\overline{X}, \overline{Y})$. We can graph the line through these two points. It should be noted that usually none of the observations in the data lie at the point of means, although this could happen by coincidence.

Third, there is zero correlation between the residuals and the explanatory variable. We already know that the average residual is zero. This third property assures us that for observations with X above $\overline{X}$, there is no tendency for the residuals to average differently from zero—and similarly for $X < \overline{X}$. To see what this means, think of all the data points as lying close to some straight line. The zero correlation between e and X assures us that the OLS fit cuts through the points rather than across them.

As an example of computation, consider the five observations on Y and X in Table 5.1. These are the first five observations on *SAVING* (Y) and *INCOME* (X) from our cross-section data set, with the values rounded to one decimal position for simplicity. As discussed above, the simple regression model (5.1) can be taken as a theoretical statement describing family saving behavior. The calculations are shown in the table, and the estimated regression is reported as

$$\hat{Y}_i = -0.0386 + 0.0863X_i \tag{5.14}$$

The data are plotted in a scatter diagram in Figure 5.3, and the fitted regression line is drawn in.

TABLE 5.1 Calculation of Regression Estimates

(1) Y_i	(2) X_i	(3) $X_i - \overline{X}$	(4) $(X_i - \overline{X})^2$	(5) $(X_i - \overline{X})Y_i$
0.0	1.9	−5.04	25.4016	0.0
0.9	12.4	5.46	29.8116	4.914
0.4	6.4	−0.54	0.2916	−0.216
1.2	7.0	0.06	0.0036	0.072
0.3	7.0	0.06	0.0036	0.018
2.8	34.7	0.0	55.5120	4.788

$$\overline{X} = \Sigma X_i/n = 34.7/5 = 6.94$$

$$\overline{Y} = \Sigma Y_i/n = 2.8/5 = 0.56$$

$$\hat{\beta}_1 = [\Sigma (X_i - \overline{X})Y_i]/[\Sigma (X_i - \overline{X})^2] = 4.788/55.512 = 0.0863$$

$$\hat{\beta}_0 = \overline{Y} - \hat{\beta}_1\overline{X} = 0.56 - (0.0863)(6.94) = -0.0386$$

5.3 Interpretation of the Regression

The econometric approach to analyzing data is based on the theory that there is some stable process of economic behavior that underlies the data we have. We use the data to learn about the process. In our work, the systematic part of the process is specified by the true regression, and calculating the best-fitting line through the data yields the estimated regression.

Consider how the actual estimated regression line compares with the theoretical true regression line. Suppose that we have a set of n observations on Y and X that were generated by an economic process correctly described by the simple regression model (5.1). The technique of ordinary least squares provides

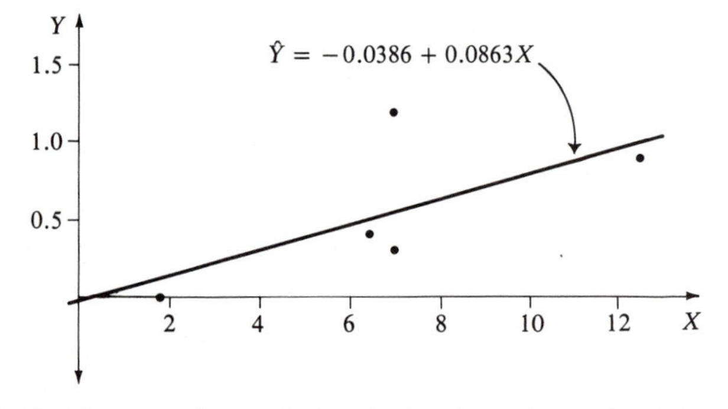

FIGURE 5.3 The scatter diagram displays the five observations on *SAVING* (Y) and *INCOME* (X) from Table 5.1, and the estimated regression line is drawn in. Compare this with Figure 3.4.

a method for calculating estimates of β_0 and β_1. Letting asterisks denote the actually computed values in a set of data, the estimates are $\hat{\beta}_0^*$ and $\hat{\beta}_1^*$.

How is $\hat{\beta}_0^*$ related to β_0 and how is $\hat{\beta}_1^*$ related to β_1? We might hope at first that $\hat{\beta}_0^* = \beta_0$ and $\hat{\beta}_1^* = \beta_1$, so that we would have discovered the true values, but this is unlikely ever to occur. We could hardly expect to learn the precise value of the coefficients in the process (5.1) on the basis of a limited set of data generated by the process.

This thinking is reflected in Figure 5.4. Suppose that we are dealing with three observations, having actual X values X_1, X_2, and X_3. We do not explore why the observations take on these values; even in theory we accept them as given. We theorize that the actual Y values for the observations are produced by the process (5.1). Accordingly, in Figure 5.4 we graph the true regression line, which is the systematic part of the process, and we also plot the actual data points resulting from the process. Given these data, the estimated regression line is determined by the OLS estimators, and it is drawn in the figure also.

For the data in Figure 5.4, the estimated line differs from the theoretical true regression line because of the particular pattern taken on by the disturbances: u_1 is large and positive, while u_2 and u_3 are relatively small. This causes the estimated regression line to have a flatter slope than the true one, in this case. Note that while $\Sigma\, e_i = 0$ for the OLS fit, it is usually true that the disturbances do not sum exactly to zero because they reflect uncontrolled random factors.

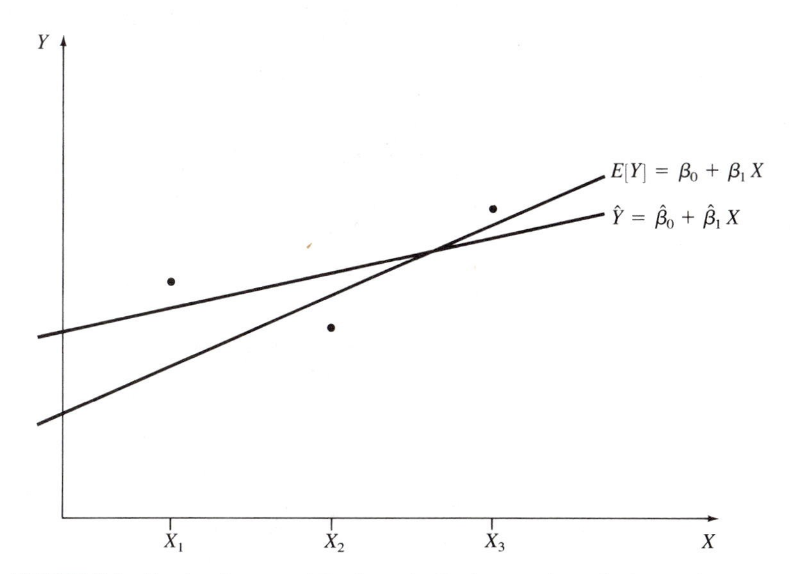

FIGURE 5.4 For the given set of X values, the Y values are determined around the true regression line as in Figure 5.1, with a large positive disturbance u_1 in this case. For these data on Y and X, the estimated regression line is also shown, as in Figure 5.2. The difference between the true regression and the estimated regression arises from the particular pattern of values taken on by the disturbances.

In general, the differences between an estimated regression line and the underlying true regression line arise from the particular pattern of values that the disturbances happen to take on. The differences between $\hat{\beta}_0^*$ and β_0, and between $\hat{\beta}_1^*$ and β_1, are estimation errors. Since the true values of β_0 and β_1 are unknown, we cannot determine the size of these estimation errors in any particular case. Of course, we always hope that the errors are small, but we can never be sure that they are. (In Part IV we develop methods of statistical inference that permit us to assess how large or small the estimation errors might be.)

What if we apply this method of estimating a regression line to data from some process that is not correctly described by the simple regression model? For example, suppose that the true process relating Y to X is

$$Y_i = X_i^{\beta_1} + u_i \tag{5.15}$$

We will explore specifications similar to this in Chapter 6, but the essential point here is that the systematic part of the relation between Y and X is not linear. If we apply our OLS methods directly to estimate a regular linear regression, our results clearly cannot provide us with a correct understanding of the true process determining Y. In general, the usefulness and validity of a regression model depends on its being an accurate specification of the process being examined.

To foreshadow the analysis that we carry out in various applications in Chapter 6, let us look at the estimated regression (5.14) of family saving on family income. Remember that as an example of computation we used only five observations, so that these numerical results can hardly be considered seriously.

Just as the theoretical true regression line underlying the data determines the systematic part of saving for a family with a given income, the estimated regression line can be used to make predictions of saving for a family with some specified income. A family whose income is 8 thousand dollars has a predicted saving of

$$\hat{Y}_i = -0.0386 + (0.0863)(8.0) = 0.6518 \tag{5.16}$$

thousand dollars (i.e., about \$652). Notice that in predicting a level for the dependent variable we are finding the height of the estimated regression line corresponding to a specific level of X.

The estimated intercept and slope values are just that: estimates of the corresponding parameters of the behavioral relation. Special attention focuses on the slope here, because the economics of the behavior is that β_1 is the marginal propensity to save. Our estimate of this parameter is 0.0863, and in a serious study we would hope that this is not too far from the true value.

Looking at the estimated slope, we see that if a family's income were to increase by 1 thousand dollars (i.e., $\Delta X = 1$) its predicted saving would increase by 0.0863 thousand dollars (i.e., \$86.30). This is a basic interpretation that we can make without any computation. By contrast, if a family's income were to increase by \$2500 (i.e., $\Delta X = 2.5$), we use our knowledge of slopes (see the appendix to Chapter 1) to compute

$$\Delta \hat{Y} = 0.0863\Delta X = (0.0863)(2.5) = 0.21575 \qquad (5.17)$$

thousand dollars. That is, predicted saving increases by about \$216. Notice that the intercept plays no role in a computation like this.

Looking at the estimated intercept, we see that if a family's income were zero, its predicted saving would be -0.0386 thousand dollars (i.e., minus \$38.60). Although this may seem odd at first, it does make economic sense: many families with temporarily low incomes do have some accumulated wealth that they would spend if their income were zero, and decreases in wealth are measured as negative saving.

5.4 Measures of Goodness of Fit

The technique of ordinary least squares guarantees that the estimated regression line is the best-fitting line that can be drawn through the data, in the sense that it has the smallest possible sum of squared residuals. Although it is the best-fitting line, whether we would judge that it fits well or not so well depends on the data: if the data in the scatterplot are widely dispersed, no line can fit very well; if the data seem to lie close to some line, a good-fitting line can be found. In this section we develop two statistics that allow us to quantify how well the regression fits.

Starting from a set of data on Y and X, the estimated regression line yields a set of fitted values, or predictions, for the actual Y_i values. These are given by

$$\hat{Y}_i = \hat{\beta}_0 + \hat{\beta}_1 X_i \qquad (5.18)$$

The associated errors of fit are given by the residuals

$$e_i = Y_i - \hat{Y}_i \qquad (5.19)$$

These residuals serve as the basis for our two measures of goodness of fit.

The Standard Error of the Regression, *SER*

The n residuals constitute a data variable, e, that can be described by the methods of Chapter 3. As noted above, it is a property of OLS estimation that the mean residual is always zero: $\bar{e} = 0$. The standard deviation of the residuals (S_e) will be some positive number, and its usual interpretation will still be valid: S_e measures the typical deviation of e from its mean, without regard to sign.

Since $\bar{e} = 0$, each deviation ($e_i - \bar{e}$) is the same as the value of the variable (e_i) itself. Hence the standard deviation can be interpreted here as the typical value of the variable, without regard to sign. Thus S_e could serve as an interesting measure of how well the regression fits the data: it answers the question, "What is the typical error of fit?" To satisfy some purposes discussed in Chapter 11, the actual measure we adopt is a slight modification of S_e.

The ***standard error of regression*** (*SER*), which is sometimes called the "standard error of estimate," is defined by

$$SER = \sqrt{\frac{\sum_{i=1}^{n} e_i^2}{n - 2}} \tag{5.20}$$

and it gives the typical error of fit. (Notice that if $n - 2$ were replaced by $n - 1$, the expression on the right-hand side would equal S_e.) As with the standard deviation, the denominator in this expression is identified as the number of degrees of freedom.

To illustrate how the *SER* is computed, we continue with the previous example dealing with family saving behavior. Table 5.2 starts with the same data on Y and X as Table 5.1. For each observation, separately, the fitted or predicted value $\hat{Y}_i$ is determined from the estimated regression (5.14) by substituting in the X_i value. Then, again for each observation separately, the residual (5.19) is calculated and squared. The sum of these squares is $\Sigma\, e_i^2$. The *SER* is then calculated directly according to (5.20).

The units of measurement of the *SER* are always the same as those of the dependent variable Y because each residual is equal to the actual value Y_i minus the fitted value $\hat{Y}_i$. In this case the *SER* is 0.416 thousand dollars, because family saving (Y) is measured in those units. We interpret this value as the typical error of fit for the regression of family saving on family income.

To assess its magnitude, the *SER* is usually compared with some value of the dependent variable. If we are making one prediction, a comparison of the *SER* with the $\hat{Y}_i$ value suggests how much in error the prediction might be. (A probabilistic analysis of prediction errors is given in Chapter 13.) If we are

TABLE 5.2 Calculation of R^2 and *SER*

(1) Y_i	(2) X_i	(3) $\hat{Y}_i$	(4) e_i	(5) e_i^2	(6) $Y_i - \bar{Y}$	(7) $(Y_i - \bar{Y})^2$
0.0	1.9	0.126	−0.125	0.0157	−0.56	0.3136
0.9	12.4	1.031	−0.131	0.0171	0.34	0.1156
0.4	6.4	0.513	−0.113	0.0129	−0.16	0.0256
1.2	7.0	0.565	0.635	0.4030	0.64	0.4096
0.3	7.0	0.565	−0.265	0.0703	−0.26	0.0676
			0.0	0.5190	0.0	0.9320

$$R^2 = 1 - \Sigma\, e_i^2/[\Sigma\,(Y_i - \bar{Y})^2] = 1 - 0.5190/0.9320 = 0.443$$

$$SER = \sqrt{\Sigma\, e_i^2/(n - 2)} = \sqrt{0.5190/3} = 0.416$$

considering the overall goodness of fit of the regression, a comparison of the *SER* with the mean of Y is useful. For example, in the saving regression, the *SER* is \$416 and $\overline{Y}$ is \$560; the typical error of fit is relatively large, indicating a fairly poor fit. However, simple comparisons with the mean are sometimes misleading; if the same residuals had been obtained with families having greater income and saving levels (and thus greater $\overline{Y}$), the *SER* would not seem so "relatively large."

As evident in Table 5.2 and Figure 5.3, one of the residuals is much larger than the others in absolute value. This, combined with the small size of the sample, leads the *SER* to be larger than four of the five residuals. (The mean absolute residual is 0.254 here, which is substantially smaller than the *SER*.) Because of this, we might hesitate before accepting the conclusion that the fit is "fairly poor."

In practical applications of simple regression, we sometimes end up estimating two or more regressions having the same dependent variable. For example, if we have two sets of data on the same variables, we might estimate two separate regressions of Y on X. Or if we have three variables in single data set, we might estimate one regression of Y on X and another regression of Y on Z. In comparing the two regressions, it makes sense to say that the one with the smaller *SER* has the better fit.

By contrast, it generally does not make sense to compare the *SER*s of two regressions when they have different dependent variables. This can lead to illogical comparisons. For example, if we have three variables in a single data set, we might estimate one regression of Y on X and another regression of Z on X. In each regression the computed *SER* measures the typical error of fit. But if Y is measured in dollars and Z is measured in percentage points, we clearly cannot compare the *SER*s to judge which regression has the better fit.

The Coefficient of Determination, R^2

The second measure we develop to quantify how well the estimated regression line fits the data yields a pure, dimensionless number. Like the magnitude of the correlation, this measure varies between zero and 1, with a higher value indicating a better fit. We go through several steps of a formal development because these are useful for understanding the interpretation we use.

Making reference to Figure 5.5, it should be clear that for any observation i,

$$(Y_i - \overline{Y}) = (\hat{Y}_i - \overline{Y}) + (Y_i - \hat{Y}_i) = (\hat{Y}_i - \overline{Y}) + e_i \qquad (5.21)$$

This is a decomposition, showing that for each observation the total deviation of Y_i from the mean is equal to the deviation of the regression's predicted value from the mean plus the regression's error of fit. In other words, the ***total deviation*** of Y_i from $\overline{Y}$ is equal to the ***deviation explained by the regression*** plus the ***unexplained deviation.***

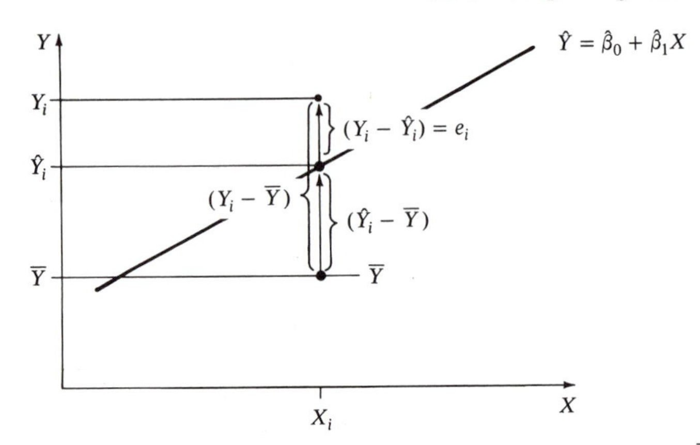

FIGURE 5.5 For any observation, the total deviation of Y_i from the mean, $(Y_i - \overline{Y})$, can be decomposed into two parts: the unexplained deviation $(Y_i - \hat{Y}_i)$, which is just the residual e_i, and the explained deviation $(\hat{Y}_i - \overline{Y})$. The coefficient of determination, R^2, is defined as the ratio of an overall measure of the explained deviations to an overall measure of the total deviations. The standard error of the regression, *SER*, is a measure of the typical unexplained deviation—that is, it is a measure of the typical error of fit, e_i.

If we square the leftmost and rightmost portions of (5.21), we get

$$(Y_i - \overline{Y})^2 = (\hat{Y}_i - \overline{Y})^2 + e_i^2 + 2e_i(\hat{Y}_i - \overline{Y}) \tag{5.22}$$

and if we add up the n equations like (5.22) that hold for each i, we get

$$\sum (Y_i - \overline{Y})^2 = \sum (\hat{Y}_i - \overline{Y})^2 + \sum e_i^2 + 2 \sum e_i(\hat{Y}_i - \overline{Y}) \tag{5.23}$$

It can be shown that the last term on the right-hand side of (5.23) is equal to zero, so that what remains is

$$\sum (Y_i - \overline{Y})^2 = \sum (\hat{Y}_i - \overline{Y})^2 + \sum e_i^2 \tag{5.24}$$

All the terms in this equation are positive or zero because they are sums of squares. The expression on the left-hand side is called the ***total variation*** of Y, because it is the sum of squares of the total deviations identified above. Similarly, the first expression on the right-hand side is called the ***explained variation*** and the second is called the ***unexplained variation.*** Thus, as a consequence of the original decomposition illustrated in Figure 5.5, we can say that the total variation of Y is equal to the variation that is explained by the regression plus the unexplained variation.

Rearranging and dividing by $\Sigma (Y_i - \overline{Y})^2$, we get

$$1 - \frac{\sum e_i^2}{\sum (Y_i - \overline{Y})^2} = \frac{\sum (\hat{Y}_i - \overline{Y})^2}{\sum (Y_i - \overline{Y})^2} \tag{5.25}$$

The right-hand side of this equation is the ratio of the explained variation to the total variation. The numerator and the denominator are positive, and since the numerator is a component of the denominator the ratio can take on values only between zero and 1. This ratio could serve as a measure of goodness of fit, and we adopt it for that purpose. The left-hand side of (5.25) is often easier to compute. In part for that reason, we use the left-hand side in the definition but give it the interpretation that applies directly to the right-hand side.

The ***coefficient of determination,*** which is usually denoted by R^2 and read as "R-squared," is defined by

$$R^2 = 1 - \frac{\sum_{i=1}^{n} e_i^2}{\sum_{i=1}^{n} (Y_i - \bar{Y})^2} \tag{5.26}$$

Based on the analysis of (5.25), the general form of our interpretation of R^2 is that it measures "the proportion of the total variation of Y that is explained by the regression." Since the regression model explains, or predicts, the values of Y on the basis of the given values of X, we might say instead that R^2 measures "the proportion of the total variation of Y that is explained by X."

The computation of R^2 is illustrated in Table 5.2. In addition to calculating $\sum e_i^2$, which was done for the *SER*, we need to calculate the total variation of Y, $\sum (Y_i - \bar{Y})^2$, as an intermediate step. Finding $R^2 = .443$, we say that the regression explains about 44 percent of the variation of Y.

In the appendix to this chapter it is proved that R^2 is equal to the square of r, the correlation between X and Y. Thus these two measures serve in the same way to quantify how well the estimated regression line fits the data. In econometric analysis we usually talk in terms of R^2 rather than r, partly because of its useful interpretation in terms of explaining the total variation of Y.

The R^2 is a zero-to-1 dimensionless measure of goodness of fit, and knowing the numerical value in any case helps us conjure up an image of the scatterplot in our minds. An R^2 of .443 is usually associated with a scatterplot in which the data are fairly dispersed around the estimated regression. In our example, we might qualify this description by noting that one residual is much larger than the others.

Our interpretation of the magnitude of R^2 also depends on the nature of the economic process being analyzed. The R^2 is often high in time-series work because Y and X often have a common trend. By contrast, the R^2 tends to be lower in cross-section work because there is no trend and because of the substantial natural variation in individual behavior. If $R^2 = .443$ were reported for a macroeconomic time-series saving function, it would be judged quite low: experienced researchers expect a regression of aggregate saving on income to have an R^2 of .95 or higher. However, an R^2 of .443 for a large-sample, cross-section saving function would be quite high compared with similar studies, and the regression would be judged to have a relatively good fit.

The R^2 can be used in a limited way to compare the fits of two or more regressions. In comparing regressions, it is natural to note which one is the most successful in explaining the variation in its dependent variable. However, these comparisons must be made with care, especially when the regressions have different dependent variables. A better fit does not necessarily mean a better regression.

For any given set of data and the corresponding estimated regression, R^2 and *SER* are useful measures of goodness of fit. It is common to report both along with the estimated regression in a display like

$$\hat{Y}_i = -0.0386 + 0.0863X_i$$
$$R^2 = .443 \qquad SER = 0.416 \tag{5.27}$$

The R^2 tells us that about 44 percent of the variation in family saving is explained by the regression. The *SER* tells us that the typical error of fit for saving is 0.416 thousand dollars (i.e., $416). In Chapter 11 we develop additional calculations in regression estimation that will be added to this reporting style.

5.5 The Effects of Linear Transformation

The units in which we measure variables are arbitrarily chosen. For example, *GNP* for 1980 might be measured as 1,480,700,000,000 dollars, 1,480.7 billion dollars, or 1.4807 trillion dollars. What impact does this choice have on our regression results? We can show that the economic sense of the fitted regression does not depend on the units of measurement, even though the actual regression coefficients do.

Suppose that we start with data on Y and X. The most common form of units adjustment involves multiplying a variable by a constant. We view this as transforming the original variables, Y and X, into new variables, y and x, according to

$$y_i = b_y Y_i \quad \text{and} \quad x_i = b_x X_i \tag{5.28}$$

For example, if Y is the unemployment rate expressed as a proportion of the labor force and if b_y equals 100, y is the unemployment rate expressed in percentage points.

Using the original data, let the estimated regression of Y on X be denoted by

$$\hat{Y}_i = \hat{\beta}_0 + \hat{\beta}_1 X_i \tag{5.29}$$

And using the transformed data, let the estimated regression of y on x be denoted by

$$\hat{y}_i = \hat{\gamma}_0 + \hat{\gamma}_1 x_i \tag{5.30}$$

The choice of notation here makes it clear that the estimated intercepts may be different in the two regressions and that the estimated slopes may be different also. (Note that γ is lowercase "gamma," the Greek "g.")

The question is: what are the relations between $\hat{\gamma}_0$ and $\hat{\beta}_0$ and between

$\hat{\gamma}_1$ and $\hat{\beta}_1$? In other words, how do the new coefficients compare with the old ones? To answer this, we start with the formulas for the OLS estimators giving $\hat{\gamma}_0$ and $\hat{\gamma}_1$ in terms of y and x, then substitute for y and x what they are in terms of Y and X, and finally rearrange until a useful statement is found. Our results from Chapter 3 regarding transformations of variables are important here. We see that

$$\begin{aligned}\hat{\gamma}_1 &= \frac{\sum (x_i - \bar{x})y_i}{\sum (x_i - \bar{x})^2} \\ &= \frac{\sum (b_x X_i - b_x \bar{X}) b_y Y}{\sum (b_x X_i - b_x \bar{X})^2} \\ &= \frac{b_x b_y \sum (X_i - \bar{X}) Y_i}{(b_x)^2 \sum (X_i - \bar{X})^2} \\ &= \frac{b_y}{b_x} \hat{\beta}_1\end{aligned} \tag{5.31}$$

and

$$\begin{aligned}\hat{\gamma}_0 &= \bar{y} - \hat{\gamma}_1 \bar{x} \\ &= b_y \bar{Y} - \left(\frac{b_y}{b_x} \hat{\beta}_1\right)(b_x \bar{X}) \\ &= b_y (\bar{Y} - \hat{\beta}_1 \bar{X}) \\ &= b_y \hat{\beta}_0\end{aligned} \tag{5.32}$$

Collecting these results, we see that when Y and X are transformed by multiplicative constants as specified in (5.28), the new coefficient estimates are related to the original ones by (5.31) and (5.32).

For example, the saving (Y) and income (X) data used in this chapter are measured in thousands of dollars. If both variables were transformed to express the amounts in dollars, the new variables would be

$$y_i = 1000 Y_i \quad \text{and} \quad x_i = 1000 X_i \tag{5.33}$$

From (5.27), (5.31), and (5.32), we see quickly that the new regression of saving on income would be

$$\hat{y}_i = -38.6 + 0.0863 x_i \tag{5.34}$$

That is, the slope is unchanged but the new intercept is 1000 times greater than the old one. By contrast, if saving were transformed to dollars while income remained in the original form, the regression of saving on income would be

$$\hat{y}_i = -38.6 + 86.3 X_i \tag{5.35}$$

The R^2 in these regressions would be the same as in (5.27), but in this case the original *SER* would be multiplied by a factor of 1000.

The question of units choice can be generalized somewhat to the following problem. Suppose that we have data on Y and X and have estimated the regression

$$\hat{Y}_i = \hat{\beta}_0 + \hat{\beta}_1 X_i \tag{5.36}$$

Suppose now that we create a new variable, y, from Y according to the linear transformation

$$y_i = a_y + b_y Y_i \tag{5.37}$$

and a new variable, x, according to

$$x_i = a_x + b_x X_i \tag{5.38}$$

(Note that a_y and b_y need not be related to a_x and b_x at all; the notation is used just to economize on symbols.) If we estimate an OLS regression of y on x, we get

$$\hat{y}_i = \hat{\gamma}_0 + \hat{\gamma}_1 x_i \tag{5.39}$$

and we ask: How do the new coefficients compare with the original ones? Derivations similar to those above lead to

$$\hat{\gamma}_1 = \frac{b_y}{b_x} \hat{\beta}_1 \tag{5.40}$$

and

$$\hat{\gamma}_0 = a_y + b_y \hat{\beta}_0 - \frac{a_x b_y}{b_x} \hat{\beta}_1 \tag{5.41}$$

For example, suppose that after-tax income (x) is related to before-tax income (X), both expressed in thousands of dollars, by

$$x_i = 2 + 0.75 X_i \tag{5.42}$$

Comparing this to (5.38), we have $a_x = 2$ and $b_x = 0.75$. Suppose now that we are interested in the relation between saving and after-tax income. The saving concept is unchanged: $y_i = Y_i$, so $a_y = 0$ and $b_y = 1$. Without estimating another regression, we can determine from (5.27) that the regression of saving on after-tax income is

$$\begin{aligned} \hat{y}_i &= \left[0 + (1)(-0.0386) - \frac{(2)(1)}{0.75}(0.0863)\right] \\ &\quad + \left[\frac{1}{0.75}(0.0863)\right] x_i \\ &= -0.269 + 0.115 x_i \end{aligned} \tag{5.43}$$

The R^2 of the original and new regressions would be the same; in this case the *SER* is also unchanged because the units of measurement for saving, the dependent variable, are unchanged.

Understanding the effects of units adjustment and linear transformation is useful for us in two ways. First, given an estimated regression, we are able to restate it to make more sensible reports without actually computing new estimates. For example, it generally does not make sense to report a regression of saving on income with one variable measured in dollars and the other in thousands of dollars.

Second, this analysis makes us aware that the magnitude of estimated regression coefficients depend in part on the units of measurement that happen to have been used in the data. Without knowing what the units are in any particular regression, finding a coefficient equal to 3400.0 is no more interesting or important than finding 0.00034 as the estimated value.

Problems

Section 5.1

5.1 In a simple linear regression model of market demand involving the quantity demanded and the price of a product, which variable is the dependent one and which is the explanatory one? Draw a figure showing the true regression line and plot some of the observations that might be observed.

⋆ **5.2** In the simple regression model illustrated in Figure 5.1, is it possible that all the actual Y_i values would lie above the true regression line? Explain.

5.3 Suppose that the quantity demanded in a market depends on the price of the product and the income of consumers. Would a simple regression model explaining the quantity demanded be appropriate? Could income be considered part of the disturbance?

5.4 Suppose that the true regression line is $E[Y] = 2 + 3X$. Determine the value of the disturbance for an observation having (X_i, Y_i) values (3, 8). Determine the disturbance for an observation (6, 21).

Section 5.2

5.5 In an OLS regression estimation, is it possible that all the actual Y_i values would lie above the estimated regression line? Explain.

⋆ **5.6** Construct a diagram to show a case for which $\Sigma\, e_i = 0$ around a line that is clearly not the best-fitting line.

5.7 In the following table Y stands for *EARNS* and X stands for *ED* from the cross-section data set, and the data are for observations 26 through

30 (with *EARNS* rounded). Plot Y and X in a scatter diagram, with Y on the vertical axis.

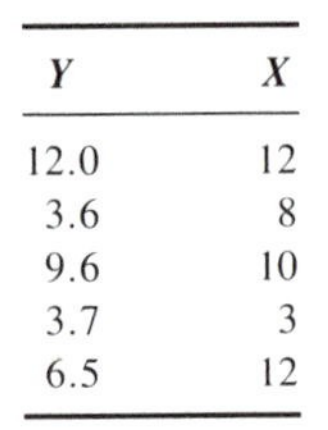

Y	X
12.0	12
3.6	8
9.6	10
3.7	3
6.5	12

5.8 Based on the data of Problem 5.7, estimate the coefficients of the OLS regression of Y on X and graph the estimated regression line through the scatter diagram of Problem 5.7. (Use a table format to organize your calculations.)

5.9 For a data set of three observations whose (X_i, Y_i) values are (10, 5), (8, 7), and (12, 9), estimate the regression of Y on X. Calculate the sum of the residuals and the covariance between e and X. Verify that the regression goes through the point of means.

5.10 Calculate the sum of squared residuals for the estimated regression in Problem 5.9. Now, add 0.5 to the intercept to get another line through the data, and calculate the sum of squared residuals for this line. Which of the two calculated *SSR*s is smaller? Why?

Section 5.3

⋆ **5.11** Draw a scatter diagram of the first five observations on *EARNS* and *ED* from the cross-section data set, and roughly draw in the best-fitting line. How does comparing this to the graph from Problem 5.8 illustrate the existence of estimation errors?

5.12 Based on the estimated regression (5.14), determine the predicted saving of a family whose income is 25 thousand dollars. For what level of income would predicted saving be zero?

Section 5.4

⋆ **5.13** Based on Problem 5.8, compute the values of R^2 and *SER*.

5.14 OLS finds the best-fitting line, while R^2 measures the goodness of fit. Does this mean that R^2 will always be high if the OLS technique is used?

5.15 Prove that the third term in Equation (5.23) is equal to zero. [*Hint:* At one stage use the fact that $\Sigma\, e_i X_i = 0$, which stems from Equation (5.48).]

5.16 Compute the values of R^2 and *SER* for the regression estimated in Problem 5.9.

5.17 In Figure 5.3 and Table 5.2, the observation with the large residual could be called an "outlier." Recompute the estimated regression, the

R^2, and the *SER* after deleting this outlier (i.e., using only four observations) and compare the results to those gathered in Equation (5.27).

Section 5.5

★ **5.18** Based on Problems 5.7 and 5.8, suppose that *EARNS* is transformed into dollars (from thousands of dollars). What would be the new estimated regression of earnings on *ED*?

5.19 Based on the saving function presented in Equation (5.27), what would be the estimated regression of saving on income if income were transformed to dollars but saving remained in the original form?

5.20 Derive the relations presented as Equations (5.40) and (5.41).

Appendix

5.21 Prove that Equations (5.54) and (5.53) are equivalent. [*Hint:* Start from the numerator in (5.53).]

5.22 Show why Equation (5.48) is equivalent to stating that e and X are uncorrelated.

★ **5.23** Consider a regression model in which β_0 is specified to be zero: $Y_i = \beta_1 X_i + u_i$. Derive the OLS estimator for β_1.

APPENDIX*

Derivation of the OLS Estimators

The derivation of the OLS estimators is a standard calculus minimization problem in which the objective function to be minimized is the sum of squared residuals [see (5.9)]. Given a set of data, all the values like Y_i and X_i are fixed numbers. Our task is to find the values of $\hat{\beta}_0$ and $\hat{\beta}_1$ that make *SSR* as small as possible. In this context, $\hat{\beta}_0$ and $\hat{\beta}_1$ are variables (arguments of the function) while the Y's and X's are constants.

The sum of squared residuals is given by

$$\sum_{i=1}^{n} e_i^2 = \sum_{i=1}^{n} (Y_i - \hat{\beta}_0 - \hat{\beta}_1 X_i)^2 \tag{5.44}$$

To find the values of $\hat{\beta}_0$ and $\hat{\beta}_1$ that minimize this expression, we take the partial derivatives of (5.44) with respect to $\hat{\beta}_0$ and $\hat{\beta}_1$ and set them equal to zero:

$$\frac{\partial}{\partial \hat{\beta}_0}\left[\sum (Y_i - \hat{\beta}_0 - \hat{\beta}_1 X_i)^2\right] = 0 \tag{5.45}$$

*This appendix is relatively difficult and can be skipped without loss of continuity.

and

$$\frac{\partial}{\partial \hat{\beta}_1}\left[\sum (Y_i - \hat{\beta}_0 - \hat{\beta}_1 X_i)^2\right] = 0 \tag{5.46}$$

Evaluating these partial derivatives gives us

$$-2 \sum (Y_i - \hat{\beta}_0 - \hat{\beta}_1 X_i) = 0 \tag{5.47}$$

$$-2 \sum (Y_i - \hat{\beta}_0 - \hat{\beta}_1 X_i) X_i = 0 \tag{5.48}$$

Next, we divide each equation by -2, leaving each side equal to zero, and then rearrange terms to get

$$\sum Y_i = n\hat{\beta}_0 + \hat{\beta}_1 \sum X_i \tag{5.49}$$

$$\sum Y_i X_i = \hat{\beta}_0 \sum X_i + \hat{\beta}_1 \sum X_i^2 \tag{5.50}$$

This pair of equations is a set of two simultaneous equations in which $\hat{\beta}_0$ and $\hat{\beta}_1$ are unknown and all the X_i and Y_i values are known.

A convenient way to solve these equations is to solve (5.49) for

$$\hat{\beta}_0 = \left\{\sum Y_i - \hat{\beta}_1 \sum X_i\right\} / n \tag{5.51}$$

and substitute this for $\hat{\beta}_0$ in (5.50). That equation then can be solved to yield

$$\hat{\beta}_1 = \frac{n \sum X_i Y_i - \sum X_i \sum Y_i}{n \sum X_i^2 - \left(\sum X_i\right)^2} \tag{5.52}$$

For theoretical and computational convenience, we note that this expression can be arranged in various ways, including

$$\hat{\beta}_1 = \frac{\sum (X_i - \overline{X})(Y_i - \overline{Y})}{\sum (X_i - \overline{X})^2} \tag{5.53}$$

and

$$\hat{\beta}_1 = \frac{\sum (X_i - \overline{X}) Y_i}{\sum (X_i - \overline{X})^2} \tag{5.54}$$

which is the same as (5.12). Now (5.51) can be manipulated to yield

$$\hat{\beta}_0 = \overline{Y} - \hat{\beta}_1 \overline{X} \tag{5.55}$$

which is the same as (5.13). We could find instead an expression that gives $\hat{\beta}_0$ solely in terms of X and Y values, but this is not useful.

The property that $\Sigma\, e_i = 0$ follows from (5.47) and that e and X are uncorrelated follows from (5.48).

Equivalence of R^2 and $(r)^2$

The value of R^2 is exactly equal to the square of r, the correlation between X and Y. To see this, first note that

$$
\begin{aligned}
\sum_{i=1}^{n} (\hat{Y}_i - \overline{Y})^2 &= \sum_{i=1}^{n} (\hat{\beta}_0 + \hat{\beta}_1 X_i - \hat{\beta}_0 - \hat{\beta}_1 \overline{X})^2 \\
&= \sum_{i=1}^{n} [\hat{\beta}_1 (X_i - \overline{X})]^2 \qquad (5.56) \\
&= (\hat{\beta}_1)^2 \sum_{i=1}^{n} (X_i - \overline{X})^2
\end{aligned}
$$

Now,

$$
\begin{aligned}
R^2 &= \frac{\sum_{i=1}^{n} (\hat{Y}_i - \overline{Y})^2}{\sum_{i=1}^{n} (Y_i - \overline{Y})^2} \\
&= \frac{(\hat{\beta}_1)^2 \sum (X_i - \overline{X})^2}{\sum (Y_i - \overline{Y})^2} = (\hat{\beta}_1)^2 \frac{\sum (X_i - \overline{X})^2}{\sum (Y_i - \overline{Y})^2} \\
&= \left[\frac{\sum (X_i - \overline{X})(Y_i - \overline{Y})}{\sum (X_i - \overline{X})^2} \right]^2 \frac{\sum (X_i - \overline{X})^2}{\sum (Y_i - \overline{Y})^2} \qquad (5.57) \\
&= \frac{\left[\sum (X_i - \overline{X})(Y_i - \overline{Y}) \right]^2}{\sum (X_i - \overline{X})^2 \sum (Y_i - \overline{Y})^2} \\
&= \left[\frac{\sum (X_i - \overline{X})(Y_i - \overline{Y})}{\sqrt{\sum (X_i - \overline{X})^2} \sqrt{\sum (Y_i - \overline{Y})^2}} \right]^2 \\
&= (r)^2
\end{aligned}
$$

Since we showed earlier that the maximum value of R^2 is 1 and the minimum is 0, the equivalence of R^2 and $(r)^2$ proves that the maximum value of r is 1 and the minimum is -1.

Simple Regression: Application

The simple regression model developed in Chapter 5 can be used to estimate the structural (behavioral) relation between two variables whenever we believe that the model accurately describes the process by which Y is determined. Although this model is too simple to describe most economic relations, it has many useful applications. In all these, the task of estimating the unknown coefficients is the same as in the general case, and the technique needs no further elaboration. What remains is to interpret the estimated model, and the first two sections of this chapter focus on that problem.

When we believe that the true process is more complex than the basic model, there are several paths open to us. If the true relation involves two variables in a nonlinear fashion, it is sometimes possible to transform the relation into a linear one and then apply the basic techniques already developed. Much of this chapter is devoted to exploring these possibilities. If more than one variable plays a systematic role in the determination of Y, we are led to multiple regression (Chapter 7).

6.1 Interpretation of the Coefficients

As presented in Chapter 5, the simple regression model

$$Y_i = \beta_0 + \beta_1 X_i + u_i \tag{6.1}$$

is a theoretical statement of the process by which the value of Y for each observation is determined on the basis of the given value of X for that observation. The systematic part of the relation is specified by the true regression line

$$E[Y] = \beta_0 + \beta_1 X \tag{6.2}$$

where $E[Y]$ denotes the expected value of Y associated with any particular value of X. This is graphed in Figure 6.1, which illustrates that $E[Y]$ is a linear function of X.

When we have data on Y and X, an estimated regression line

$$\hat{Y} = \hat{\beta}_0 + \hat{\beta}_1 X \tag{6.3}$$

can be determined by the method of ordinary least squares. This line is the empirical counterpart of the true regression line, and it serves as an estimated model of the systematic relation between Y and X. We should keep in mind, however, that even in the best of circumstances this model is only an estimate of the true relation—in the sense that $\hat{\beta}_0$ and $\hat{\beta}_1$ are only estimates of β_0 and β_1. Figure 6.1 illustrates an estimated regression line that (hypothetically) is based on data produced by the process described by the true regression line in the same figure. As explained in Section 5.3, the differences between the estimated and true regressions arise from the particular pattern of disturbances in the data set.

After estimation, our interest moves on to interpreting and using the estimated model. For this we blend together economic reasoning and some mathematical analysis. Since (6.3) is the equation of a straight line, the mathematics is easy.

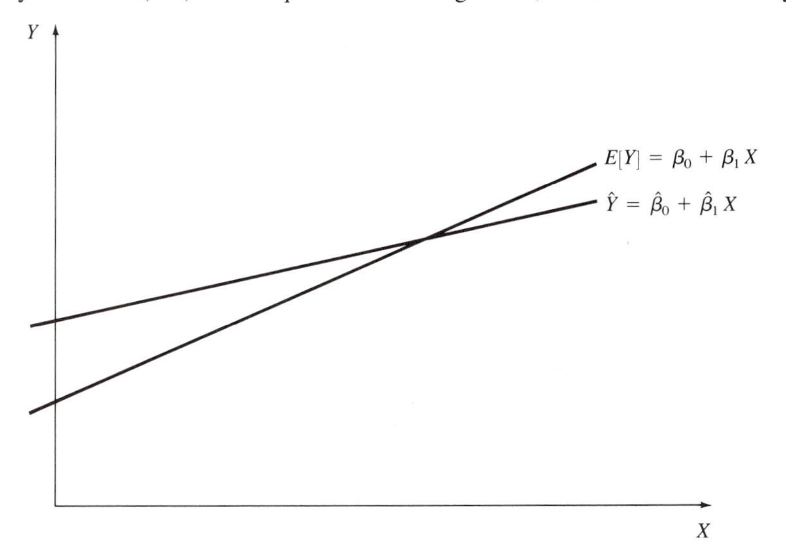

FIGURE 6.1 The systematic part of the simple regression model is specified by the true regression line, whose intercept is β_0 and whose slope is β_1. An estimated regression line fit to data produced by this process has an intercept of $\hat{\beta}_0$ and a slope of $\hat{\beta}_1$.

The parameter $\hat{\beta}_0$ in the estimated regression is called the ***intercept*** because it is the intercept of the estimated regression line with the vertical axis drawn through $X = 0$. Strictly speaking, the intercept gives the predicted value of Y for an observation with $X = 0$. In some cases this interpretation is of special interest. However, in many cases such an interpretation does not make economic sense. For example, with an aggregate consumption function it would be absurd to use a real-world model to predict what consumption would be if income were equal to zero: in all probability, either everyone would starve or chaos would reign as inventories of food and products were consumed.

The parameter $\hat{\beta}_1$ is called the ***slope coefficient.*** Based on the simple analytics of a straight line (see the appendix to Chapter 1), we know that the change in predicted Y that would be associated with any particular change in X is given by

$$\Delta\hat{Y} = \hat{\beta}_1 \, \Delta X \tag{6.4}$$

In this context, a "change in X" that is economically interesting might be the difference between the values of X for two observations or it might be a hypothetical change in the value of X for a particular observation. In either case, we are considering moving from one point to another along the estimated regression line.

Based on (6.4), we can see that if X changes by 1, $\hat{Y}$ will change by $\hat{\beta}_1$. That is,

$$\text{if} \quad \Delta X = 1, \quad \text{then} \quad \Delta\hat{Y} = \hat{\beta}_1 \tag{6.5}$$

This simple relation is the basis of our usual interpretation of a slope coefficient: $\hat{\beta}_1$ is the change in $\hat{Y}$ that results from a unit change in X. Meaning this, we say that $\hat{\beta}_1$ is the ***impact*** of X on $\hat{Y}$. (Note that a "unit change in X" means $\Delta X = 1$; this is different from a change in the units of measurement for X.)

When we believe that the form of the model correctly specifies the true process relating Y to X, which is a belief that we usually hold, we can recast our interpretations in terms of the true parameters. For example, since $\hat{\beta}_1$ is an estimate of the true β_1, we might say that $\hat{\beta}_1$ is an estimate of the impact of X on $E[Y]$.

A semantic problem arises in discussing the impact of the explanatory variable in a regression. It is tempting to use the word "significant" when the impact is big or noteworthy in an economic sense. Unfortunately, this use of "significant" in statistical discussion has been preempted by a technical meaning that is different from ordinary usage. As set out in Chapter 12, the meaning of "significant" in a basic hypothesis test is simply "not zero," and to use the word to discuss the importance of the impact of one variable on another can lead only to confusion.

Finally, since $\hat{\beta}_1$ estimates the impact of X on Y, it is tempting to jump to the conclusion that the larger is the $\hat{\beta}_1$ value in a regression the more important is X in explaining Y. This conclusion is valid when thinking about possible values for the coefficient in a particular regression, but it is invalid when thinking about a comparison of coefficients in different regressions. Indeed, as shown in Section 5.5, the magnitude of a coefficient is affected by the units of measurement of the variables in the data. Therefore, the magnitude by itself does not indicate the importance of the coefficient.

6.2 The Earnings Function and the Consumption Function

In this section we present two examples of regular simple regression that serve as small case studies of econometric research. In both cases we start with a theory or belief about the form of the relation. The actual calculations are not shown, because they are standard applications of the methods of Chapter 5 and because we nearly always rely on a computer to carry them out anyway. The emphasis is on interpreting and using the estimated model.

The Earnings Function

We start with a simple theory that labor markets serve to determine persons' earnings according to their educational attainment as specified in

$$EARNS_i = \beta_0 + \beta_1 ED_i + u_i \tag{6.6}$$

using mnemonic variable names instead of Y and X. The theory might be based on the idea that education enhances productivity, which is rewarded in labor markets with greater earnings. Alternatively, it might be based on the ideas that educational attainment mainly certifies the existence of potential abilities, and that earnings are based on this certification. In either case, (6.6) is an appropriate model.

Using the 100 observations in our cross-section data set, in which *EARNS* measures earnings in thousands of dollars and *ED* measures years of schooling, and applying (5.12) and (5.13) to calculate the OLS coefficients, we find that

$$\begin{aligned} \widehat{EARNS}_i &= -1.315 + 0.797 ED_i \\ R^2 &= .285 \qquad SER = 4.361 \end{aligned} \tag{6.7}$$

The estimated coefficient on *ED* is $\hat{\beta}_1^* = 0.797$, which means that the estimated labor market reward for an extra year of schooling is 0.797 thousand dollars greater annual earnings. Put more simply, we have estimated that the impact of education on earnings is \$797 greater annual earnings for each additional year of schooling. Is this an important effect? Certainly it is not trivial. Given the relative ease of acquiring another year of education and taking into account the magnitude of this impact in comparison with the variation in earnings that exists among the observations, most economists (and educators) would say that education has a fairly important effect on earnings in this model. (We point out that the fact that even though 0.797 looks like a small number, this alone tells us nothing about the importance of the coefficient.)

The intercept $\hat{\beta}_1^* = -1.315$ tells us that the predicted earnings of a person with no schooling is negative $1315. Since labor markets do not offer negative earnings, this unrealistic finding needs careful attention. One possibility is that a worker with no schooling would indeed have negative productivity in a job, and so he would not be hired. In this case the estimated model is in accord with the true behavior of labor markets, but the naive interpretation of the intercept is misleading because it attempts to apply the model for a value of *ED* that is not appropriate. Another possibility is that workers with no schooling do indeed have positive earnings in labor markets, but that the particular outcomes for the disturbances in our data have led to a negative estimate for the intercept even though the true value is positive. Yet another possibility is that our model (6.6) is incorrect—the true form might be nonlinear, with positive earnings for workers with no education—and that the attempt to fit a straight line to the data has led to an unrealistic result. It turns out that in our data the minimum value for *ED* is 2 years, for which the estimated model predicts positive earnings. Thus we might proceed with some faith in the model and its estimation. (However, the issue unresolved, and we note that our attempt to make a careful interpretation of an estimated coefficient has led us to recognize the need for more research.)

The R^2 indicates that only about 30 percent of the variation in earnings among the observations is explained by the level of educational attainment. This may seem low, but it is similar to R^2 values found in other cross-section studies of earnings.

The relatively poor fit is also evidenced by the *SER*, which is 4.361 thousand dollars. To interpret this, the magnitude must be compared to values of *EARNS*, which is the dependent variable. In the data, *EARNS* ranges from 0.750 to 30.000 and has a mean of 7.911. Taking the mean value as a standard for comparison, the typical error of fit indeed seems quite large.

Using (6.7) we can predict that the earnings in 1963 of a college graduate ($ED = 16$) who was a male head of family aged 25–54 would have been

$$\widehat{EARNS} = -1.315 + (0.797)(16) = 11.437 \tag{6.8}$$

thousand dollars. The typical error associated with an out-of-sample prediction such as this turns out to be even larger than the standard error of the regression, *SER*. (In Chapter 13 we see how these typical prediction errors are calculated.)

It should be noted that our data pertain only to male heads of families in the 25–54 age range. Making predictions or assessing the impact of education on the basis of this estimated earnings function is appropriate only with reference to this particular group. We do not expect that it will adequately predict women's earnings, for example. Also, the dollar magnitudes are based on 1963 labor market conditions.

The Consumption Function

As discussed in Chapter 1, an aggregate consumption function based on simplified Keynesian ideas can be specified as

$$CON_i = \beta_0 + \beta_1 DPI_i + u_i \tag{6.9}$$

where CON_i is aggregate personal consumption expenditure in year i, and DPI_i is aggregate disposable personal income in the same year. β_1 is interpreted as the marginal propensity to consume, because (6.9) implies that if DPI increases by 1 dollar, then $E[CON]$ will increase by β_1 dollars (assuming that both variables are measured in the same units).

Using the 25 observations for the years 1956–1980 in our time-series data set, in which CON and DPI are both measured in billions of 1972 dollars, and calculating the ordinary least squares estimates, we find that

$$\begin{aligned} \widehat{CON}_i &= 0.568 + 0.907 DPI_i \\ R^2 &= .997 \qquad SER = 8.935 \end{aligned} \tag{6.10}$$

The estimated marginal propensity to consume ($\hat{\beta}_1^*$) is 0.907, which is consistent with Keynes' conjecture. The fit is extraordinarily good: the R^2 of .997 means that nearly 100 percent of the variation in CON over this period is explained by the regression (i.e., is explained by variation in DPI). Although it seems that we might have discovered some fundamental economic law, judgment must be reserved on this question. Very high R^2 values are common in time-series studies because most variables tend to increase over time, and therefore high correlations will exist among them even if cause-and-effect relations are absent or weak. Also, other specifications of the process determining consumption behavior may be preferred in economic research.

We can use (6.10) to predict how high CON will be when DPI is 1.2 trillion dollars:

$$\widehat{CON} = 0.568 + (0.907)(1200) = 1089 \tag{6.11}$$

billion dollars. Also, if DPI were to decrease by 20 billion dollars (from whatever level it might be at), the estimated model predicts a change in consumption of

$$\Delta\widehat{CON} = (0.907)(-20) = -18.14 \tag{6.12}$$

billion dollars. Note that the estimated intercept plays no role in a calculation like this.

6.3 Alternative Model Specifications

As already noted, the appropriateness of using the regular simple regression model is contingent on its being an accurate description of the particular process being studied. Most important, the model requires that for each observation the expected value of Y be a linear function of the value of X. In the earnings function and consumption function, this was taken to be an appropriate specification of the relations.

In other cases, however, theory and evidence may lead us to believe that the relation between two variables is definitely not linear, or that it is not contemporaneous. Thus the regular model will not be an accurate description of the process, and using it will be inappropriate. (This statement might be tempered with the notion that if the relation is *approximately* described by the regular model, its use can be considered appropriate.)

In this section and the next we see how some structural relations that do not conform to the regular model can be respecified in such a way that OLS can be applied. The key to this is realizing that the Y and X in the regular model (6.1) need not be the interesting variables themselves, but that they can be variables that are constructed from the interesting variables. To reduce confusion, it is useful to refer to Y as the ***regressand*** rather than the dependent variable and to X as the ***regressor*** rather than the explanatory (or independent) variable.

Ratios of Variables

Part of the controversy surrounding the Keynesian consumption theory focused on the average propensity to consume (APC), which is defined as

$$APC_i = \frac{CON_i}{DPI_i} \tag{6.13}$$

One tradition and body of evidence viewed the consumption–income ratio as an economic constant that did not change over time. By contrast, Keynes conjectured that the APC would decline over time.

One way to look at the data and provide evidence on this question begins by assuming that the average propensity to consume is a simple linear function of time. The variables CON and DPI from the time-series data set are used to construct a new variable, APC, defined by (6.13). A regression model is specified and then estimated using APC as the regressand and the time trend T as the regressor. With $n = 25$, the results are

$$\begin{aligned} \widehat{APC_i} &= 0.911 - 0.000213T_i \\ R^2 &= .021 \qquad SER = 0.011 \end{aligned} \tag{6.14}$$

Sometimes regressions like this one are reported as

$$\widehat{\left(\frac{CON_i}{DPI_i}\right)} = 0.911 - 0.000213T_i \tag{6.15}$$

to emphasize that the regressand is constructed as a ratio of two variables.

These results show that the average propensity to consume decreases over time, seemingly in conformity with the Keynesian view. Each year, the *APC* is estimated to decrease by about 0.000213. Is this a lot or a little? In the middle year of the sample (1968, with $T = 13$) the predicted *APC* is $0.911 - (0.000213)(13) = 0.908$, and the annual decrease (0.000213) is very small compared with this. The annual decrease seems quite unimportant. Hence some persons might be inclined to say that for all practical purposes the average propensity to consume is constant.

The Reciprocal Specification

Often economic theory predicts that the systematic relation between two variables is nonlinear. If the nonlinearity is judged to be not too severe, it might be reasonable to proceed with the regular simple regression model as an approximation. However, this is not usually advisable because it inhibits our investigating the nonlinear features. An alternative approach involves finding a nonlinear mathematical form to specify a relation that is appropriate for the economic process and that is transformable into a linear relation.

One possibility is the ***reciprocal*** relation

$$Y = \beta_0 + \frac{\beta_1}{X} \tag{6.16}$$

The geometry of the reciprocal relation is illustrated in the left side of Figure 6.2. Y may be positive or negative. We focus only on cases in which all the X values are positive; although the reciprocal relation is defined for negative X values, it is rarely used in these cases. If β_1 is negative, the slope of the relation between Y and X is positive and it becomes flatter as X increases. Y never rises above the value $Y = \beta_0$, and in fact it never quite reaches it. If β_1 is zero, Y is constant. If β_1 is positive, the slope of the relation between Y and X is negative and becomes flatter as X increases. Y never reaches or falls below the value $Y = \beta_0$.

Now consider a new variable, *XINV*, that is equal to the reciprocal (i.e., inverse) of X: $XINV = 1/X$. It follows from (6.16) that Y is a linear function of *XINV*:

$$Y = \beta_0 + \beta_1 XINV \tag{6.17}$$

The relation between Y and *XINV* is illustrated on the right side of Figure 6.2, with three cases depending on the value of β_1. We see that the specific form of the nonlinear relation (6.16) implies a linear relation between Y and *XINV*.

The potential for applying this bit of mathematical analysis to econometric regression modeling should be clear. If we have a theory or belief that the systematic part of the relation between Y and X is reciprocal, like (6.16), this theory can be reexpressed to state that Y is linearly related to *XINV*. Further,

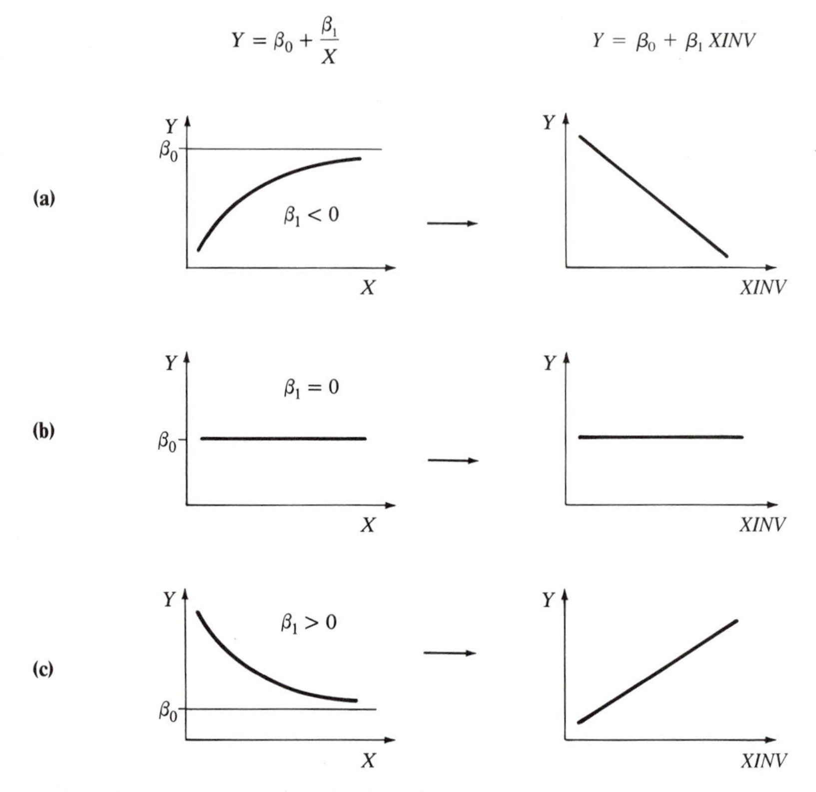

FIGURE 6.2 The geometry of the reciprocal relation depends on the sign of β_1. Except when $\beta_1 = 0$, Y is a nonlinear function of X; as X increases, Y increases ($\beta_1 < 0$) or decreases ($\beta_1 > 0$) and approaches the asymptotic limit β_0. Although Y is a nonlinear function of X, it is a linear function of the inverse of X—which is denoted by *XINV*.

the parameters β_0 and β_1 of the linear relation (between Y and *XINV*) are the same as the parameters of the reciprocal relation (between Y and X). If we add a disturbance term to (6.17), we have the specification of a regression model.

For example, before the combined high inflation and unemployment of the 1970s, the systematic part of the Phillips curve relation between the rate of inflation and the unemployment rate was considered to be nonlinear and resemble the left side of Figure 6.2c. Hence a reciprocal relation between inflation and unemployment was considered to be an appropriate form, and we use 15 annual observations (1956–1970) to estimate the model.

The regressand is simply *RINF1*. The regressor is a new variable, equal to the inverse of *UPCT*:

$$UINV_i = \frac{1}{UPCT_i} \tag{6.18}$$

The estimated regression is

$$\widehat{RINF1}_i = -1.984 + 22.234UINV_i \qquad R^2 = .549 \qquad SER = 0.956 \tag{6.19}$$

Commonly, results such as these are reported simply with $1/UPCT$ in place of $UINV$, but we wish to stress that the actual regressor is a transformed variable.

The results of regressions involving transformed variables require care in interpretation. Since $\hat{\beta}_1^*$ is positive, the actual estimated regression resembles the right side of Figure 6.2c. However, the economic interpretation is best carried out in terms of the implied relation between predicted *RINF1* and *UPCT*, which resembles the left side of Figure 6.2c. Looking first at the estimated intercept, we see that as *UPCT* increases, *RINF1* decreases and approaches a lower limit of -1.984 percent. The sign of the estimated slope tells us the general shape (here, like the left side of Figure 6.2c), but the magnitude is difficult to interpret. To understand the meaning of this coefficient, it is useful to make predictions for the rate of inflation at various levels of *UPCT*. For example, if $UPCT = 6$ percent, the predicted *RINF1* is $-1.984 + (22.234)(1/6) = 1.72$ percent; if $UPCT = 3$ percent, the predicted *RINF1* is $-1.984 + (22.234)(1/3) = 5.43$ percent. The estimated regression is consistent with earlier findings on the Phillips curve.

The restriction of the data to observations from the period 1956–1970 reflects the fundamental econometric requirement that all the data used to estimate a model must have been generated from the same economic process. We believe that after 1970 expectations of continuing inflation began to get built into the behavior of the economy in a way that they had not been in the earlier period. In other words, the behavioral pattern changed. This means that although the model (6.19) may be quite appropriate for the earlier period, it cannot be used to make predictions after 1970. For this later period, a more complicated model is required.

Lagged Variables and First Differences

The regular simple regression model is specified so that Y_i is related to X_i. In a time-series context, this means that the values of Y and X are to be measured in the same time period. However, economic behavior is dynamic, and an effect may occur substantially later than its cause. For example, a firm's investment decision may be made at one point in time but the machines might not be produced and delivered until a year or more later. Similarly, people may budget their consumption expenditures on the basis of last year's income rather than its current level. In these cases the explanatory variable is said to determine the dependent variable with a ***lag.***

If there is a one-period lag between cause and effect, it is natural to formulate a simple regression model as

$$Y_i = \beta_0 + \beta_1 X_{i-1} + u_i \tag{6.20}$$

where X_{i-1} is the value of X one period before i. In other words, the current (ith period) value of Y depends on the one-period lagged value of X and the current disturbance. This formulation appears to complicate the estimation of the coefficients, because the paired Y_i, X_{i-1} values are no longer a row in a rectangular data matrix and the OLS estimators are no longer appropriately defined.

However, the creation of a new regressor straightens out these difficulties. Table 6.1 illustrates the procedure. The table shows some data on two variables Y and X. A new regressor is created in such a way that each of its values is equal to the previous period's value of X. Appropriately, this new regressor is called *XLAG*. In period i, the value $XLAG_i$ is equal to X_{i-1}, and

$$Y_i = \beta_0 + \beta_1 XLAG_i + u_i \tag{6.21}$$

has the same meaning as (6.20). In this specification the earlier difficulties disappear, and the coefficients can be estimated in the ordinary way. It should be noted that in the construction of *XLAG*, no value could be assigned to $XLAG_1$ because X_0 is not in the data set. Hence in estimating (6.21) we must ignore the first observation in the data, and use only 2 through n.

For example, using observations 2 through 25 of the time-series data set, an aggregate consumption function embodying the theory that *DPI* affects *CON* with a one-period lag is estimated as

$$\begin{aligned} \widehat{CON}_i &= 10.913 + 0.923 DPILAG_i \\ R^2 &= .993 \qquad SER = 14.953 \end{aligned} \tag{6.22}$$

The results differ slightly from the contemporaneous model (6.10). The standard error of regression, which measures the typical error of fit, is about 50 percent greater here, but the R^2 is only slightly lower. Although the original specification provides a better fit, we do not have a statistical basis yet for choosing between

TABLE 6.1 Construction of the Lag Regressor

i	Y	X	$XLAG$
1	405.4	446.2	—
2	413.8	455.5	446.2
3	418.0	460.7	455.5
⋮			
$i-1$	Y_{i-1}	X_{i-1}	$XLAG_{i-1}$
i	Y_i	X_i	$XLAG_i$
⋮			

the two. This lag specification is so common and easy to understand that reports of equations like (6.22) often are written with DPI_{i-1} rather than $DPILAG_i$ on the right-hand side, because no confusion is likely to occur.

Another common specification in time-series modeling involves letting the regressand or regressor, or both, be the ***first difference*** of a variable. The first difference is a constructed variable whose value in any period is equal to the value of the original variable in that period minus its value in the previous period. For example, starting with data on *GNP*, the new variable ΔGNP is defined by

$$\Delta GNP_i = GNP_i - GNP_{i-1} \qquad (6.23)$$

In other words, the first difference for *GNP* is equal to *GNP* minus the lagged value of *GNP*, for each and every observation. (Note that if we have n observations on *GNP*, the values of ΔGNP are defined only for observations 2 through n.)

The accelerator theory of aggregate investment behavior is based on the idea that changes in the level of *GNP* are the main determinant of the level of investment spending by business. This leads to a simple regression model

$$INV_i = \beta_0 + \beta_1 \, \Delta GNP_i + u_i \qquad (6.24)$$

where *INV* is real gross private domestic investment. Using observations 2 through 25 from the time-series data set in Chapter 2, the estimated model is

$$\begin{aligned} \widehat{INV}_i &= 136.6 + 0.691 \, \Delta GNP_i \\ R^2 &= .175 \qquad SER = 40.4 \end{aligned} \qquad (6.25)$$

The positive intercept gives the estimated amount of investment (136.6 billion dollars) that would occur if *GNP* were not growing (i.e., if $\Delta GNP = 0$); this might reflect investment to replace depreciated assets. The estimated slope is substantially smaller than predicted by simple accelerator theory, which suggests that (6.24) might not be an appropriate model of investment behavior.

6.4 Logarithmic Functional Forms*

As seen in Section 6.3, in some cases a transformation of a nonlinear relation leads to an equivalent linear relation involving newly created variables. The benefit of this is that a linear regression model can be used to estimate the parameters. The focus of our interest and interpretation, however, remains with the original relation.

A special class of nonlinear relations become linear when they are transformed with logarithms. The wide range of nonlinearities that can be captured

*This section is relatively difficult and can be skipped without loss of continuity.

and the associated ease of interpretation make these specifications very popular with applied econometricians. These relations and their transformations are explored in this section, and the appendix to this chapter contains the derivations of some of the interpretations given here.

The Log-Linear Specification

Suppose that we think the exact relation between Y and X is

$$Y = e^{\beta_0} X^{\beta_1} \tag{6.26}$$

where e^{β_0} stands for any positive constant. If we take the natural logarithm of both sides of the equation, we obtain

$$\ln Y = \beta_0 + \beta_1 \ln X \tag{6.27}$$

This is known as the ***log-linear*** relation between Y and X because it is linear in the logarithms of the original variables. Since (6.27) involves the logarithm of Y and X, the relation is applicable only if all the values of Y and X are positive: none can be zero, none can be negative.

The log-linear relation is particularly useful because of the variety of graphical shapes that it can represent. Figure 6.3 shows the geometry of (6.26) and its logarithmic transformation (6.27) when β_1 takes on different values. If β_1 is negative, the relation between Y and X is downward sloping and its slope becomes flatter as X increases. If β_1 is zero, then Y is just a constant. If β_1 is between 0 and 1, then the relation between Y and X extends out from the origin and slopes upward, but the slope becomes flatter as X increases. If β_1 is equal to 1, the relation between Y and X passes through the origin and is linear; Y is proportional to X. If β_1 is greater than 1, the relation between Y and X extends out from the origin and slopes upward, but the slope becomes steeper as X increases. In all cases, β_0 is a factor that affects all possible observations equiproportionally.

Perhaps the most attractive feature of this model is that β_1 can be directly interpreted as the ***elasticity*** of Y with respect to X. In economics, this elasticity is equal to the proportional change in Y divided by the proportional change in X resulting from a movement along the relation between Y and X. As shown in the appendix to this chapter, the specification underlying (6.26) and (6.27) is such that the elasticity is the same everywhere along the whole function when the considered changes in Y and X are small. That is, the point elasticity is constant, and it is exactly equal to β_1:

$$\beta_1 = \frac{dY/Y}{dX/X} \tag{6.28}$$

where dY and dX can be thought of as small changes (Δ's) in Y and X, respectively. Hence the log-linear relation is sometimes called the ***constant-elasticity*** relation.

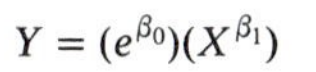

$Y = (e^{\beta_0})(X^{\beta_1})$

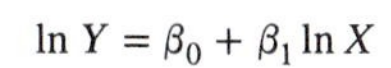

$\ln Y = \beta_0 + \beta_1 \ln X$

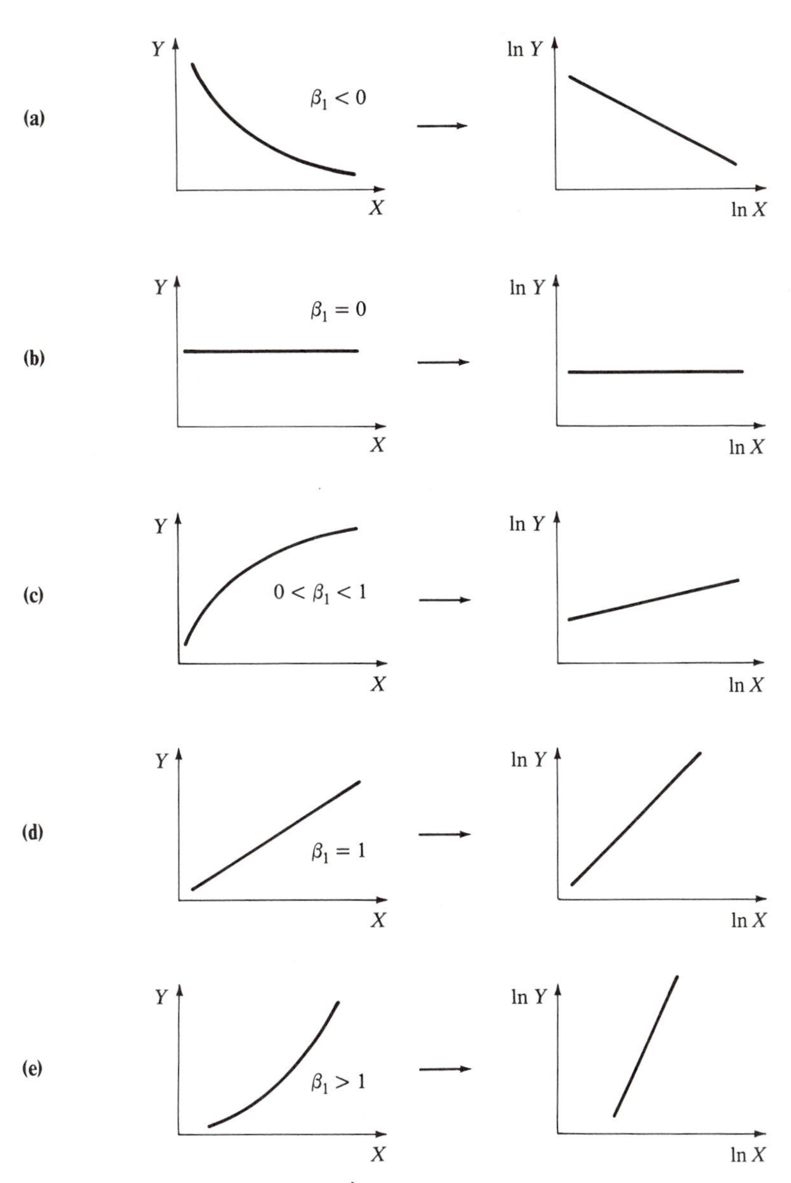

FIGURE 6.3 The geometry of the log-linear relation depends on the sign of β_1. When Y decreases as X increases [case (a), with $\beta_1 < 0$], it is concave upward. When Y increases with X (with $\beta_1 > 0$), the concavity may be upward or downward, depending on the magnitude of β_1. Although Y is a nonlinear function of X, $\ln Y$ is a linear function of $\ln X$; the slope of that line is the same β_1 as in the original formulation. The parameter β_1 is the elasticity of Y with respect to X.

When the elasticity of the relation is known, it provides the basis for calculating the effect on Y of changes in X. Let "pc of Y" be the proportionate change in Y, which might otherwise be denoted by $\Delta Y/Y$. The definition of elasticity implies that

$$\text{pc of } Y \approx (\beta_1)(\text{pc of } X) \tag{6.29}$$

For example, if the elasticity (β_1) equals 1.2, a 5 percent change in X will lead to a 6 percent change in Y. [The "pc" should be entered into (6.29) as a regular proportion—not in percentage points; that is, it should be entered like .05, not 5. However, it is conventional to verbalize a regular proportion like .05 as "5 percent." It might be noted that entering the "pc" in percentage points will not lead to an error in (6.29), but doing so in similar formulas will.]

So far we have focused on an exact theoretical relation between variables Y and X. It is easy to see that the log-linear relation can be taken as the basis of a simple regression model

$$\ln Y_i = \beta_0 + \beta_1 \ln X_i + u_i \tag{6.30}$$

where the regressand is $\ln Y$ and the regressor is $\ln X$. This econometric model is appropriate if the process determining Y is such that the expected value of $\ln Y$ is a linear function of $\ln X$.

For example, consider the aggregate demand for money in the United States. Good economics leads us to specify the model in real terms, so we construct the variable M to be the real quantity of money (see Section 2.3), based on variables in our data set:

$$M_i = \frac{M1_i}{PGNP_i}(100) \tag{6.31}$$

Economic theory suggests that it is reasonable to specify the constant-elasticity form

$$\ln M_i = \beta_0 + \beta_1 \ln GNP_i + u_i \tag{6.32}$$

as the regression model. It is important to realize that the regressand and regressor of the model are transformed variables: they are the logarithms of M and GNP, respectively. For notational convenience we let the names of the new variables be LNM and $LNGNP$.

Using the 25 observations in the time-series data set, the estimated regression is

$$\begin{aligned} \widehat{LNM}_i &= 3.948 + 0.215 LNGNP_i \\ R^2 &= .780 \qquad SER = 0.0305 \end{aligned} \tag{6.33}$$

Since $0 < \hat{\beta}_1^* < 1$, we see that the implied relation between M and GNP resembles the left side of Figure 6.3c. The estimated income elasticity of the demand for money is 0.215; in other words, if GNP increases by 1 percent, we predict that the demand for money will increase by 0.215 percent. Using (6.29) we see that a 5 percent increase in *GNP* leads to a 1.075 percent increase in predicted M. However, when we start to consider large changes in *GNP* or M, the calculation based on the point elasticity holds only approximately.

Now, suppose that we wish to predict the demand for money when $GNP = 1000$. First, we calculate $\ln(1000) = 6.908$. Then we determine the predicted value of $\ln M$: $3.948 + (0.215)(6.908) = 5.433$. Finally, we take the antilog of this number, yielding a predicted demand for money of 228.8 billion dollars. (A refinement of this procedure is suggested in Section 16.2.)

The constant elasticity model is probably the second most useful specification for a simple regression, after the regular linear form. This is because economic theory often leads us to characterize relations in elasticity terms, and the model allows for a simple approach to estimating "the elasticity." It should be noted that the presumption that the elasticity is constant throughout the relation is a strong one—but so is the corresponding assumption that the slope is constant in a regular linear model.

The Semilog Specification

Suppose that we think the exact relation between Y and X is

$$Y = e^{\beta_0} e^{\beta_1 X} \tag{6.34}$$

If we take the natural logarithm of both sides of the equation, we obtain

$$\ln Y = \beta_0 + \beta_1 X \tag{6.35}$$

This is known as the ***semilog*** relation because only part of it is specified in logarithmic form. (Another variant, which we do not consider, specifies Y as a function of $\ln X$.)

The geometry of (6.34) and (6.35) is illustrated in Figure 6.4. X may take on positive or negative values, but Y must be positive if $\ln Y$ is to be defined. If β_1 is negative, Y decreases as X increases and its slope becomes flatter. If $\beta_1 = 0$, Y is constant. If β_1 is positive, Y increases as X increases and its slope becomes steeper.

Part of the usefulness of the semilog relation derives from the ease of interpretation of β_1. Thinking of dY and dX as small changes resulting from movement along the relation between Y and X, it is shown in the appendix to this chapter that

$$\beta_1 = \frac{dY/Y}{dX} \tag{6.36}$$

That is, β_1 can be interpreted as the proportional change in Y that results from a unit change in X. This implies that

$$\text{pc of } Y \approx \beta_1 \, \Delta X \tag{6.37}$$

when the changes in X and Y are small.

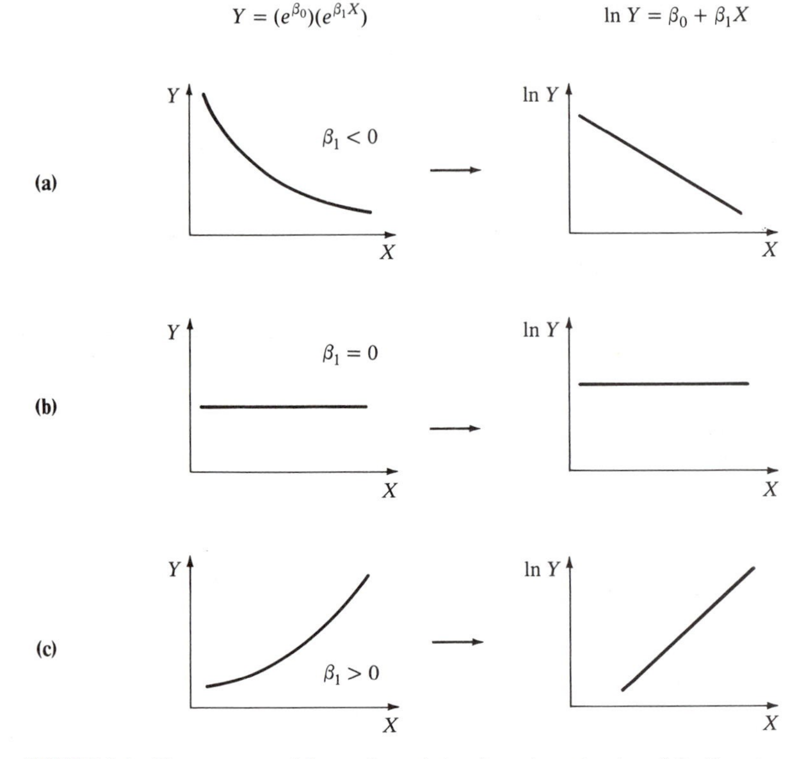

FIGURE 6.4 The geometry of the semilog relation depends on the sign of β_1. Except when $\beta_1 = 0$, Y is a nonlinear function of X. As X increases, Y decreases ($\beta_1 < 0$) or increases ($\beta_1 > 0$), and the relation is concave upward in both cases. Although Y is a nonlinear function of X, $\ln Y$ is a linear function of X; the slope of that line is the same β_1 as in the original formulation. The parameter β_1 can be interpreted as the proportional change in Y that results from a unit change in X.

Moving toward econometrics, it is easy to see that the exact relation (6.35) can be taken as the basis of a simple regression model

$$\ln Y_i = \beta_0 + \beta_1 X_i + u_i \tag{6.38}$$

where the regressand is $\ln Y$ and the regressor is simply X.

One application of the semilog specification is to the human capital theory of earnings determination. In one derivation, the theory states that the systematic relation between earnings and educational attainment is

$$\ln EARNS = \beta_0 + \beta_1 ED \tag{6.39}$$

Adding a disturbance term yields a semilog regression model. The estimate of this earnings function, based on the 100 observations in our cross-section data set, is

$$\widehat{LNEARNS}_i = 0.673 + 0.107ED_i \qquad (6.40)$$
$$R^2 = .405 \qquad SER = 0.446$$

Since $\hat{\beta}_1^* > 0$, we see that the implied relation between *EARNS* and *ED* resembles the left side of Figure 6.4c. On the basis of this estimation we predict that each additional year of schooling increases earnings approximately by the proportion 0.107, or 10.7 percent. In other words, the new level of earnings will be 1.107 times the original level. Using (6.37), a three-year increase in schooling would lead earnings to increase approximately by the proportion 0.321, or 32.1 percent.

It is interesting to compare this with the result of (6.7), which estimates that the impact of each additional year of schooling is to increase annual earnings by $797. Which model is more appropriate for the earnings function? On the basis of general principles, we might surmise that the semilog model is appropriate when theory suggests that equal-sized increases in X lead to equal proportionate increases in Y for different observations, whereas the linear model is appropriate when theory suggests that equal-sized increases in X lead to equal absolute increases in Y for different observations. In practice, one way to compare (6.7) with (6.40) is to graph the residuals against the explanatory variable, *ED* (a computer can do this easily). If the residuals appear to be unrelated to *ED* in one case, but related to *ED* in a $\cup$ or $\cap$ shape in the other, then the case with the unrelated residuals may be judged better because it is in greater conformity with regression theory. Note, however, that there will be no linear relation between the residuals and *ED* because OLS always makes this correlation equal to zero.

Another common application of the semilog specification is to estimate the trend rate of growth or shrinkage in a time-series variable. Consider a simple model of growth

$$Y_i = Y_0(1 + r)^i \qquad (6.41)$$

where i indicates the number of years since period 0. Since our time variable (T) conforms to our observation numbering, we can write

$$Y_i = Y_0(1 + r)^{T_i} \qquad (6.42)$$

Taking logarithms yields

$$\ln Y_i = \ln Y_0 + T_i \ln (1 + r) \qquad (6.43)$$

Now, $\ln Y_0$ and $\ln (1 + r)$ are both constants, which we can rename β_0 and β_1, respectively. Adding a disturbance for econometric reality, (6.43) becomes

$$\ln Y_i = \beta_0 + \beta_1 T_i + u_i \qquad (6.44)$$

which is suitable for OLS estimation using the transformed variable $\ln Y$ as the regressand. Since $\beta_1 = \ln (1 + r) \approx r$, $\hat{\beta}_1$ is approximately the estimated annual rate of growth [see (2.12)].

For example, choosing *GNP* as a variable of interest, we find that

$$\widehat{LNGNP_i} = 6.456 + 0.0354T_i \qquad (6.45)$$
$$R^2 = .988 \qquad SER = 0.0297$$

for the 25 observations in the time-series data set. The annual rate of growth of *GNP* is estimated to be 3.54 percent based on the assumption behind model (6.44) that the systematic rate of growth was constant over this period. The disturbance term allows for random fluctuation of *GNP* from its trend each year.

Problems

Section 6.1

⋆ **6.1** If the expected value of Y for the ith observation is 35, how could it be that the actual value is 33?

6.2 Suppose that an appropriately estimated regression is $\hat{Y}_i = 3 + 4X_i$.
(a) Determine the change in $\hat{Y}$ associated with $\Delta X = 2$.
(b) Comparing two particular observations in the data, it turns out that $\Delta Y = 9$ while $\Delta X = 2$. What accounts for the difference between this ΔY and your answer to part (a)?

6.3 In regressions of saving on income reported in Section 5.5, the slope coefficient was 86.3 in one case and 0.0863 in another. In which case is income more important in explaining saving? Why?

Section 6.2

6.4 Based on the regression reported as Equation (6.7):
(a) Determine the predicted value of *EARNS* corresponding to $ED = 12$.
(b) Determine the change in predicted *EARNS* associated with a change from $ED = 12$ to $ED = 16$.
(c) Thinking of $\Delta ED = 1$, is the effect on earnings of the senior year in college the same as for the sophomore year?

⋆ **6.5** Suppose that Equation (6.6) is the correct specification of the relation between *EARNS* and *ED*. If the regression (6.7) explains 28.5 percent of the variation in earnings in the sample, what accounts for the rest of the variation?

⋆ **6.6** Suppose that in 1980 all earnings levels had been inflated by a factor of 100 percent compared with the levels of 1963. If the relation between *EARNS* and *ED* otherwise remained unchanged, what would be the impact of an additional year of schooling on predicted earnings in 1980?

6.7 Based on Equation (6.10), how much of an increase in *DPI* is required to increase predicted *CON* by 1 billion dollars?

6.8 Interpret the estimated intercept in Equation (6.10).

Section 6.3

6.9 Suppose that instead of T in Equation (6.14) we had used the actual year number, t, as the regressor (i.e., $t_1 = 1956$, $t_2 = 1957$, etc.). What would be the estimated regression of APC on t?

⋆ **6.10** From Table 2.3, find the actual values of CON and DPI for 1974. What is the actual value for the APC? What is the predicted value of the APC based on Equation (6.14)?

6.11 Assuming that DPI increases steadily over time, what does Equation (6.10) imply about what happens to the average propensity to consume over time?

6.12 Suppose that in normal times the yield on bonds increases with the time to maturity, but that it increases at a decreasing rate and that it never goes above some natural level. Explain how a regression model can best be used to estimate the relation between yield and time to maturity.

6.13 Based on the estimated Phillips curve (6.19), what level of unemployment would have been required to bring the predicted rate of inflation down to zero?

⋆ **6.14** Based on Equation (6.22), what is the predicted value of consumption for 1974? What is the predicted value based on Equation (6.10)?

6.15 Based on Equation (6.25), determine the predicted level of investment for 1976.

Section 6.4

6.16 With reference to Equation (6.26), what happens to the value of all possible observations on Y if β_0 is increased by a constant amount?

6.17 Based on the slope coefficient in Equation (6.33), what would be the proportional impact on predicted M of the actual increase in GNP that occurred between 1967 and 1968? What was the actual proportional change in M? (See Table 2.3.)

⋆ **6.18** Suppose that $\ln Y = 1.0 + 0.25 \ln X$. Compute the predicted values of Y for these four values of X: 100, 500, 1000, 1500.

6.19 Plot the four Y, X points determined in Problem 6.18. How does the shape of this graph compare with Figure 6.3?

⋆ **6.20** Suppose that X increases from 500 to 1000. Based on your calculations for Problem 6.18, what is the percentage increase in predicted Y? How does this compare with the estimated elasticity?

6.21 Based on Equation (6.40), what is the proportional change in predicted earnings that would result from gaining a college education rather than stopping after completing high school?

⋆ **6.22** Consider the specification of a demand curve: $Q = AP^b$. Interpret the meaning of b. Based on examination of Figure 6.3 and knowledge of economics, is b likely to be positive or negative? How could you use simple regression to estimate the value of b?

6.23 Suppose that the labor force has been growing at a steady rate over time. Explain how a regression model can be used to estimate the rate of growth.

Appendix

⋆ **6.24** Based on the relation in Problem 6.18, follow the procedure in Equation (6.48) to determine the exact predicted percentage increase in Y associated with a 100 percent increase in X. Compare this result with that in Problem 6.20.

6.25 Based on Equation (6.40), what is the exact proportionate change for Problem 6.21? Compare this with the simpler calculation.

APPENDIX: The Coefficients in Logarithmic Models*

The slope coefficients in log-linear and semilog models have straightforward interpretations when ΔX and the accompanying ΔY are small. These are given in Section 6.4 and are derived here. For large changes these interpretations are poor approximations, and to determine the effect ΔY resulting from a given ΔX we must work through the mathematical specification of the model, as also shown here.

For the log-linear functional form, the point elasticity interpretation is derived using calculus:

$$\begin{aligned} \ln Y &= \beta_0 + \beta_1 \ln X \\ d(\ln Y) &= d\beta_0 + d(\beta_1 \ln X) \\ dY/Y &= 0 + \beta_1(dX/X) \\ \beta_1 &= \frac{dY/Y}{dX/X} = \frac{dY}{dX} \cdot \frac{X}{Y} \end{aligned} \tag{6.46}$$

The middle expression in the last line of (6.46) is the elasticity, and the derivation shows that this is equal to β_1, a constant. Roughly speaking, we may say that a 1 percent change in X leads to a β_1 percent change in Y.

For discrete changes in X and Y we consider the movement from p' to p'' along the relation between Y and X illustrated in Figure 6.5. Letting X_1 and Y_1 correspond to p', and X_2 and Y_2 correspond to p'', we find that

*This appendix is relatively difficult and can be skipped without loss of continuity.

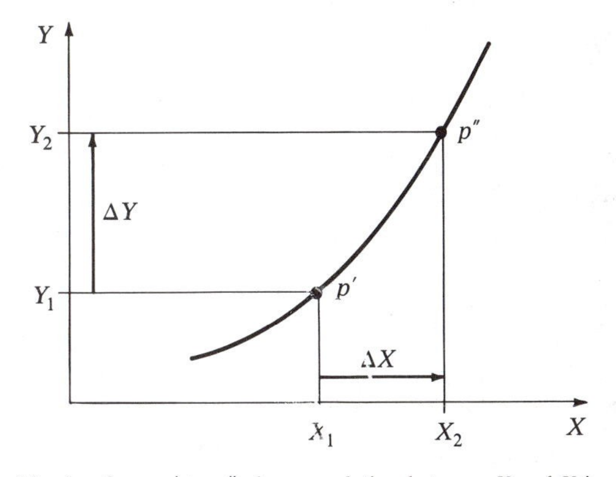

FIGURE 6.5 Moving from p' to p'' along a relation between Y and X is viewed as a change in Y (i.e., ΔY) resulting from a given change in X (i.e., ΔX). In the linear form (not illustrated here), Y and X are related in such a way that ΔY is a constant multiple of ΔX. In the log-linear form, Y and X are related in such a way that $\Delta Y/Y$ is a constant multiple of $\Delta X/X$, approximately; that is, the elasticity is constant, and therefore it does not vary with the level of X. In the semilog form, Y and X are related in such a way that $\Delta Y/Y$ is a constant multiple of ΔX, approximately; that is, the proportionate change in Y resulting from a given ΔX does not depend on the level of X.

$$\begin{aligned}
\ln Y_2 &= \beta_0 + \beta_1 \ln X_2 \\
\ln Y_1 &= \beta_0 + \beta_1 \ln X_1 \\
\ln Y_2 - \ln Y_1 &= \beta_0 - \beta_0 + \beta_1 \ln X_2 - \beta_1 \ln X_1 \\
\ln (Y_2/Y_1) &= \beta_1 \ln (X_2/X_1) \\
\beta_1 &= \frac{\ln (Y_2/Y_1)}{\ln (X_2/X_1)} = \frac{\ln (1 + \text{pc of } Y)}{\ln (1 + \text{pc of } X)} \\
&\approx \frac{\text{pc of } Y}{\text{pc of } X}
\end{aligned} \tag{6.47}$$

where "pc of Y" denotes the proportionate change in Y. The first two lines are statements of the relation holding at p'' and p', and the third line results from subtracting the second line from the first. The fourth line is obtained by applying the rules of logarithms. The fifth line isolates β_1 and shows what it is equal to in terms of proportionate changes of Y and X. When the proportionate changes are small, the final approximation is good; this yields the elasticity.

We can use this result to calculate the effect on Y resulting from any given change in X. For example, $\hat{\beta}_1^*$ in the demand for money equation (6.33) is 0.215. If GNP were to increase by 100 percent (pc = 1.00), as it does over the 1956–1980 sample period, the proportionate effect on M could be determined as follows. First (6.47) is rewritten as

$$\begin{aligned}
\ln (1 + \text{pc of } Y) &= \beta_1 \ln (1 + \text{pc of } X) = \beta_1 \ln (1 + 1) \\
&= (0.215)(0.693) = 0.149
\end{aligned} \tag{6.48}$$

Taking antilogs yields

$$1 + \text{pc of } Y = 1.161, \quad \text{so} \quad \text{pc of } Y = 0.161 \tag{6.49}$$

Hence the correct predicted proportionate change in Y is 16.1 percent, which is considerably smaller than the 21.5 percent change that would be predicted by simply applying the point elasticity.

For the semilog form,

$$\begin{aligned} \ln Y &= \beta_0 + \beta_1 X \\ d(\ln Y) &= d\beta_0 + d\beta_1 X = \beta_1\, dX \\ \beta_1 &= \frac{d(\ln Y)}{dX} = \frac{dY/Y}{dX} \end{aligned} \tag{6.50}$$

Roughly speaking, a unit change in X leads to a β_1 proportionate change in Y.

For discrete changes in X and Y we consider moving from p' to p'' in Figure 6.5 again:

$$\begin{aligned} \ln Y_2 &= \beta_0 + \beta_1 X_2 \\ \ln Y_1 &= \beta_0 + \beta_1 X_1 \\ \ln Y_2 - \ln Y_1 &= \beta_0 - \beta_0 + \beta_1 X_2 - \beta_1 X_1 \\ \ln (Y_2/Y_1) &= \beta_1\, \Delta X \\ \beta_1 &= \frac{\ln (1 + \text{pc of } Y)}{\Delta X} \approx \frac{\text{pc of } Y}{\Delta X} \end{aligned} \tag{6.51}$$

When the changes are small, the final approximation is good; this is our easy interpretation.

The effect on Y resulting from any given change in X may be derived as follows. Restating (6.51) leads to

$$\begin{aligned} \ln (1 + \text{pc of } Y) &= \beta_1\, \Delta X \\ 1 + \text{pc of } Y &= e^{\beta_1 \Delta X} \\ \text{pc of } Y &= e^{\beta_1 \Delta X} - 1 \end{aligned} \tag{6.52}$$

For example, in the earnings function (6.40) the estimated slope coefficient, 0.107, was interpreted as indicating that each additional year of schooling increases earnings by approximately the multiplicative factor 0.107, or 10.7 percent. The actual computed increase is $e^{0.107} - 1 = 0.113$, or 11.3 percent. A three-year increase in schooling increases earnings by the factor $e^{0.321} - 1 = 0.379$, or 37.9 percent. This is substantially larger than the 32.1 percent computed by the approximation (6.37).

Multiple Regression: Theory and Application

Most economic relations, and the processes they describe, involve more than one determinant of some particular dependent variable. For example, the earnings function examined in Chapter 6 presumes that education is the only identifiable variable that affects a person's earnings. Surely many more variables are directly relevant, and in this chapter we will go on to examine the roles of experience, demographic characteristics, and other variables.

This chapter covers both the theory and application of multiple regression, which involves more than one explanatory variable in a single regression equation. Most of the ideas regarding simple regression carry over, so there are relatively few new concepts to learn.

7.1 Two Explanatory Variables

As a first step in multiple regression, we consider an economic process in which the variable Y is determined by two given variables, X_1 and X_2. Much of what needs to be said about the theory and estimation of the corresponding model is a direct extension of the case of simple regression.

Our thinking is that n observations are subjected to this process, one at a time. For each observation separately, the values of X_1 and X_2 are fed in, and the value of Y is determined. The n values of X_1, X_2, and Y are all observable, and they can be collected in a data set. Graphically, the observations can be plotted in a three-dimensional scatter diagram, in which the explanatory variables X_1 and X_2 are measured along the two axes defining the base and the dependent variable Y is measured vertically. Each observation is graphed as a single point, and together the observations resemble a cloud.

Two aspects of this process should be noted. First, X_1 and X_2 are the only identifiable and observable variables that affect Y. By explicit exclusion, other variables that might be measured for each observation are understood to play no direct role in the determination of Y. Second, the values of X_1 and X_2 are taken as given. These are determined outside the process under consideration, and we make no effort to understand why they take on whatever values they do.

We move toward statistical analysis by making a more concrete specification of the process by which Y is determined. We theorize that the value of variable Y for each observation is determined by the ***linear multiple regression model***

$$Y_i = \beta_0 + \beta_1 X_{1i} + \beta_2 X_{2i} + u_i \tag{7.1}$$

Variable Y is the dependent variable (or regressand), and X_1 and X_2 are the explanatory variables (or regressors). The disturbance u is considered to be a random term that represents pure chance factors in the determination of Y. When a particular observation is referred to, a second subscript is used for the explanatory variables.

Based on (7.1), we can decompose each Y_i value into a systematic component

$$E[Y_i] = \beta_0 + \beta_1 X_{1i} + \beta_2 X_{2i} \tag{7.2}$$

and a random component, u_i. The systematic part of the relation between Y, X_1, and X_2 is given by an equation of the form

$$E[Y] = \beta_0 + \beta_1 X_1 + \beta_2 X_2 \tag{7.3}$$

which we call the true regression. This describes a plane graphed in the three-dimensional scatter diagram, but it is also known as a linear equation. With Y measured in the vertical direction, the observations (points) lie above or below the true regression, each at a vertical distance u_i.

Given this theoretical statement of how observable data are generated, our econometric task is to estimate the coefficients β_0, β_1, and β_2. As with simple regression, our technique is based on the method of ordinary least squares. When we fit a plane to the three-dimensional scatter of data points, it has an equation of the form

$$\hat{Y} = \hat{\beta}_0 + \hat{\beta}_1 X_1 + \hat{\beta}_2 X_2 \tag{7.4}$$

This plane is called the estimated regression. It decomposes each actual Y_i value into its fitted (or predicted) value

$$\hat{Y}_i = \hat{\beta}_0 + \hat{\beta}_1 X_{1i} + \hat{\beta}_2 X_{2i} \tag{7.5}$$

and its residual

$$e_i = Y_i - \hat{Y}_i = Y_i - \hat{\beta}_0 - \hat{\beta}_1 X_{1i} - \hat{\beta}_2 X_{2i} \tag{7.6}$$

The OLS technique calculates $\hat{\beta}_0$, $\hat{\beta}_1$, and $\hat{\beta}_2$ so as to make the sum of squared residuals as small as possible, and a derivation similar to that given in the appendix to Chapter 5 leads to estimators for the coefficients (i.e., formulas for calculating them). To simplify the presentation, we adopt the notation here that a lowercase letter stands for the deviation of a variable from its mean. Thus

$$y = Y - \overline{Y}, \quad x_1 = X_1 - \overline{X}_1, \quad \text{and} \quad x_2 = X_2 - \overline{X}_2 \tag{7.7}$$

The estimators are

$$\hat{\beta}_2 = \frac{\left(\sum x_2 y\right)\left(\sum x_1^2\right) - \left(\sum x_1 y\right)\left(\sum x_1 x_2\right)}{\left(\sum x_1^2\right)\left(\sum x_2^2\right) - \left(\sum x_1 x_2\right)^2} \tag{7.8}$$

$$\hat{\beta}_1 = \frac{\left(\sum x_1 y\right)\left(\sum x_2^2\right) - \left(\sum x_2 y\right)\left(\sum x_1 x_2\right)}{\left(\sum x_1^2\right)\left(\sum x_2^2\right) - \left(\sum x_1 x_2\right)^2} \tag{7.9}$$

$$\hat{\beta}_0 = \overline{Y} - \hat{\beta}_1 \overline{X}_1 - \hat{\beta}_2 \overline{X}_2 \tag{7.10}$$

Notice that the estimators for all three coefficients involve all the values for all the variables. For example, $\hat{\beta}_2$ depends on the values of X_1 as well as those of Y and X_2. Hence these estimators are not equivalent to the coefficient estimators from the simple regression model applied twice. In other words, the multiple regression coefficients cannot be obtained by estimating two simple regressions, one of Y on X_1 and another of Y on X_2. [The exception to this is the special case when the correlation (and covariance) between X_1 and X_2 is zero; this would imply that $\Sigma\, x_1 x_2 = 0$.]

Our interpretations of the coefficients of the estimated regression differ in an important way from the case of simple regression. First, if we compare one possible observation with another, they may differ with regard to the values of X_1 and X_2 and also with regard to the predicted value $\hat{Y}$. These differences are linked by

$$\Delta \hat{Y} = \hat{\beta}_1\, \Delta X_1 + \hat{\beta}_2\, \Delta X_2 \tag{7.11}$$

This is derived by subtracting an equation like (7.5) specified for one point from the corresponding equation specified for another. This equation determines how changes in the values of the explanatory variables affect the predicted value of dependent variable. Graphically, this equation compares two points on the estimated regression plane and shows how differences in the three dimensions are related.

Now, if only one explanatory variable changes in value while the other remains the same, then

$$\Delta\hat{Y} = \hat{\beta}_1\, \Delta X_1 \qquad \text{when} \quad \Delta X_2 = 0 \tag{7.12}$$

and

$$\Delta\hat{Y} = \hat{\beta}_2\, \Delta X_2 \qquad \text{when} \quad \Delta X_1 = 0 \tag{7.13}$$

These provide the basis for interpreting the slope coefficients: $\hat{\beta}_1$ gives the impact on the predicted value of Y of a unit increase in X_1, holding constant the value of X_2. Note that this phrasing corresponds closely to the economic concept of ***ceteris paribus***. Similarly, $\hat{\beta}_2$ gives the impact on $\hat{Y}$ of a unit increase in X_2, holding X_1 constant. These interpretations will be illustrated and enhanced in later examples and discussion.

As with simple regression, the *SER* and R^2 serve as measures of goodness of fit, and their interpretations are the same. The general definitions of these measures are given in the next section.

The Earnings Function

As an example, we extend our analysis of the earnings function explored in Chapter 6. Suppose that labor market theory suggests that in addition to formal education, experience working in the labor force has a direct effect on workers' earnings. This may be because experience represents on-the-job training and thereby increases a person's productivity and wage, or it may be because of some other considerations. If this theory is correct and if the relation is linear, it is appropriate to formulate the multiple regression model

$$EARNS_i = \beta_0 + \beta_1 ED_i + \beta_2 EXP_i + u_i \tag{7.14}$$

Turning to estimation, the 100 observations in the cross-section data set in Chapter 2 are used to obtain the estimated regression

$$\begin{aligned} \widehat{EARNS}_i &= -6.179 + 0.978ED_i + 0.124EXP_i \\ R^2 &= .315 \qquad SER = 4.288 \end{aligned} \tag{7.15}$$

Holding constant the level of education, each year of experience is estimated to increase expected earnings by \$124. The *ceteris paribus* qualification "holding constant the level of education" is an interpretation based on (7.13); it does not mean that we or the computer hold some values specially fixed during estimation or make comparisons only among observations with common values for *ED*. Considering the 30-year age difference between the youngest and the oldest men in the sample, the estimated coefficient implies a (0.124)(30) = 3.720 thousand dollar annual earnings difference. Thus experience might be judged to be a moderately important factor in the determination of earnings.

The estimated impact of education on earnings is increased substantially from the finding in the simple regression (0.797 to 0.978). Of course, the true impact of education has not changed—we did nothing capable of altering that—but the change in specification has affected our estimated impact. We return to this later.

We can use the estimated model to predict earnings for given values of *ED* and *EXP* in the usual way. For example, the predicted earnings (in 1963) for a college graduate with five years of experience is

$$\widehat{EARNS_i} = -6.179 + (0.978)(16) + (0.124)(5) = 10.089 \quad (7.16)$$

thousand dollars. In addition, the method of constructing *EXP* provides auxiliary information that permits us to answer a question like this: taking into account that for a specific individual an increase in *ED* of one year means a decrease in *EXP* of one year, what is the total economic effect on earnings of going to school for one more year? That is, we are seeking the joint impact of $\Delta ED = 1$ and $\Delta EXP = -1$. Based on (7.11),

$$\Delta\widehat{EARNS} = (0.978)(1) + (0.124)(-1) = 0.854 \quad (7.17)$$

thousand dollars per year. Note that the estimated intercept plays no role in a calculation like this.

The R^2 in the multiple regression is .315, which is somewhat higher than the .285 in the simple regression. Does the fact that R^2 increases by only .030 mean that the impact of experience is marginal, compared with that of education? No, not necessarily. *ED* could seem to get most of the credit simply because we considered it first. The R^2 of a simple regression of *EARNS* on *EXP* in these data could be fairly high, although in fact it is not in this case.

7.2 The General Case

The ***general linear multiple regression model*** is a particular specification of some economic process in which the values of a dependent variable (or regressand) are determined by several explanatory variables (or regressors). In general we may say that Y depends on k explanatory variables:

$$Y_i = \beta_0 + \beta_1 X_{1i} + \beta_2 X_{2i} + \cdots + \beta_k X_{ki} + u_i \quad (7.18)$$

where the names of the variables are $X_1, X_2, \ldots, X_k$. A typical variable is denoted by X_j, and a typical coefficient by β_j. When a particular observation is referred to, a second subscript is used, so the ith observation on the jth variable is X_{ji}.

As with simple regression, our notion is that this model accurately reflects the way some process works. Indeed, this notion is more acceptable with multiple regression because this technique allows us to take into account all the important variables that help determine the value of the dependent variable.

Since most economic processes involve multiple causes of a single effect, this feature is especially important.

If it happens that $k = 1$, this model reduces to that of simple regression

$$Y_i = \beta_0 + \beta_1 X_{1i} + u_i \tag{7.19}$$

in which X_1 is the same as X in (5.1). We have treated the case of simple regression separately because it readily permits graphical interpretation and because the algebra of the derivations is relatively simple. If $k = 2$, the model reduces to the case with two explanatory variables, treated in the preceding section.

The ordinary least squares (OLS) technique for estimating the coefficients $\beta_0, \beta_1, \beta_2, \ldots, \beta_k$ is an extension of that for the simpler cases, and the estimates of these parameters are denoted by $\hat{\beta}_0, \hat{\beta}_1, \hat{\beta}_2, \ldots, \hat{\beta}_k$. For any observation the true value of Y_i can be decomposed into the fitted value and the residual:

$$Y_i = \hat{Y}_i + e_i = \hat{\beta}_0 + \hat{\beta}_1 X_{1i} + \cdots + \hat{\beta}_k X_{ki} + e_i \tag{7.20}$$

Recall that with simple regression ($k = 1$) the observations can be plotted in a two-dimensional graph, and the estimated regression is the line that provides the best possible fit to the scattered points. When $k = 2$, the observations can be plotted in a three-dimensional graph, and the estimated regression is the plane that provides the best possible fit to the scattered points. By extension, in the general case the observations can be thought of as plotted in a $(k + 1)$-dimensional graph, and the estimated regression is a hyperplane fit through the scattered points.

The OLS technique calculates the $\hat{\beta}_j$ so as to make the sum of squared residuals as small as possible:

$$\text{OLS criterion:} \quad \text{minimize} \quad SSR = \sum_i e_i^2 \tag{7.21}$$

Suffice it to say that we end up with a set of estimators analogous to those for the simpler cases. All the data are used together to solve simultaneously for the $k + 1$ coefficients; it is not the case that we just estimate k simple regressions. A computer can carry out the necessary calculations, and we rely on this facility.

The discussion in Section 5.3 regarding the comparison of the estimated regression coefficients with the corresponding true regression coefficients applies fully to multiple regression. Although we might wish that each estimated coefficient were equal to the corresponding true coefficient, so that $\hat{\beta}_j^* = \beta_j$, this is unlikely ever to occur. (As before, the asterisk indicates the computed values in a set of data.) The difference between $\hat{\beta}_j^*$ and β_j arises from the particular pattern of values taken on by the disturbances. (Just how the value of $\hat{\beta}_j$ might be related to β_j is the subject of sampling theory, which is discussed in Chapter 11.) Overall, we recognize that the estimated regression will be different from the true regression, but we hope that the differences are not too great.

To interpret and apply the estimated regression model we need to see how changes in the regressors affect the predicted value of the regressand. The estimated regression is given by

$$\hat{Y} = \hat{\beta}_0 + \hat{\beta}_1 X_1 + \hat{\beta}_2 X_2 + \cdots + \hat{\beta}_k X_k \tag{7.22}$$

For any given set of changes in the explanatory variables,

$$\Delta\hat{Y} = \hat{\beta}_1\,\Delta X_1 + \hat{\beta}_2\,\Delta X_2 + \cdots + \hat{\beta}_k\,\Delta X_k \tag{7.23}$$

provides the method for calculating the effect on the predicted value of Y of specified changes in the X_j's.

If only one explanatory variable changes in value while all the others remain the same, then

$$\Delta\hat{Y} = \hat{\beta}_j\,\Delta X_j \quad \text{, holding all other regressors constant} \tag{7.24}$$

This provides the basis for interpreting the value of any single coefficient: $\hat{\beta}_j$ gives the impact on the predicted value of Y of a unit increase in X_j, holding constant the values of all the other variables. The latter qualification is important, and it corresponds closely to the economic concept of ***ceteris paribus.*** This concept does not require or imply that there are no relations among the explanatory variables, but it ignores them in assessing the effect of each variable.

If a change in X_j does cause a change in other explanatory variables, we recognize that the total economic effect on $\hat{Y}$ includes the *ceteris paribus* effect $\hat{\beta}_j$ plus the effects of the consequent changes in the other variables, through (7.23). Additional information about how the explanatory variables affect each other would be needed to make this calculation, but usually it is not available. Sometimes such information is developed in multiequation models, which are beyond the scope of our present concern. In working with single-equation models, the only effects revealed are the *ceteris paribus* effects given by the individual regression coefficients, and these are what we are interested in.

Some of the properties of the estimated regression, which is usually called a line even though it is not one, are the same as for simple regression. First, it turns out that $\Sigma\, e_i = 0$; that is, the positive and negative residuals cancel out in summation, so the average error is zero. Second, the point $\overline{Y}, \overline{X}_1, \overline{X}_2, \ldots, \overline{X}_k$ lies on the fitted line; that is, the regression goes through the point of means. Third, the correlation between the residuals and any explanatory variable is zero.

The standard error of regression is given by

$$SER = \sqrt{\frac{\sum_{i=1}^{n} e_i^2}{n - k - 1}} \tag{7.25}$$

and it measures the typical error of fit. Notice that the denominator is equal to n minus the number of coefficients (including $\hat{\beta}_0$) that are estimated; this count is the number of degrees of freedom for the estimation.

How well the estimated regression fits the data is also measured by the coefficient of determination, R^2, which is calculated as in the case of simple regression

$$R^2 = 1 - \frac{\sum_{i=1}^{n} e_i^2}{\sum_{i=1}^{n} (Y_i - \overline{Y})^2} \tag{7.26}$$

Since each e_i is the part of Y_i that *is not* explained by the regression, R^2 is again interpreted as the proportion of the variation in Y that *is* explained by the regression (i.e., that is explained by the variation in the X_j's). A perfect fit, in which each $e_i = 0$, yields $R^2 = 1$, and if the fit is not very good at all, R^2 is close to zero.

In practical econometrics we often estimate more than one regression involving the same dependent variable. For example, so far we have estimated two earnings functions: one involving *ED* as the only explanatory variable, and a second involving both *ED* and *EXP* as explanatory variables. In comparing regressions like these it is useful to have a statistic or an indicator that tells us which regression fits better. Our natural candidates are the *SER* and R^2.

The R^2 is a more popular measure of fit for any single regression. However, for comparing regressions like our earnings functions, it always gives the same answer: the regression with additional variables included fits better. This is because the addition of an explanatory variable to an original regression model cannot raise the sum of squared residuals, *SSR*. (Since OLS is acting to minimize this sum, it need not allow an additional specified variable to increase the *SSR*; it could effectively ignore the new variable rather than let it worsen the *SSR*.) This sum, which is $\sum e_i^2$, appears in the numerator of the ratio in (7.26). Thus for a given set of data on a dependent variable, Y, the addition of an explanatory variable to a regression model cannot decrease R^2, and in practice it always increases it at least a bit.

However, the increase in R^2 is obtained at a statistical "cost": the inclusion of another variable. An indicator of whether the new equation "really" fits better should assess whether the decrease in *SSR* achieved by including a new regressor is substantial enough to outweigh the cost of doing so. A statistic that does this is known as the ***adjusted*** or ***corrected*** R^2, which is denoted by $\overline{R}^2$ and defined as

$$\overline{R}^2 = 1 - \frac{\sum_{i=1}^{n} e_i^2/(n - k - 1)}{\sum_{i=1}^{n} (Y_i - \overline{Y})^2/(n - 1)} \tag{7.27}$$

The symbol $\overline{R}^2$ is conventionally read as "*R* bar squared." (The overbar notation does not signify a mean here.)

To examine what happens to $\bar{R}^2$ as another variable is added to a regression, we need only look at the numerator of the ratio in (7.27) because the denominator stays fixed. The numerator itself is a ratio. If adding a variable causes Σe_i^2 to decrease proportionately more than $n - k - 1$ decreases, $\bar{R}^2$ will increase; if adding the variable causes Σe_i^2 to decrease only slightly, by proportionately less than the decrease in $n - k - 1$, $\bar{R}^2$ will decrease. Note that the expression $n - k - 1$ is the number of degrees of freedom, and its decrease is the link to the "statistical cost" of adding another variable.

For computational purposes, it is possible to determine $\bar{R}^2$ from R^2 and readily available parameters:

$$\bar{R}^2 = R^2 - \frac{k}{n - k - 1}(1 - R^2) \tag{7.28}$$

From this we see that $\bar{R}^2$ is less than R^2, except if $R^2 = 1$. Unfortunately, $\bar{R}^2$ does not have as straightforward an interpretation as R^2 does, and sometimes it can be negative.

Whether $\bar{R}^2$ would be higher or lower in comparable regressions can be determined by examining the *SER*s. To see this, note that the numerator of the ratio in (7.27) is the square of *SER*. Thus a decrease in the *SER* occurs whenever $\bar{R}^2$ increases, and an increase in the *SER* occurs whenever $\bar{R}^2$ decreases. In other words, the *SER* functions exactly the same as $\bar{R}^2$ as an indicator of whether the addition of an explanatory variable "really" improves the fit. For this reason, we make little use of the $\bar{R}^2$.

An implication of this is that our two basic measures of fit, *SER* and R^2, can sometimes give conflicting signals. Whenever a variable is added to the specification of a regression, R^2 will increase and our basic interpretation is that the "overall fit" is improved. However, when the improvement in the fit is relatively small, *SER* will increase and our interpretation is that the typical error of fit got worse.

The Consumption Function

The original Keynesian idea that aggregate consumption is determined primarily by income can be expanded to include other potentially important explanatory variables. We might think that the rate of interest available on savings and the rate of inflation in consumer prices are important. Economic theory does not specify strictly whether these effects would be positive or negative, so we approach the data with an exploratory frame of mind.

Using the 25 annual observations from our time-series data set, we estimate

$$\widehat{CON}_i = -2.370 + 0.910DPI_i + 0.500RAAA_i - 0.562RINF2_i \tag{7.29}$$

$$R^2 = .997 \qquad SER = 9.309$$

The estimated marginal propensty to consume, which gives the impact on *CON* of a unit increase in *DPI* while holding *RAAA* and *RINF2* constant, is 0.910. We see also that a one-percentage-point increase in the long-term interest rate is estimated to increase consumption (i.e., decrease saving) by 0.500 billion dollars, and a one-point increase in the rate of inflation decreases consumption by 0.562 billion dollars.

In comparison with the simple regression of *CON* on *DPI* (6.10), we see that the estimated marginal propensity to consume is changed only slightly. The R^2 is slightly higher (in the fifth decimal position, not shown) and also the *SER* is higher here than in (6.10). As noted above, this situation leads to a lower $\bar{R}^2$ in the present model.

We see that the addition of *RAAA* and *RINF2* to the simple form has not led to a model with appreciably more explanatory power. This, in itself, does not mean that the new variables do not belong in the model. If *DPI* were treated as a "new" variable, and compared with a simple regression of *CON* or *RAAA* or *RINF2*, then it might appear to have little additional explanatory power. (We see from Table 3.4 that the correlation between *CON* and *RAAA* is .952.) In Chapter 12 we present formal tests that also bear on this question.

7.3 Dummy Variables

In all the regression models considered so far, every variable has been a cardinal measure of some economic characteristic. For example, *CON* measures aggregate consumption in billions of constant dollars and *ED* measures the years of schooling completed by individual persons. These variables are included in a regression model in a natural way, so that changes in the numerical value of an explanatory variable have consistent numerical effects on the dependent variable.

Another type of data variable introduced in Chapter 2 carries information that is essentially categorical, such as a person's race, sex, or region of residence. This information can be used to classify or categorize observations or to separate them in some way, but the characteristic cannot be measured in any meaningful way. For example, the variable *REG* in the cross-section data set is equal to 1 if the worker lives in the Northeast, 2 if he lives in the North Central region, and so on. The variable *REG* indicates where the worker lives, but the values 1, 2, 3, and 4 do not result from measuring or counting anything. Hence the information contained in *REG* cannot be included directly into a regression model.

However, when the numerical outcome of an economic process depends in part on some categorical characteristic of the observation, this information must be brought into the regression specification somehow in order for the model to describe the process correctly. The technique for doing this involves constructing new regressors known as ***dummy variables*** and treating them exactly like other regressors in the multiple regression framework.

To start with a simple case, suppose that we are focusing on the relation between Y and X and that some such relation occurs both for men and women. One possibility is that the process determining Y is quite different for men and women. In this case, where there are essentially two separate processes occurring, we would separate the data according to sex and carry out separate statistical analyses. That is, we would have two separate models, one estimated with data for men and the other estimated with data for women.

Another possibility, which leads to the use of dummy variables, is that we think the process is such that the effect on Y of a change in X is the same for both sexes, but that there may be a systematic difference between men and women in the levels of Y associated with each particular value of X. These ideas are represented in Figure 7.1, which shows that the expected value of Y is a separate linear function of X for each sex, with equal slopes but different intercepts. Let the common slope value be β_1 and let the intercept for men be β_0. The intercept for women could be given a separate symbol, but instead we let β_2 be the difference between the women's and men's intercepts, so that the intercept for women is $\beta_0 + \beta_2$. Since the two functions have the same slope, β_2 also is the difference in $E[Y]$ between women and men having any particular value of X. As drawn in the figure, β_2 is positive so that the relation for women lies above that for men. In our general thinking the sign of β_2 is not specified, so that the relation for women may lie above or below that for men.

These three parameters can be estimated by constructing a dummy variable, S, that is equal to 0 for every observation that is a man and is equal to 1 for every observation that is a woman. We can formulate our specification so far as

$$E[Y] = \beta_0 + \beta_1 X + \beta_2 S \tag{7.30}$$

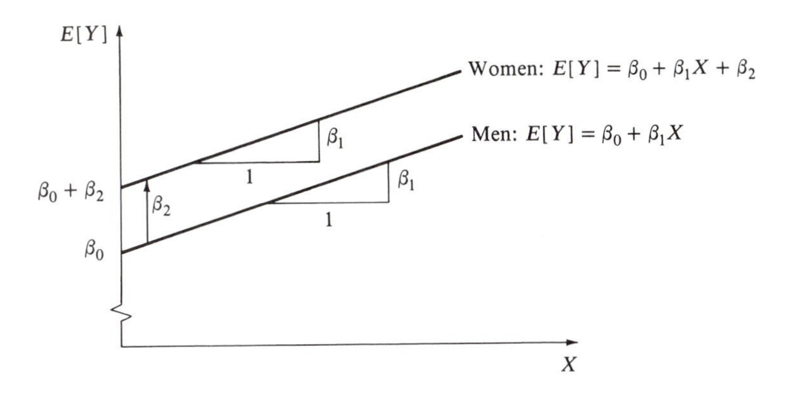

FIGURE 7.1 This illustrates an economic process in which the impact on Y of a change in X is the same for men and women, but in which men and women having the same value for X have different expected values for Y. Assuming linearity, the parameterization in the diagram leads naturally to the dummy variable formulation of a regression model.

Regardless of the value of S, the slope in the relation between $E[Y]$ and X is β_1. For men, with $S = 0$, the right-hand side of (7.30) equals $\beta_0 + \beta_1 X$, so the intercept is β_0. For women, with $S = 1$, the right-hand side of (7.30) can be rearranged as $(\beta_0 + \beta_2) + \beta_1 X$, so the intercept is $\beta_0 + \beta_2$. Adding a disturbance for econometric reality, we have a multiple regression model

$$Y_i = \beta_0 + \beta_1 X_i + \beta_2 S_i + u_i \tag{7.31}$$

which can be estimated by OLS.

The essential property of a dummy variable is that it identifies each of the observations as being in one of two groups. For all the observations in the first group the variable is set equal to zero, and for all the observations in the second it is set to 1. When the dummy variable is included as a regressor in a multiple regression, its coefficient represents the difference in the intercept between the second group and the first. Therefore, it also measures the impact on the expected value of the regressand of an observation's being in the second group rather than the first, holding all the other regressors constant. Sometimes the dummy is called a shift variable, because it simply causes a shift in the relation between $E[Y]$ and the other regressors; it does not otherwise alter that relation.

When the regression model is specified, the group given a dummy value of zero is called the ***excluded group,*** and the group given a dummy value of 1 is called the ***included group.*** With a single dummy variable in the regression model, the intercept for the excluded group is simply β_0 and the intercept for the included group is $\beta_0 + \beta_2$. That is, the intercept for the included group is equal to the intercept for the excluded group plus the coefficient on the dummy variable. Given two groups, either may be taken as the included one; this choice will affect the values of particular coefficients but not the overall interpretation of the regression.

For example, suppose theory suggests that earnings depends linearly on educational attainment but that there is a shiftlike difference between races. Our cross-section data set includes the variable *RACE*, which was coded in the interview as 1 for whites and 2 for blacks; other races were ignored in the data selection. This variable is unsatisfactory for our purposes. We construct instead a new regressor, *DRACE*, which is a dummy variable taking the value 0 for whites and 1 for blacks. Algebraically,

$$DRACE_i = RACE_i - 1 \tag{7.32}$$

The earnings function theory leads to a multiple regression, which is estimated as

$$\begin{aligned} \widehat{EARNS}_i &= -0.778 + 0.762 ED_i - 1.926 DRACE_i \\ R^2 &= .293 \qquad SER = 4.356 \end{aligned} \tag{7.33}$$

The coefficient $\hat{\beta}_1^* = 0.762$ estimates that the impact on earnings of an additional year of schooling is \$762 for both blacks and whites. The estimated impact

TABLE 7.1 First 10 Observations For Estimating Equation (7.33)

Obs.	*EARNS*	*ED*	*DRACE*
1	1.920	2	1
2	12.403	9	0
3	5.926	17	0
4	7.000	9	0
5	6.990	12	0
6	6.500	13	0
7	26.000	17	0
8	15.000	16	0
9	5.699	9	1
10	8.820	16	0

of race is that the earnings of blacks are $1926 less than those of whites, holding constant the level of education.

The first 10 observations used for estimating (7.33) are shown in Table 7.1. The variables *EARNS* and *ED* are taken directly from Table 2.2. The variable *DRACE* is constructed according to (7.32). Like any dummy variable, *DRACE* takes on only the values 0 and 1.

In time-series regressions it may be that for some periods the relation between a dependent variable and a set of explanatory variables is shifted by a constant amount. For example, during war years the consumption function might shift down at all levels of income, because of the decrease in production of consumer goods and the special incentives given to saving. In this case, a model of the form

$$CON_i = \beta_0 + \beta_1 DPI_i + \beta_2 WAR_i + u_i \tag{7.34}$$

would be appropriate. The dummy variable *WAR* takes on the value 1 for wartime observations and zero for others; the coefficient β_2 is the shift in the consumption function, and we expect that $\hat{\beta}_2$ will be negative.

Another example is provided by the Phillips curve, which represents the trade-off between inflation and unemployment. It is conjectured that the relation shifted up in the middle of the 1960s because of the Viet Nam War and structural changes in the economy. To examine this conjecture, we add a dummy variable to the model earlier specified (6.19):

$$RINF1_i = \beta_0 + \beta_1 \left(\frac{1}{UPCT_i}\right) + \beta_2 D_i + u_i \tag{7.35}$$

where $D_i = 0$ for the observations 1956–1964 and $D_i = 1$ for 1965–1970. With 15 observations in total, we find that

$$\widehat{RINF1}_i = -0.803 + 15.030 \left(\frac{1}{UPCT_i}\right) + 0.894 D_i$$
$$R^2 = .600 \qquad SER = 0.936 \tag{7.36}$$

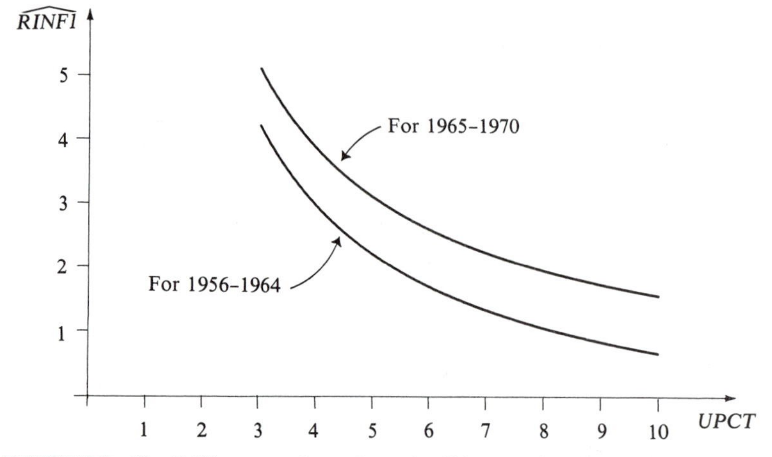

FIGURE 7.2 The Phillips curve shows the trade-off between inflation and unemployment. Using a dummy variable specification, Equation (7.36) finds evidence that is consistent with the conjecture that the curve shifted upward in the middle of the 1960s, as compared with its position earlier. The reciprocal specification of the effect of unemployment on inflation leads to the nonlinear relation shown here.

(Note that the first regressor is actually *UINV* as defined previously.) The coefficient on the dummy variable indicates that the Phillips curve shifted up by 0.894 percentage point in the latter part of the sample period as compared with where it was earlier. The two estimated Phillips curves are shown in Figure 7.2.

Using categorical information with the dummy variable technique is more complicated when there are more than two categories involved. For example, the variable *REG* in our cross-section data set carries information about the region of the country in which the person lives. The information is coded 1 for Northeast, 2 for North Central, 3 for South, and 4 for West. Suppose that we are focusing on a linear relation between Y and X, but want to take region into account. Would the specification

$$E[Y] = \beta_0 + \beta_1 X + \beta_2 REG \tag{7.37}$$

make sense? No. This specification does show a common impact of X on $E[Y]$ among the regions, but holding X constant it specifies that $E[Y]$ is β_2 greater in the North Central region than in the Northeast, $2\beta_2$ greater in the South, and $3\beta_2$ greater in the West. There is no reason why the actual differences should be so ordered or why they should be multiples of one another.

To bring in multiple-category information like region, we must construct a set of dummy variables. One way to think about doing this is to create a separate dummy variable for each category (region), taking on the value 1 if the observation belongs to that category and 0 if it does not. From *REG*, we can create four dummy variables: *DNEAST*, *DNCENT*, *DSOUTH*, and *DWEST*, to use mnemonic names. As before, each dummy variable serves to identify each obervation as being in one of two groups (i.e., in the specified region or not). In formulating the regression model, one of the dummy variables must be excluded. Then, the coefficient on each of the included dummy variables represents the difference between the intercepts of that category and the excluded category. The intercept for the excluded category is simply β_0. Were all the categories' dummy variables included, we would have one more coefficient than we could interpret logically; such a redundancy is symptomatic of a major mistake in specification, which we review in Section 7.6.

Revising (7.37) we have

$$E[Y] = \beta_0 + \beta_1 X + \beta_2 DNCENT + \beta_3\, DSOUTH + \beta_4 DWEST \tag{7.38}$$

If a person lives in the excluded Northeast, the value of each of the included dummy variables is zero and

$$E[Y] = \beta_0 + \beta_1 X \tag{7.39}$$

If a person lives in one of the other regions, the corresponding dummy variable is 1 but the other two are 0; thus

$$E[Y] = \beta_0 + \beta_1 X + \beta_j \qquad (j = 2,\ 3,\ \text{or } 4) \tag{7.40}$$

The interpretation of each dummy variable coefficient (β_j) is the difference in $E[Y]$ of living in that region rather than the excluded Northeast region, holding X constant.

For example, we reconsider the simplest earnings function. Now taking region into account, the estimated regression is

$$\widehat{EARNS_i} = -0.803 + 0.794 ED_i + 0.288 DNCENT_i - 0.828 DSOUTH_i - 1.992 DWEST_i \tag{7.41}$$

$$R^2 = .310 \qquad SER = 4.351$$

which is graphed in Figure 7.3. The coefficients on the dummy variables estimate shiftlike differences among the predicted earnings for persons living in different regions. It does not matter for the ultimate interpretation which of the region dummies is excluded, but it does affect the constant and dummy coefficients that are actually found.

The dummy variable technique can be extended to cases with more than one categorizing variable. For example, if theory specifies that earnings is a function of education, race, and region, then the regression model would include *DRACE* and the three dummy variables for region. For men in the sample who are white and live in the Northeast, the values of all the dummy variables are zero; this is the reference group to which other observations are compared.

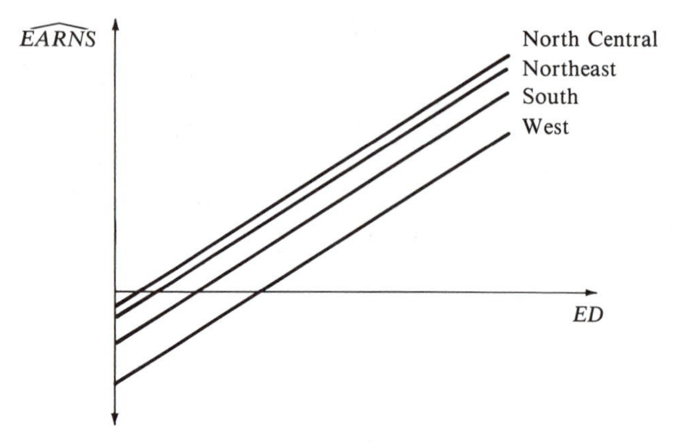

FIGURE 7.3 The simplest earnings function taking region into account uses three dummy variables to estimate the shiftlike differences among the regions. In Equation (7.41), the dummy for the Northeast region is excluded, and the other dummy-variable coefficients give the shift between each of the other regions and the Northeast. If a different region were excluded, the resulting graphical representation would be exactly the same as shown here, but the dummy variable coefficients and the regression intercept would be different because of the change of reference.

7.4 Polynomial Specifications

In Chapter 6 a variety of functional forms were introduced that allow the linear regression model to be used even when the basic behavioral relation is essentially nonlinear. The idea is that transformations of variables create new variables that can be examined in a regression framework. In the multiple regression extension of this idea, the same transformation can be applied to all the original variables, or different transformations can be applied to different variables. The important requirement is that a linear relation be specified between the ultimate regressand and the ultimate regressors.

In this section we look at another functional form that enhances the flexibility of the regression model. In the example provided, one of the original explanatory variables is left unchanged while the other is treated in the new way.

Thinking purely in mathematical terms, suppose that the exact relation between Y and X is like that in Figure 7.4a or b. These relations are clearly nonlinear, but they cannot be characterized by any of the functional forms examined in Chapter 6. The shape is that of a parabola, whose equation is written as

$$Y = \beta_0 + \beta_1 X + \beta_2 X^2 \tag{7.42}$$

If $\beta_2 > 0$, the parabola is concave upward, and if $\beta_2 < 0$, the parabola is concave downward. The slope of a parabola is given by

$$\text{slope of parabola} = \beta_1 + 2\beta_2 X \tag{7.43}$$

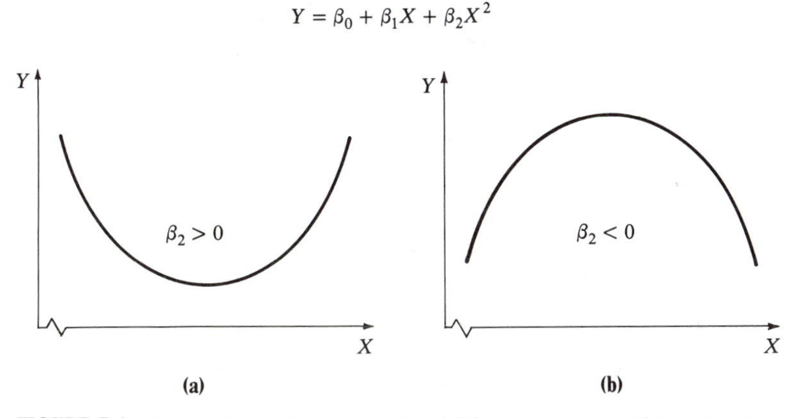

FIGURE 7.4 A parabolic relation between Y and X is concave upward if $\beta_2 > 0$ and concave downward if $\beta_2 < 0$. Although Y is a nonlinear function of X, Y also can be viewed as a linear function of X and X^2 together. This allows a multiple regression model to specify a parabolic form for the systematic relation between Y and X.

And as with any mathematical function, we know that when $\Delta X = 1$, the corresponding ΔY is approximately equal to the slope.

Returning to econometrics, if the systematic part of the relation between two variables, Y and X, is parabolic (or quadratic), we can use a linear multiple regression model of the form

$$Y_i = \beta_0 + \beta_1 X_{1i} + \beta_2 X_{2i} + u_i \tag{7.44}$$

to describe the process. To do this, we let the regressor X_1 be equal to the original variable X and we let the regressor X_2 be equal to X^2. That is, we make two regressors out of a single original variable.

With this understanding, it is conventional to rewrite (7.44) as

$$Y_i = \beta_0 + \beta_1 X_i + \beta_2 X_i^2 + u_i \tag{7.45}$$

Although the regression model specifies that Y is a linear function of X and X^2, our interest is in the implied parabolic relation between Y and X. We satisfy this interest by recognizing that X^2 must change if X does. Thus it does not make sense to say that β_1 alone measures the impact of X on Y. Rather, the impact on $E[Y]$ of a unit change in X is given approximately by the slope of the implicit quadratic equation: $\beta_1 + 2\beta_2 X$.

In studies of earnings functions it sometimes is suggested that the impact of experience diminishes and perhaps even becomes negative as the amount of experience increases. This is based on notions of physical and mental aging, diminishing returns, and optimal investment in human capital. Holding constant the level of education, the impact of experience on earnings is theorized to be like a hill-shaped parabola. Letting *EXPSQ* be the name of the regressor that is equal to the square of *EXP*, we estimate an earnings function of the form

$$\widehat{EARNS}_i = -9.791 + 0.995ED_i + 0.471EXP_i$$
$$-0.00751EXPSQ_i \tag{7.46}$$
$$R^2 = .329 \qquad SER = 4.267$$

In reporting regressions like this, the symbol EXP^2 is sometimes used to denote the second regressor.

To assess the impact of experience, we see first that the coefficient on *EXPSQ* is negative. Therefore, the estimated relation between *EARNS* and *EXP* (holding *ED* constant at any level) is a hill-shaped parabola, as in Figure 7.5. The slope of this relation is

$$\text{slope} = 0.471 + (2)(-0.00751)EXP \tag{7.47}$$

Using this, we find that a man with five years of experience will have his earnings increased by about \$396 after gaining another year, but a man with 20 years of experience will have his earnings increased by about only \$171 after gaining another year. To find the level of experience that corresponds to the peak of earnings, we solve (7.47) for the *EXP* value associated with a zero slope. Here, peak earnings occurs after about 31 years of experience, and beyond that negative returns to experience set in. All this holds true regardless of what the level of education is.

To assess the effect on predicted earnings of any particular change in experience two approaches are reasonable. The first, which makes sense for small changes in experience, makes use of the slope at the original point to make an approximation:

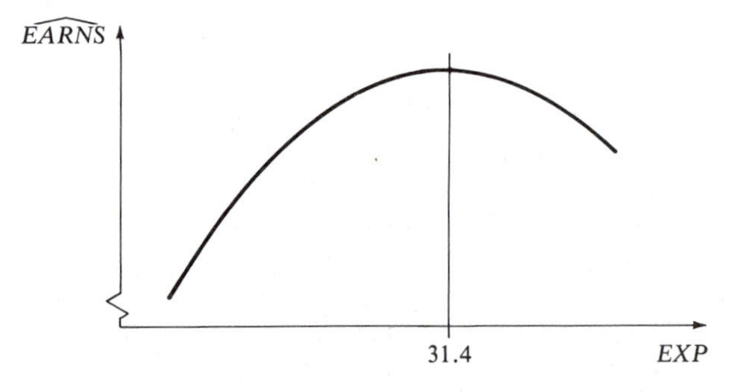

FIGURE 7.5 The earnings function (7.46) includes a parabolic (quadratic) specification for the partial relation between earnings and experience. Holding education constant at any level, the relation between predicted earnings and the level of experience is illustrated here. The slope, which is given by Equation (7.47), gives the impact on predicted earnings of a one-year increase in experience. For any given level of education, predicted earnings reach a peak at 31.4 years of experience.

$$\Delta EARNS \approx [0.471 + (2)(-0.00751)EXP_i]\ \Delta EXP \tag{7.48}$$

The second approach makes use of (7.23) to find the exact change along the estimated regression:

$$\Delta EARNS = 0.471\Delta EXP - 0.00751\Delta EXPSQ \tag{7.49}$$

In this computation, note that $\Delta EXPSQ$ is not equal to the square of ΔEXP.

It sometimes is theorized that the relation between Y and X is specified by a higher-order polynomial, such as

$$Y = \beta_0 + \beta_1 X + \beta_2 X^2 + \beta_3 X^3 \tag{7.50}$$

Clearly, our technique can be extended to bring in the values of X, X^2, and X^3 as three separate regressors in order to make estimates of the unknown parameters. These applications are rare, however, in contrast to the more common quadratic specification.

7.5 Logarithmic Specifications*

If the regressand in a multiple regression is the logarithm of a regular variable, then the regressors might be any combination of logarithmic, regular, and dummy variable forms. The model might be pure log-linear, pure semilog, or a hybrid of types.

The Demand for Money

Although the demand for money depends importantly on the level of income because money holdings are used to finance the transactions that generate income, it probably also depends on the rate of interest because money is held as an asset, as part of wealth. In addition, even the money balances held for transactions purposes may be interest sensitive. For both reasons, the interest impact is theorized to be negative. Thus a multiple regression in which the amount of money demanded is related to both the level of income and the rate of interest seems more appropriate than the simple model proposed in Chapter 6.

The most common specification in money demand regressions is the pure log-linear form, because it yields constant elasticity estimates. Using the 25 annual observations from 1956 through 1980, we find

$$\begin{aligned} \widehat{LNM}_i &= 3.759 + 0.246LNGNP_i - 0.0205LNRTB_i \\ R^2 &= .785 \qquad SER = 0.0309 \end{aligned} \tag{7.51}$$

*This section is relatively difficult and can be skipped without loss of continuity.

where M is the real quantity of money [see (6.31)], GNP is real national income, and RTB is the interest rate on Treasury bills. As theory predicts, the income elasticity is positive and the interest elasticity is negative. Comparing the multiple with the simple regression (6.33), we see that the estimated income elasticity is somewhat greater (0.246 versus 0.215) as is the R^2 (.785 versus .780).

How important is the interest rate relative to the level of income? We may directly compare the regression coefficients, because they are elasticities: a 1 percent increase in income leads to a 0.246 percent increase in the predicted demand for money, while a 1 percent increase in the interest rate leads to a 0.0205 percent decrease. Thus the interest rate appears to be much less important than income, but this comparison is misleading. In the short run the interest rate is proportionately much more variable than income: year-to-year changes of 25 percent (not percentage points!) or more often occur in the interest rate, while changes of only 5 percent or more in income occur with roughly the same frequency. Comparing these two hypothetical changes in RTB and GNP, we find the resulting impact of the interest rate to be about half as much as that of the level of income. Thus the rate of interest should be viewed as having a moderately important impact on the demand for money.

The Earnings Function

In applied labor market research, the preferred form for earnings functions specifies the regressand to be the logarithm of earnings. When education is the only explanatory variable, we have a pure semilog form as in (6.39). As more variables are taken into account the form of the function may become mixed.

For example, we consider a model in which the logarithm of earnings depends on the level of education, the amount of experience (entered quadratically), the logarithm of the number of months worked (to yield an elasticity), and the person's race and region of residence. Based on our cross-section data set, the estimated regression is

$$\begin{aligned}\widehat{LNEARNS_i} = {} & -2.031 + 0.106ED_i + 0.0501EXP_i \\ & -0.000930EXPSQ_i + 0.908LNMONTHS_i \\ & -0.239DRACE_i - 0.00468DNCENT_i \\ & -0.193DSOUTH_i - 0.162DWEST_i\end{aligned} \tag{7.52}$$

$$R^2 = .511 \qquad SER = 0.420$$

The impact of each coefficient is interpreted in the same way as in simpler formulations.

An additional year of schooling increases the level of earnings by approximately 10.6 percent, which is quite close to the finding in the simple semilog model (6.40). The level of earnings increases with the amount of experience up to a peak of about 27 years of experience and decreases thereafter. The elasticity of *EARNS* with respect to *MONTHS* worked is 0.908; if the elasticity were 1.0, then earnings would be proportional to months worked, which might be expected on the basis of simple reasoning.

The dummy variable coefficients show the impact on the regressand of being in the included category. We find that *LNEARNS* is 0.239 lower for blacks than for whites. Being black rather than white corresponds to a unit change in *DRACE* (i.e., $\Delta DRACE = 1$), so the effect of being black is to change earnings by approximately -23.9 percent. In other words, earnings are about 23.9 percent less for blacks than for whites. Note that this estimate of the racial difference in earnings pertains to a comparison in which all the other explanatory factors (education, experience, months worked, and region) are held constant; it is not an estimate of the difference between the average earnings of blacks and whites. The coefficients on the regional dummies show that earnings are lower in all these regions than in the Northeast, holding constant the other explanatory factors. Earnings are approximately 1/2 of 1 percent lower in the North Central region, 19 percent lower in the South, and 16 percent lower in the West.

The R^2 value indicates that more than half of the variation in observed *LNEARNS* has been explained by the regression, which is fairly good for this type of regression.

7.6 Specification Questions

In our development and use of multiple regression, we have followed a consistent approach in all cases. We start from the presumption that there is some stable process determining the values of some variable. Next we set up a multiple regression model that correctly describes the process in mathematical terms. Finally, we use data to estimate the unknown parameters and interpret or apply the estimated model according to our needs.

One of the difficulties with this approach is that we must correctly describe the process in terms of a regression model. This is a very demanding requirement. To meet it, or at least to come close to meeting it, we need to know more about how to specify regression models. In this section we limit ourselves to questions relating to the selection of variables for a linear model.

The Causal Nexus

In thinking about how several variables together affect or determine another one, it is natural to make a distinction between direct and indirect effects. Figure 7.6 sketches the causal linkages in a hypothetical case, with arrows indicating the paths and directions of causation. Variable X_1 is directly affected by X_3 and unlabeled variables. Variable X_2 is directly affected by X_1 and an unlabeled variable, and it is indirectly affected by X_3 and the other unlabeled variables.

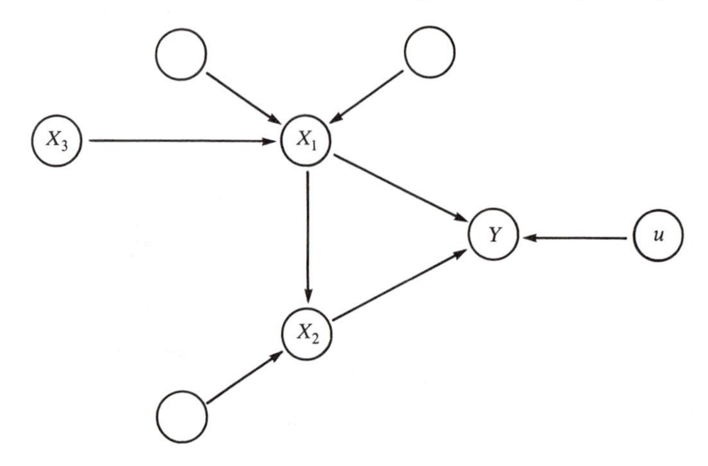

FIGURE 7.6 The causal nexus determining Y in a hypothetical case is illustrated in this schematic diagram. Arrows indicate the paths and directions of causation, and unlabeled circles contain other variables. This behavioral process is properly specified by Equation (7.53), assuming linearity. A regression model that mistakenly excludes X_2 or one that mistakenly includes X_3 misspecifies this process, but these two types of misspecification have very different consequences.

Variable Y is directly affected by X_1 and X_2, and it is indirectly affected by X_1, X_3, and the unlabeled variables. Note that X_1 has both direct and indirect effects on Y. Also, note that the disturbance u has a direct effect on Y and no linkage at all to any of the other variables.

For example, in thinking about the aggregate demand for money in macroeconomics, theory suggests that it depends directly on the level of GNP and on some interest rate and indirectly on the Federal Reserve discount rate. Through bank behavior and the action of financial markets, the discount rate affects the general level of the interest rate. Through expenditure decisions, the level of the interest rate affects GNP. And through the behavior of firms and individuals, the interest rate and the level of GNP affect the demand for money. In a commonsense way, this logic is represented in Figure 7.6, with Y being the demand for money, X_1 being the interest rate, X_2 being the level of GNP, and X_3 being the Federal Reserve discount rate.

Suppose that the correct form of the model determining Y is linear. Which variables should be included? The answer here is that only X_1 and X_2 should be included as explanatory variables, so that the proper model is

$$Y_i = \beta_0 + \beta_1 X_{1i} + \beta_2 X_{2i} + u_i \tag{7.53}$$

That is, in the specification of a multiple regression model, all the variables that have direct effects should be included, and variables that have only indirect effects (or no effects at all) should not be included.

The validity of this general rule depends on specific notions of what "direct" and "indirect" mean. For our purposes, a variable has a direct effect on Y only if in our general thinking about the process determining Y it is true that a change in the variable would lead to a change in Y while all other variables affecting Y are held constant. In other words, if a variable truly has a *ceteris paribus* effect on Y, then it has a direct effect. This leads to the general rule for specifying the regression model, because each coefficient is only a *ceteris paribus* effect.

In the causal nexus illustrated in Figure 7.6, variable X_3 has only an indirect effect. We recognize its effect as working this way: a change in the value of X_3 affects X_1, and any change in X_1 affects Y. There is no other path along which X_3 affects Y. Thus if X_1 somehow remains constant there is no way for X_3 to affect Y. Since X_3 has no *ceteris paribus* effect, it does not have a direct effect and it should not be included in the regression model. In the demand for money example, neither firms nor individuals care about the Federal Reserve discount rate. It affects the demand for money only through the general interest rate. If the general interest rate somehow remains constant, changes in the discount rate will not affect the demand for money.

The distinction between direct and indirect effects means that we must be especially careful in interpreting regression coefficients for variables like X_1, which has both kinds. The proper interpretation of β_1 is that it gives the impact on Y of a change in X_1, holding constant the value of X_2. This is the direct effect. The indirect effect of X_1 works through consequent changes in X_2. If we knew the amount of change in X_2 that would be associated with the initial change in X_1, we would have a basis for calculating the magnitude of the indirect effect. However, a regression model does not provide any information about the linkages between or among its explanatory variables, so generally it does not provide enough information to determine the magnitude of an indirect effect like that of X_1. Thus a regression model always tells us about the partial effects of the explanatory variables, but it is silent on the question of total effects. In applied research it is often true that there are some linkages between or among explanatory variables, and therefore careful interpretation is usually needed.

In thinking about what determines some economic outcome, sometimes it is natural to consider a variable that has only an indirect effect, and sometimes this indirect effect may be of special interest to us. For example, we may want to know what effect the Federal Reserve discount rate has on the demand for money. If the variable, like X_3, is not included in the model, how can we learn about its effect? What is needed is a second regression model that describes how X_1 is determined by X_3 and other variables. This model and (7.53) taken together form a recursive system of equations that can be used to analyze the effect X_3 has on Y. Such a system is a special case of ***simultaneous-equation models,*** which are discussed in Chapter 17. Until then our attention will be focused on single-equation models, which we realize are necessarily limited in scope.

Consequences of Misspecification

Sometimes it is difficult to know whether a variable under consideration for inclusion in a regression model has a direct effect, only an indirect effect, or no effect at all. Hence it is difficult to decide which variables should be included in the model, and it is likely that some models we see or create will be mis-specified.

Let us return to the causal nexus illustrated in Figure 7.6. The process is correctly described by (7.53), as explained above. Variables X_1 and X_2 are known as ***relevant variables*** because they should be included in the regression, and all other variables (such as X_3) are known as ***irrelevant variables*** because they should be excluded. Suppose that we have data on all the necessary variables. The estimated form of the correct model is

$$\hat{Y}_i = \hat{\beta}_0 + \hat{\beta}_1 X_{1i} + \hat{\beta}_2 X_{2i} \tag{7.54}$$

In carrying out our research we might make either of two mistakes, unfortunately: we might exclude a relevant variable from the regression, or we might include an irrelevant one. What are the consequences of these mistakes?

Suppose first that we leave out a relevant variable, say X_2, from the equation and estimate just the simple regression

$$\hat{Y}_i = \hat{\gamma}_0 + \hat{\gamma}_1 X_{1i} \tag{7.55}$$

(γ is lowercase gamma, the Greek "g"). Here $\hat{\gamma}_1$ denotes the estimated coefficient on X_1. In general, the value of $\hat{\gamma}_1$ in (7.55) will be different from the value of $\hat{\beta}_1$ in (7.54) when the regressions are estimated from the same set of data. This is because the formulas used are different: the simple regression slope coefficient is calculated by (5.12), and it depends only on the values of Y and X_1; the multiple regression slope coefficient on X_1 is calculated by (7.9), and it depends on the values of X_2 as well as on the values of Y and X_1.

In general we believe that the $\hat{\beta}_1$ in the estimated multiple regression provides the best estimate of the true *ceteris paribus* effect of X_1 on Y, which is denoted by β_1 in the true regression model (7.53). This idea is explored further in Chapter 13. Since $\hat{\gamma}_1$ is different from $\hat{\beta}_1$, it makes sense to say that it should be considered a not-so-good estimate of the true effect. Mathematical analysis of the estimating formulas shows that $\hat{\gamma}_1$ differs from $\hat{\beta}_1$ by factors that represent the relation between X_1 and X_2 and the relation between Y and X_2. Loosely speaking, the calculation of $\hat{\gamma}_1$ captures both the direct effect of X_1 on Y and some part of the direct effect of X_2 on Y. The latter effect is captured because of the correlation between X_2 and X_1 in the data. Hence $\hat{\gamma}_1$ is systematically distorted from the true value β_1, which is only the direct effect of X_1 on Y.

For example, consider the earnings functions (7.15) and (6.7). In the estimated multiple regression the coefficient on *ED* is 0.978. In the estimated simple regression the coefficient on *ED* is 0.797, which is substantially smaller. This difference is consistent with what would be expected if the multiple regression were the true model. To see this we note that *ED* and *EXP* are negatively correlated ($r = -.57$) in the data. This means that higher-than-average *ED* values tend to be accompanied by lower-than-average *EXP* values among the observations. Since experience is estimated to have a positive direct effect [$\hat{\beta}_2 = 0.124$ in (7.15)], observations with higher-than-average values of *ED* tend to have their earnings diminished by their lower-than-average values of *EXP*. Thus it would be expected that $\hat{\gamma}_1$ would be smaller than $\hat{\beta}_1$.

Suppose now that we make the mistake of including an irrelevant variable, say X_3, in the regression specification and that we end up with

$$\hat{Y}_i = \hat{\delta}_0 + \hat{\delta}_1 X_{1i} + \hat{\delta}_2 X_{2i} + \hat{\delta}_3 X_{3i} \tag{7.56}$$

(δ is lowercase delta, the Greek "d"). In comparison with (7.54), the inclusion of the irrelevant variable affects the estimates of the other coefficients in the sense that $\hat{\delta}_2 \neq \hat{\beta}_2$, $\hat{\delta}_1 \neq \hat{\beta}_1$, and $\hat{\delta}_0 \neq \hat{\beta}_0$. However, the effects on these estimates are rather random and usually mild. The coefficient $\hat{\delta}_3$ serves to estimate the true partial impact of X_3 on Y, which is zero (because X_3 is irrelevant). However, because of the randomness introduced into the data by the disturbances, the actual estimated value is not likely to be zero. Hence inclusion of the irrelevant variable can lead to misinterpretation of the true economic process.

In deciding whether or not to include a variable in a regression specification, the consequences of these two types of mistakes must be compared. In most cases the introduction of additional randomness into the estimation process is less serious than the introduction of systematic distortion. Hence, including an irrelevant variable is usually considered to be less of a problem than excluding a relevant one. If one has good theoretical reasons for including a variable, it is best to do so. However, this should not be taken as a suggestion to hunt for variables with the hope that some might turn out to look good. In practice, most researchers estimate more than one specification of the process they are studying and then try to determine which of them is best. This judgment must be based on a blending of economic and econometric analysis, including considerations covered in Chapters 11 through 13.

Finally, it is often the case that our interest is in determining just the effect of one variable on another. For example, we might want to estimate the effect of education on earnings but we might be unconcerned with the role of experience. It is tempting to estimate just a simple regression of earnings on education and look at the slope coefficient. However, this is not what we are interested in. The analysis above shows that the estimate of the slope in the simple regression is not a good estimate of the *ceteris paribus* effect of education on earnings. Also, it is not a good estimate of what was identified in the preceding section as the total effect of education on earnings, because to calculate that we would need a simultaneous-equation model.

To generalize this, even when we are interested in the effect of a particular variable, it is necessary to specify and estimate a regression model that fully reflects the behavior of the economic process at work. Sometimes the variables that we are not interested in are called ***statistical controls.*** For example, we might say that the multiple regression (7.15) estimates the effect of education on earnings, controlling for experience. Similarly, the coefficient on *DRACE* in (7.33) estimates the effect of race (i.e., the effect of being black rather than white) on earnings, controlling for education. What all this amounts to saying is that the only effects that can be estimated in a single-equation, multiple regression model are the *ceteris paribus* (direct) effects and that the proper way to estimate these effects is with a correctly specified model.

Multicollinearity

In the data used to estimate a multiple regression model, it is usually the case that there is some correlation or (more technically) some degree of linear dependence among the explanatory variables. For example, in Figure 7.6 a case is illustrated in which there is a direct relation between X_1 and X_2. In other cases there may be a correlation between explanatory variables even when they are not connected by a behavioral relation. The general model and our method of estimation accept this situation as valid; indeed, the absence of any relations among the explanatory variables is a very special case that we rarely encounter in econometrics. The main consequence of this situation, so far, has been that we must be careful to interpret regression coefficients as direct effects and to recognize that there may be indirect effects as well.

In addition, when two explanatory variables have a very high correlation or when there are some other special relations among the explanatory variables, the situation has some unfortunate consequences for statistical inference. We will examine these in Chapter 13. Loosely speaking, it becomes very difficult to disentangle the separate effects of the explanatory variables on the dependent variable. For example, suppose that aggregate consumption depends on aggregate income, the Treasury bill rate, and the interest rate on consumer debt. Because of the behavior of financial markets, there is likely to be a high correlation between the two interest rates over time. One might guess that it would be difficult to determine the effects of each interest rate separately with any great precision because the two variables might be nearly linear transformations of each other. A common consequence of this is that if we happen to add a few new observations to the data set, or drop a few from it, the new regression coefficients may be very different from the original ones.

This situation is known as ***multicollinearity.*** From the point of view of specifying the model, it does not indicate any mistake. Rather, multicollinearity arises from the nature of the data, and usually we have to accept it as part of reality. Multicollinearity is common in time-series regressions, because several of the explanatory variables may increase over time and therefore be highly correlated.

However, consider the possibility that there is a perfect correlation ($r = 1$) between a pair of explanatory variables, or (more generally) that there is a perfect linear dependence among the explanatory variables. Technically, this is a limiting case of multicollinearity, and indeed it is called ***perfect multicollinearity.*** In this situation, the OLS method no longer can produce estimates. The point of difficulty may be seen in the case with two explanatory variables: the denominators in (7.9) and (7.8), which define the estimators for $\hat{\beta}_1$ and $\hat{\beta}_2$, become equal to zero.

Although this would seem to complicate matters for us immensely, it turns out not to be much of a problem. In contrast to regular multicollinearity, which is a situation that occurs naturally in data, perfect multicollinearity nearly always is the result of making a mistake in the specification of the model. The remedy is simple: respecify the model appropriately. To understand this, we consider several cases in which perfect linear dependence can arise in a regression model. These mistakes share the characteristic that they include in the specification some variable that is not really needed or that does not bring new information to the model; in this sense, the mistaken specification is redundant.

A very special case of perfect linear dependence occurs if an explanatory variable, X_j, is a constant. In this case, the coefficient β_j plays the same role as the intercept β_0 in the regression specification: β_0 and the term $\beta_j X_j$ are constants that are just added in during the determination of Y. There is no unique way for any statistical technique to assign some of the constancy to β_0 and the rest to $\beta_j X_j$.

Perfect linear dependence also occurs if one variable is simply a multiple of another. For example, suppose that we are trying to explain the exports of cars from Japan to the United States and that we include both the price of these cars in Japan (measured in yen) and the price in the United States (measured in dollars) among the explanatory variables. If all our observations are from a period of fixed exchange rates during which all the dollar prices were the same multiple of the yen prices, then the two price variables measure exactly the same set of economic facts. It does not make sense to include them both, and because of the linear dependence we could not.

Perfect linear dependence also occurs if some set of the explanatory variables satisfy an additive identity. For example, suppose that we are interested in estimating the marginal propensities to consume (mpc's) out of labor income, property income, and total income. We might think of regressing consumption on these three income variables in one equation. However, since total income equals labor income plus property income, it must be that the mpc out of total income equals the sum of the two type-specific mpc's. Trying to estimate three mpc's is redundant and therefore not necessary; since it involves a linear dependence among the explanatory variables, it is also impossible.

A final case of perfect linear dependence occurs if dummy variables for all the groups of a categorical variable are included in the regression. For example, in our treatment of region in Section 7.3 we distinguished four groups but included only three in the specified earnings function. This was adequate to specify the theory behind the model, because each dummy variable coefficient specified the difference between the estimated intercept for that group and the intercept (β_0) for the excluded group. Including the fourth dummy variable would be redundant and would introduce a linear dependence.

As might be realized from these cases, it is quite possible to specify a model with perfect multicollinearity if the work is done with insufficient thought. Usually, a computer program will detect the situation and give some kind of error message. However, because of either imprecision in the data or design of the computational algorithm, it is possible that a computer program might not detect the situation and it would produce some calculations. In this case, the user would think he has an estimated regression when in fact he has nonsense.

Problems

Section 7.1

7.1 Formulate a multiple regression model showing how the quantity demanded of a certain product depends on both the price of the product and the income of consumers. What are the anticipated signs of the coefficients?

7.2 What is the graphical interpretation of the demand model estimated from the specification in Problem 7.1?

⋆ **7.3** Continuing Problem 7.2, if income is fixed at a certain amount, what is the graphical interpretation of the relation between predicted demand and price? How does this graph illustrate the *ceteris paribus* concept?

7.4 Based on Equation (7.15), what is the impact on predicted earnings of gaining a college education ($ED = 16$) rather than stopping after completing high school ($ED = 12$)? Assume that EXP is held constant.

⋆ **7.5** Consider two men of age 35. Suppose that the first has four more years of schooling than the second and therefore has four fewer years of working experience. Based on Equation (7.15), who has greater predicted earnings? By how much?

7.6 Show that if the covariance between X_1 and X_2 is zero, the estimator given in Equation (7.8) is identical to the slope estimator in the simple regression of Y on X_2.

Section 7.2

7.7 Based on Equation (7.29), what is the impact on predicted consumption of an increase in the interest rate from 5 percent to 7 percent?

7.8 Suppose that inflationary forces increase both the interest rate and the rate of inflation by three percentage points. Based on Equation (7.29), determine the effect of these changes on predicted consumption.

7.9 Suppose that a one-percentage-point increase in the rate of interest causes DPI to decrease by 200 million dollars. Based on Equation (7.29), determine the total effect of this change in the rate of interest.

7.10 Determine the values of $\overline{R}^2$ for Equations (7.29) and (6.10).

7.11 Derive the relation in Equation (7.28).

Section 7.3

★ **7.12** Based on Equation (7.33), what is the predicted level of earnings for a black man with 16 years of schooling? For a white man with 12 years of schooling?

7.13 Graph the estimated regression (7.33), clearly labeling all its features.

7.14 Based on Equation (7.41), determine the predicted levels of earnings for high school graduates ($ED = 12$) in each of the four regions of the country.

★ **7.15** Suppose that you want to estimate the impact of education and marital status on the earnings of women. If the data show three marital status categories (single, married, and divorced), how would you set up a regression model?

7.16 Based on Equation (7.41) and Figure 7.3, determine the coefficients of the estimated regression of *EARNS* on *ED* and the regional dummies if the West is the excluded region.

7.17 Does the estimated coefficient on *DRACE* in Equation (7.33) give the total effect of race on earnings? Explain.

7.18 Based on Equation (7.36), determine the predicted rate of inflation in 1960 and compare this with the actual rate. Now, try to predict the rate of inflation for 1980, and compare whatever prediction you make with the actual rate. Explain.

Section 7.4

7.19 Based on Equation (7.46), what is the effect on the predicted earnings of a person with 25 years of experience gaining one more year? What about a person with 35 years of experience?

7.20 Based on Equation (7.46), determine the effect on predicted earnings of a five-year increase in experience for a worker already having 20 years of experience. Do this first using an approximation based on the slope of the implied relation and then using an exact calculation in the estimated regression.

7.21 Verify that the peak in earnings in Figure 7.5 occurs at about 31.4 years of experience.

7.22 Suppose that we have data on a factory's average cost of production and the amount of output in different periods. Specify a regression model that could estimate a U-shaped average cost curve.

Section 7.5

7.23 Suppose that the quantity demanded of a certain product depends on its price and consumers' income. Formulate a constant-elasticity regression model for estimating this demand relation. What are the anticipated signs of the estimated elasticities?

⋆ **7.24** In a cross-section context, suppose that output depends on labor and capital inputs. Formulate a regression model for estimating the output elasticities of labor and capital.

⋆ **7.25** In a time-series context, suppose that "disembodied technical progress" leads the output yielded by all combinations of capital and labor inputs to grow at a fixed rate per year. Assuming constant output elasticities for labor and capital, formulate a regression model for estimating the rate of disembodied technical progress.

7.26 Modify the model formulated in Problem 7.25 to take account of the effect of "energy restrictions" that prevailed during three years. Explain clearly the econometric assumption underlying this modification.

7.27 In the earnings function estimated as Equation (7.52), would it make sense to have all the regressors in logarithmic form? Explain.

Section 7.6

7.28 Supposing that the rate of inflation affects the interest rate, and making any other economic assumptions that seem appropriate, illustrate the causal nexus determining aggregate consumption as estimated in Equation (7.29). Where could the Federal Reserve discount rate enter?

7.29 Suppose that a properly specified earnings function includes *ED*, *EXP*, and *DRACE* as explanatory variables. Illustrate the causal nexus determining earnings and explain the linkages.

7.30 Compare the estimated effect of education in Equations (7.33) and (6.7) from the point of view of possible misspecification.

7.31 Suppose that we wish to estimate the effect of being a union member on workers' earnings, using cross-section data. Specify a regression model that would be appropriate for estimating this effect. What theory or assumptions are required to make it appropriate?

7.32 Consider the Phillips curve model in Equation (7.36). Could this be estimated using data just for 1965 through 1970? Explain.

7.33 Suppose that the fourth dummy variable *DNEAST* were added to the regression specification in Equation (7.41) and that a computer provided "estimates." Try to interpret all the coefficients.

Probability Distributions

Probability Theory

The theory of statistical inference, which allows us to assess the relations between the estimated coefficients in a regression model and the true values of these parameters, is based on the discussion of random varibles and probability distributions in Chapters 9 and 10. As helpful background for understanding that material, this chapter discusses some elementary probability theory. However, later chapters are written so that this one can be skipped without any loss of continuity.

8.1 Outcomes and Probabilities

Notions of probability arise in situations of uncertainty, where several possible outcomes are candidates to be the one actual outcome of an ***activity.*** (In more formal presentations, these basic activities are called ''experiments.'') Some of the best examples are gambling games, and it was in their study that the subject of probability was developed.

To provide a context for introducing some basic probability concepts, consider flipping a coin. We assume that a coin never lands on its edge, and so we identify only two possible ***outcomes:*** the coin lands on its head side (H) or it lands on its tail side (T). In general, each possible outcome is denoted by e_i. In coin flipping the two outcomes are denoted by e_1 and e_2, and they are identified as $e_1 = H$, $e_2 = T$. In general, all the possible outcomes of an activity are collectively known as the ***sample space,*** which is denoted by S. For coin flipping, $S = \{e_1, e_2\}$.

We consider two ways of arriving at the concept of probability. First, we could take a given coin and (hypothetically) flip it a great many times, recording the actual outcomes. At any stage, we could calculate the proportion of flips that resulted in heads and the proportion that resulted in tails. These two proportions are relative frequencies and they sum to 1. Now, if we think of the number of flips as approaching infinity, we could define the ***probability*** of H occurring as the proportion of flips that result in H. This probability is denoted by Pr (H) or Pr (e_1). Although this is not a practical approach to determining probabilities, it is a useful conceptual approach. Second, we could analyze the coin physically in order to determine the symmetry of its design and weight distribution. If our understanding of the natural laws of physics is correct, this analysis could lead us to proclaim knowledge of Pr (e_1) and Pr (e_2) without any experimenting. This approach focuses on characterizing the uncertainty regarding the single result of an activity that will occur, and the practicality of repeating the activity is not an issue. We rely on both approaches for interpretation.

Based on these concepts, two properties of probabilities emerge. Thinking of probabilities as relative frequencies, we realize that none can be negative and that none can be greater than 1. For convenience in describing some sample spaces, we allow a probability to be zero. Hence for any outcome

$$0 \leq \Pr(e_i) \leq 1 \tag{8.1}$$

And again thinking of probabilities as realtive frequencies, we realize that they must sum to 1:

$$\sum \Pr(e_i) = 1 \tag{8.2}$$

where the summation is understood to include all the outcomes in the sample space.

As an example, we consider a ***fair*** coin. By this we mean that if the coin were flipped a great many times, then in the limit half the actual outcomes would be H and half would be T. Hence Pr (H) $= 1/2$ and Pr (T) $= 1/2$. Alternatively viewed, we mean that the physics of the coin is such that each possible outcome is judged equally likely. Since there are two outcomes, Pr (H) $= 1/2$ and Pr (T) $= 1/2$.

It is possible to use the game of flipping a fair coin to define a variety of different activities. The first activity we consider is just the flip of one coin. As we have seen, there are two outcomes, and Pr (e_1) $= 1/2$ and Pr (e_2) $= 1/2$. The second activity is flipping a coin twice in a row; here it is important to identify the sequence as well as the result of each flip in distinguishing the outcome. There are four outcomes: $e_1 = HH$, $e_2 = HT$, $e_3 = TH$, and $e_4 = TT$. A little physics and logic lead to the conclusion that these outcomes are equally likely, so Pr (e_i) $= 1/4$ for each. The third activity is flipping a coin three times in a row. There are eight outcomes, identified in Table 8.1, and Pr (e_i) $= 1/8$ for each of them. It should be noted that these outcomes can be identified and labeled in any order; none has any special claim to being e_1.

TABLE 8.1 Outcomes for Three Flips of a Coin

Outcome		Pr (e_i)	A	B	C	D
HHH	e_1	1/8			✓	
HHT	e_2	1/8	✓			
HTH	e_3	1/8	✓			✓
THH	e_4	1/8	✓	✓		
HTT	e_5	1/8				✓
THT	e_6	1/8		✓		
TTH	e_7	1/8		✓		✓
TTT	e_8	1/8		✓		✓

8.2 Events

Often we have special interest in groups, or sets, of outcomes. For example, we might be concerned with getting exactly two heads in the activity of flipping a coin three times. Referring to Table 8.1 we see that e_2, e_3, and e_4 are all the outcomes meeting this description. It is convenient to define an ***event*** as a set of outcomes. Letting A be the event of getting exactly two heads, then

$$A = \{e_2, e_3, e_4\} \tag{8.3}$$

Our basic concepts lead us to define the probability of an event as being equal to the sum of the probabilities of the constituent outcomes. In this case,

$$\Pr(A) = \Pr(e_2) + \Pr(e_3) + \Pr(e_4) \tag{8.4}$$

The logic of this is best seen in terms of the proper treatment of relative frequencies. By Pr (A) we mean what would be the number of actual outcomes with two heads as a proportion of the number of repetitions of the three-flip activity, in the limit. A little algebra justifies (8.4) and the general definition.

Similarly, we might be concerned with getting a tail on the first flip. This event, B, is defined as

$$B = \{e_4, e_6, e_7, e_8\} \tag{8.5}$$

and Pr $(B) = 1/2$. A third event, C, is getting no tails at all. This leads to

$$C = \{e_1\} \tag{8.6}$$

with Pr $(C) = 1/8$.

Two events are said to be ***mutually exclusive*** if they have no outcomes in common. In our three-flip activity, A and C are mutually exclusive; also, B and C are mutually exclusive. A and B are not mutually exclusive because they have e_4 in common.

The basic notions used in defining events are sometimes illustrated in ***Venn diagrams.*** In this type of diagram a point is used to represent an outcome, and groups of points make up events. A Venn diagram for an undescribed activity

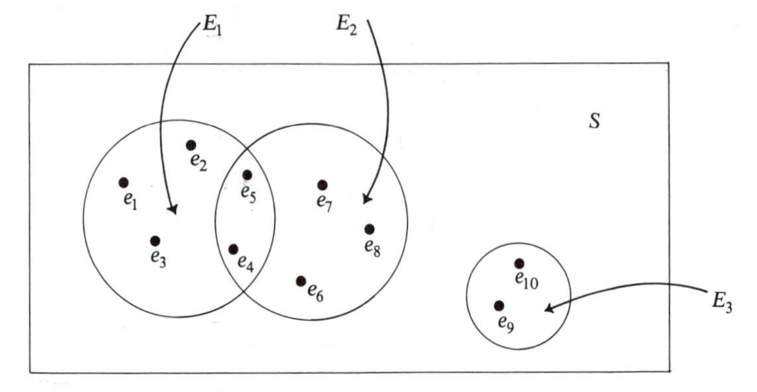

FIGURE 8.1 A Venn diagram illustrates all the outcomes of an activity and some events defined as sets of the outcomes. In this case, we see that events E_1 and E_2 are not mutually exclusive, because they have outcomes e_4 and e_5 in common. The outcomes e_4 and e_5 constitute the intersection $E_1 \cap E_2$.

is given in Figure 8.1. The points represent all the outcomes (i.e., the sample space) of the activity, and this set of points is enclosed in a boundary to indicate this. Three events are defined as

$$E_1 = \{e_1, e_2, e_3, e_4, e_5\} \tag{8.7}$$

$$E_2 = \{e_4, e_5, e_6, e_7, e_8\} \tag{8.8}$$

$$E_3 = \{e_9, e_{10}\} \tag{8.9}$$

Events E_3 and E_1 are mutually exclusive, as are E_3 and E_2. Graphically, we see that E_3 has no points in common with either E_1 or E_2. By contrast, E_1 and E_2 have e_4 and e_5 in common, so they are not mutually exclusive.

A Venn diagram can be drawn to reflect the design of the activity. For example, in the two-flip activity the outcomes can be arranged in a matrix like that in Figure 8.2. In more complicated cases such a diagram can help simplify the identification of the outcomes that constitute an event defined in terms of the procedure of the activity.

8.3 Unions and Intersections

Sometimes events are defined as special combinations of other events. One type of combination is the union of events, and another is their intersection.

The ***union*** of two events is a new event whose constituent outcomes are all those in either or both of the two events. Notationally, the union of E_1 and E_2 is $E_1 \cup E_2$, which is read as "E_1 union E_2." A specific outcome is in $E_1 \cup E_2$ if it is in E_1 only, or if it is in E_2 only, or if it is in both. Although we could

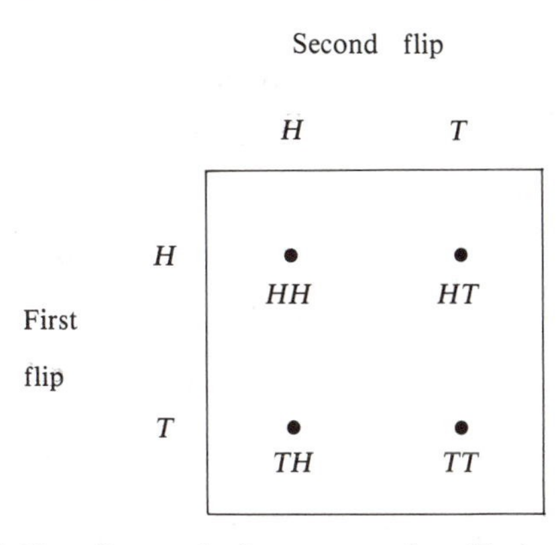

FIGURE 8.2 In this Venn diagram the four outcomes from flipping a coin twice are arranged in a matrix. The first row contains all the outcomes having a head on the first flip, and the second row contains all those having a tail on the first flip. Similarly, the first column contains all the outcomes having a head on the second flip, and the second column contains all those having a tail on the second flip.

assign a new name to this event, such as E_3 (defined as $E_3 = E_1 \cup E_2$), we usually refer to it simply as $E_1 \cup E_2$. In our three-flip example

$$A \cup B = \{e_2, e_3, e_4, e_6, e_7, e_8\} \tag{8.10}$$

$$A \cup C = \{e_1, e_2, e_3, e_4\} \tag{8.11}$$

$$B \cup C = \{e_1, e_4, e_6, e_7, e_8\} \tag{8.12}$$

Notice that since (8.10) just identifies the constituent outcomes of $A \cup B$, there is no double listing for e_4, which is in both A and B.

The ***intersection*** of two events is a new event whose constituent outcomes are only those in both of the two events. Notationally, the intersection of E_1 and E_2 is $E_1 \cap E_2$, which is read as "E_1 intersect E_2." A specific outcome is in $E_1 \cap E_2$ if it is in E_1 and if it is also in E_2. This "and" combination contrasts with the "either-or-both" combination of a union. In our three-toss example

$$A \cap B = \{e_4\} \tag{8.13}$$

$$A \cap C = \{\varnothing\} \tag{8.14}$$

$$B \cap C = \{\varnothing\} \tag{8.15}$$

where $\varnothing$ indicates an empty set, in which there are no outcomes. There are no outcomes in $A \cap C$ and no outcomes in $B \cap C$. This occurs because A and C are mutually exclusive, as are B and C.

The probability of an event that is defined either as a union or an intersection is equal to the sum of the probabilities of its constituent outcomes. This basic definition is the same as our previous definition of the probability of any event. For example, Pr $(A \cup B) = 6/8$ while Pr $(A \cap B) = 1/8$.

Moving up from the basic definition, the probability of a union also can be determined from the probabilities of the combining events according to an ***addition rule:***

$$\Pr(E_1 \cup E_2) = \Pr(E_1) + \Pr(E_2) - \Pr(E_1 \cap E_2) \qquad (8.16)$$

The logic of this stems from the basic probability definition. The outcomes in $E_1 \cap E_2$ are in E_1 and they are in E_2. Thus, the sum $\Pr(E_1) + \Pr(E_2)$ double counts the probability associated with the outcomes in $E_1 \cap E_2$, and therefore $\Pr(E_1 \cap E_2)$ is subtracted to keep the counting right. Note that if E_1 and E_2 are mutually exclusive, $\Pr(E_1 \cap E_2)$ is zero and no double counting exists. In our three-flip activity,

$$\begin{aligned} \Pr(A \cup B) &= \Pr(A) + \Pr(B) - \Pr(A \cap B) \\ &= 3/8 + 4/8 - 1/8 = 6/8 \end{aligned} \qquad (8.17)$$

$$\begin{aligned} \Pr(A \cup C) &= \Pr(A) + \Pr(C) - \Pr(A \cap C) \\ &= 3/8 + 1/8 - 0 = 4/8 \end{aligned} \qquad (8.18)$$

$$\begin{aligned} \Pr(B \cup C) &= \Pr(B) + \Pr(C) - \Pr(B \cap C) \\ &= 4/8 + 1/8 - 0 = 5/8 \end{aligned} \qquad (8.19)$$

In the Venn diagram of an undescribed activity shown in Figure 8.1, the union of E_1 and E_2 (i.e., $E_1 \cup E_2$), consists of the outcomes e_1 through e_8. The intersection of E_1 and E_2 (i.e., $E_1 \cap E_2$), consists of e_4 and e_5 only. With regard to the probability of a union such as $\Pr(E_1 \cup E_2)$, it is clear that $\Pr(e_4)$ and $\Pr(e_5)$ are double counted in the sum $\Pr(E_1) + \Pr(E_2)$. Since the outcomes e_4 and e_5 constitute $E_1 \cap E_2$, $\Pr(E_1 \cap E_2)$ is subtracted away in the addition rule.

These concepts extend easily to more than two events entering into a combination. A union of events consists of the outcomes in any or all of the combining events, whereas an intersection of events consists only of the outcomes in all the combining events.

8.4 Conditional Probability and Independence

Our basic idea of probability is that any of the possible outcomes in the sample space might occur as the actual outcome of the activity, and concern often focuses on the occurrence of some subset of the possible outcomes. For example, in the activity of flipping a coin three times, $\Pr(A)$ is "the probability of two heads occurring, given that the set of outcomes that might possibly occur is the sample space S." The condition that any of the outcomes in S may occur helps determine the calculation of the probability that two heads will occur. To highlight this, we may denote Pr (A) by the fuller statement Pr $(A|S)$. This is read as "the probability of A, given S," which concisely conveys our basic idea.

Now, it might be that we already know that the first flip resulted in a tail, or it might be that we are considering the hypothetical situation that the first flip results in a tail. In either context, suppose that we are still concerned with the occurrence of two heads, and we wish to determine "the probability of two heads occurring, given that the set of outcomes that might possibly occur is B," where B is the event of getting a tail on the first flip. This probability, which we denote by Pr $(A|B)$, is known as a ***conditional probability,*** and it might be equal to or different from the unconditional probability amount Pr (A).

Returning to Table 8.1, we see that if B occurs (tail on first), then the final result must be e_4, e_6, e_7, or e_8. Among these outcomes, only e_4 also corresponds to A (two heads). Hence our concern now is focused on the occurrence of e_4, given that e_4, e_6, e_7, or e_8 is the set of now-possible outcomes of the activity. Since each of these four outcomes is equally likely, the probability of e_4 occurring is 1/4 now. Thus we may state that Pr $(A|B) = 1/4$. This conditional probability is different from the unconditional Pr (A), which is 3/8. Notice that to determine Pr $(A|B)$ we calculate a ratio in which the numerator is based on the set of outcomes that corresponds both to A (two heads) and B (tail on first); this is the set $A \cap B$. The denominator of the ratio is based on the set of outcomes that corresponds to B; this is simply the set B.

The probability calculation in this example generalizes quite readily. For any two events, E_1 and E_2,

$$\text{Pr }(E_1|E_2) = \frac{\text{Pr }(E_1 \cap E_2)}{\text{Pr }(E_2)} \tag{8.20}$$

This expression determines the conditional probability of E_1 given E_2. In essence, (8.20) uses the conditioning information (i.e., the occurrence of E_2) to restrict our concern for the occurrence of E_1 to just those outcomes that occur in conjunction with E_2 (i.e., to just $E_1 \cap E_2$). In the previous example, where we followed more basic principles, we formed the ratio of the numbers of equally likely outcomes. Following (8.20) instead, we calculate the ratio of probabilities, yielding

$$\text{Pr }(A|B) = \frac{\text{Pr }(A \cap B)}{\text{Pr }(B)} = \frac{1/8}{4/8} = 1/4 \tag{8.21}$$

as before.

In some probability applications it is easy to determine probability amounts like Pr $(E_1|E_2)$ and Pr (E_2) but difficult to get down to the basic level to determine Pr $(E_1 \cap E_2)$. In such a case, (8.20) can be reexpressed as

$$\text{Pr }(E_1 \cap E_2) = \text{Pr }(E_1|E_2) \cdot \text{Pr }(E_2) \tag{8.22}$$

This is especially helpful in the case of independence.

Among the events of an activity, two events E_1 and E_2 are statistically ***independent*** if and only if

$$\Pr(E_1|E_2) = \Pr(E_1) \qquad \text{for } E_1, E_2 \text{ independent} \tag{8.23}$$

That is, when the two events are independent, the conditional probability of one occurring (conditional on the other) is the same as its unconditional probability of occurring. Oppositely, $\Pr(E_1|E_2) \neq \Pr(E_1)$ corresponds to E_1 and E_2 being not independent. [Note that independence is symmetric, so that $\Pr(E_2|E_1) = \Pr(E_2)$ also holds.] For example, let D be the event of getting a tail on the second flip in a three-flip activity (see Table 8.1):

$$D = \{e_3, e_5, e_7, e_8\} \tag{8.24}$$

Is getting a tail on the second coin independent of getting a tail on the first—is D independent of B? From (8.23), they would be independent if $\Pr(D|B) = \Pr(D)$. Now,

$$\Pr(D|B) = \frac{\Pr(D \cap B)}{\Pr(B)} = \frac{\Pr(e_7) + \Pr(e_8)}{\Pr(B)} = \frac{2/8}{4/8} = 1/2 \tag{8.25}$$

And by basic concepts $\Pr(D) = 4/8 = 1/2$, so D and B are independent. When two events are independent, the conditioning information that one of them did or must occur does not alter the chance that the other event might occur. In coin flipping, the occurrence of a tail on the first flip in no way affects the chance of a tail occurring on the second flip; this is consistent with the laws of physics, because a coin has no memory and is not physically altered by its history of outcomes.

For events that are independent, we may substitute $\Pr(E_1)$ for $\Pr(E_1|E_2)$ in (8.22), yielding the ***multiplication rule***

$$\Pr(E_1 \cap E_2) = \Pr(E_1) \cdot \Pr(E_2) \qquad \text{for } E_1, E_2 \text{ independent} \tag{8.26}$$

This rule holds true if and only if E_1 and E_2 are independent, and it may be taken as an alternative form of the definition (8.23).

The multiplication rule for independent events is helpful in determining the probabilities of outcomes in activities that are designed as sequences of repetitions of simpler activities. For example, suppose that we have an unfair coin for which $\Pr(H) = 2/3$ and $\Pr(T) = 1/3$. Consider the activity of flipping a coin two times, for which the Venn diagram is given in Figure 8.2. Let H_1 be the event of getting a head on the first toss and H_2 be the event of getting a head on the second toss. A physical analysis of the activity leads one to realize that H_1 and H_2 are independent. Thus we realize that $\Pr(H_1) = \Pr(H_2) = \Pr(H)$. A similar analysis holds for T_1 and T_2. Therefore, the multiplication rule can be used to determine the probabilities of the four outcomes:

$$\Pr(HH) = \Pr(H_1) \cdot \Pr(H_2) = (2/3)(2/3) = 4/9 \tag{8.27}$$

$$\Pr(TH) = \Pr(T_1) \cdot \Pr(H_2) = (1/3)(2/3) = 2/9 \tag{8.28}$$

$$\Pr(HT) = \Pr(H_1) \cdot \Pr(T_2) = (2/3)(1/3) = 2/9 \tag{8.29}$$

$$\Pr(TT) = \Pr(T_1) \cdot \Pr(T_2) = (1/3)(1/3) = 1/9 \tag{8.30}$$

8.5 A Pair of Dice

Rolling a pair of dice is an activity that nicely illustrates the probability concepts developed in this chapter. In this example we view the dice as distinguishable: one is red and the other is green.

Since any of the outcomes on the red die can occur in conjunction with any on the green die, there are 36 possible outcomes for the activity. These are enumerated in Figure 8.3, which arranges the outcomes of the activity in matrix form according to the outcome on each die. Since the laws of physics show that each of the outcomes of the activity is equally likely, the probability of each is 1/36. Alternatively, we might start by recognizing that the six outcomes for one die are equally likely events, so that the probability of each is 1/6. Then, recognizing that the outcomes on the two dice are independent, we obtain the probability of each of the 36 outcomes of the activity by using the multiplication rule—each $\Pr(e_i) = 1/36$.

Determination of the probabilities associated with events is straightforward,

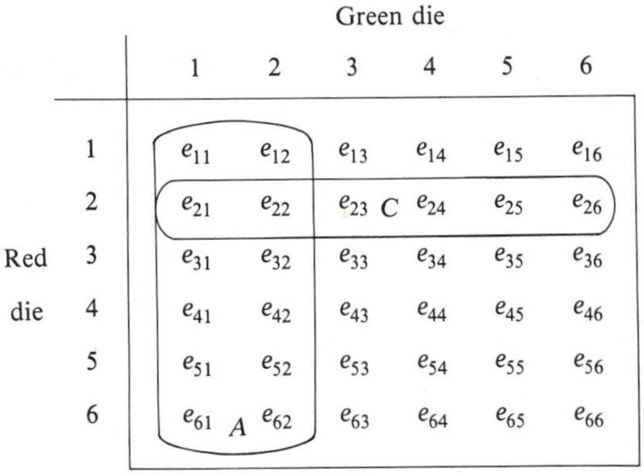

Green die

		1	2	3	4	5	6
	1	e_{11}	e_{12}	e_{13}	e_{14}	e_{15}	e_{16}
	2	e_{21}	e_{22}	e_{23} C	e_{24}	e_{25}	e_{26}
Red	3	e_{31}	e_{32}	e_{33}	e_{34}	e_{35}	e_{36}
die	4	e_{41}	e_{42}	e_{43}	e_{44}	e_{45}	e_{46}
	5	e_{51}	e_{52}	e_{53}	e_{54}	e_{55}	e_{56}
	6	e_{61} A	e_{62}	e_{63}	e_{64}	e_{65}	e_{66}

FIGURE 8.3 The 36 outcomes from rolling a pair of dice are illustrated in a Venn-like diagram that arranges the outcomes in a matrix. Each row contains all the outcomes having a particular value occurring on the red die, and each column contains all those having a particular value on the green die. This arrangement facilitates the identification of some events. For example, event A is "getting a 2 or less on the green die," and C is "getting a 2 on the red die."

		Green die					
		1	2	3	4	5	6
Red die	1	2	3	4	5	6	7
	2	3	4	5	6	7	8
	3	4	5	6	7	8	9
	4	5	6	7	8	9	10
	5	6	7	8	9	10	11
	6	7	8	9	10	11	12

FIGURE 8.4 In this diagram the outcome identifiers are replaced with the sum of the values occurring on the dice for that outcome. This facilitates the identification of events defined in terms of the sum of the dice values. Event B is "getting a sum of 5." "Getting a 2 or less on the green die" (event A in Figure 8.3) corresponds to all the outcomes in the first two columns.

but needs some care. For example, let event A be defined as "getting a 2 or less on the green die." Since this definition places no restriction on the red die, the outcomes making up A are given in the first two columns of the matrix in Figure 8.3. Hence $\Pr(A) = 12/36$.

In dice throwing, the outcomes often are characterized by the sum of the two numbers occurring. Figure 8.4 shows the matrix of Figure 8.3 rewritten with sums replacing the outcome identifiers. The event B, defined as "getting a sum of 5," is shown to correspond to four outcomes. Hence $\Pr(B) = 4/36$.

The event $A \cap B$, "getting a 2 or less on the green die and a sum of 5," corresponds to $\{e_{41}, e_{32}\}$ only, and $\Pr(A \cap B) = 2/36$. The event $A \cup B$ corresponds to all the outcomes in the first two columns, plus e_{23} and e_{14}. Thus, from the basic principles $\Pr(A \cup B) = 14/36$. Alternatively, from the addition rule,

$$\begin{aligned} \Pr(A \cup B) &= \Pr(A) + \Pr(B) - \Pr(A \cap B) \\ &= 12/36 + 4/36 - 2/36 = 14/36 \end{aligned} \tag{8.31}$$

As noted, $\Pr(B) = 4/36$. However, if we wish to determine the probability of "getting a sum of 5, given that the green die is 2 or less," then we formulate and calculate

$$\Pr(B|A) = \frac{\Pr(B \cap A)}{\Pr(A)} = \frac{2/36}{12/36} = \frac{6}{36} \tag{8.32}$$

In this case $\Pr(B|A) > \Pr(B)$, and we realize that A and B are not independent.

Since the outcome on the green die cannot affect the outcome on the red die, events defined solely in terms of one die are independent of events defined solely in terms of the other. Let C be "getting a 2 on the red die," for which Pr (C) = 6/36. Then the probability of "getting a 2 or less on the green die and a 2 on the red die" is given by

$$\Pr(A \cap C) = \Pr(A) \cdot \Pr(C) = (12/36)(6/36) = 2/36 \qquad (8.33)$$

using knowledge of independence. Alternatively, basic principles identify $A \cap C$ as $\{e_{21}, e_{22}\}$, and yield the same probability of course.

It should be noted that the toss of a pair of dice is the same as two successive occurrences of tossing a single die. Unless the die has a memory, successive occurrences will be independent. The probability analysis of events defined in terms of two successive occurrences of a die toss is exactly the same as that for a single occurrence with a pair of dice.

Problems

Section 8.1

8.1 Take a fair coin and flip it 20 times. After each flip, calculate the proportion of flips up to that point that have resulted in H and record the results in a table or graph. As the number of flips increase, does the proportion that are heads tend toward .5?

★ **8.2** Suppose that the four sides of a tetrahedron are labeled "1," "2," "3," and "4," and that the result of any flip is the number that is face down.

(a) List all the outcomes for the activity of flipping a tetrahedron twice.

(b) What is the probability of each of these outcomes?

(c) How many outcomes are there in a three-flip activity with a tetrahedron?

Section 8.2

8.3 For the three-flip coin activity, define (in terms of their constituent outcomes) the following events and determine their probabilities:

(a) Three tails

(b) No heads

(c) Head on first flip with tail on second

(d) Two tails

8.4 Draw a Venn diagram similar to Figure 8.1 to represent the outcomes of the three-flip coin activity, and identify on it the events A, B, and C defined in Table 8.1.

8.5 Devise a simple three-dimensional diagram to represent the outcomes of the three-flip coin activity, and identify on it the four events defined in Table 8.1.

★ **8.6** Suppose that there are 10 balls in a hat: one labeled "5," two "6," three "7," and four "8."

(a) What is the probability of drawing a particular ball?

(b) What is the probability of drawing a ball labeled "5"? a "6"? a "7"? a "8"?

(c) In a draw from the hat, what is the probability that the number on the ball will be less than or equal to 7?

★ **8.7** For the three-flip activity, are the two events in each of the following pairs mutually exclusive?

(a) Two tails, two heads

(b) Head on first flip, two tails

(c) Tail on first flip, tail on third flip

Section 8.3

★ **8.8** For the unions of the pairs of events in Problem 8.7, write out the set listings of the outcomes in each combination, and apply basic probability concepts to determine the probabilities of each union.

8.9 For the intersections of the pairs of events in Problem 8.7, do the same analysis as in Problem 8.8.

★ **8.10** Use the addition rule to determine the probability of the union of each of the pairs of events in Problem 8.7.

8.11 Draw a Venn diagram illustrating a case in which the intersection of three events has one outcome. Label all the two-event intersections in the diagram.

Section 8.4

8.12 For the three-flip coin activity, what is the probability of getting a tail on the third flip given that

(a) There is a tail on the first flip?

(b) There is a tail on the first two flips?

(c) There is a total of one tail?

(d) There is a total of two tails?

★ **8.13** Suppose that a worker is twice considered for a bonus, that each time the chance of getting it is 20 percent, and that getting it once does not affect the chance of getting it again. Determine the probability of the worker's

(a) Never getting a bonus

(b) Getting it once

(c) Getting it twice

Section 8.5

8.14 The game of craps is played by rolling a pair of dice. If a player rolls a sum of 7 or 11 on the first roll, he wins; if he rolls a sum of 2, 3, or 12 (craps), he loses.

(a) What is the probability of winning on the first roll?
(b) What is the probability of losing on the first roll?
(c) What is the probability of neither winning nor losing on the first roll?

* **8.15** In the game of craps, if a player neither wins nor loses on the first roll, whatever sum appeared is called the "point." After a point is established, the player keeps on rolling the dice until his point appears (an ultimate win) or a 7 appears (an ultimate loss).
(a) What is the probability of winning given that the point is 8?
(b) What is the probability that a player just approaching the table will ultimately win with a point of 8 in his first game?

Random Variables and Probability Distributions

In this chapter and the next we develop the theory of random variables and probability distributions, which is part of the general study of mathematical statistics. The topics are rather abstract, and we rely on intuition rather than formal derivation as much as possible. Our treatment of random variables and probability distributions emphasizes their similarity to data variables and frequency distributions, so it would be helpful to review Chapter 4 at this time.

9.1 Discrete Random Variables

In Chapter 2 we defined a data variable as a measurable characteristic that we can observe and for which we can collect data. We are now interested in developing a theoretical framework for thinking about the process that generated some data we have. In contrast to the economic processes that we thought about in connection with regression theory, these chance processes have no inputs.

Our general notion is that underlying any data is some process for producing numbers (i.e., for generating the data). Especially at the beginning of our study, it is useful to describe or imagine some physical activity that is going on inside a box: there might be a woman throwing dice, a man picking numbers out of a hat, or a computer churning away; all that we can see is that numbers are generated. The numbers that get produced are called the ***outcomes*** of the process.

Two important ideas about the processes we consider will hold in general. First, the process can produce an unlimited number of outcomes; that is, it goes on forever, so in practice we could never observe all the outcomes that are produced. Second, the outcomes of the process are ***independent*** of each other, in the sense that the numerical value of one outcome is unaffected by whatever values might have occurred for previous outcomes.

The term ***random variable*** is used to refer to a chance process like this. If we observe a set of outcomes from the random variable, this series of numbers is a ***sample*** taken from the random variable. The sample can be analyzed by the methods of descriptive statistics presented in Chapters 3 and 4. Hence we think of a sample drawn from a random variable as being the same as a regular data variable.

To describe or characterize a random variable, we focus our attention on the next outcome that is about to occur. Letting X denote the name of the random variable, our possible knowledge about the next outcome of X has two components, which together are called the ***probability distribution*** of X. The first component is a list or statement of the possible values that X might ever take on, and we denote this generally by $\{X\}$. For example, if X can take on only the values 3, 4, and 5, then the list $\{X\}$ is $\{3, 4, 5\}$. The second component is a list of values or a function that gives information about the probability or likelihood that the next outcome of X will be equal to some particular value or fall in some range of values. This second component is denoted by $p(X)$ generally.

In working with data in Chapters 2 through 4, a distinction was made between discrete and continuous data variables. The same distinction is made for random variables, and we treat the discrete case first.

A ***discrete random variable*** is one that has a countable number of possible outcomes. For the random variable X, the list of all its possible outcomes is denoted by $\{X\}$. Letting X_k denote a particular possible outcome, we let $\Pr(X = X_k)$ be a shorthand way of writing "the probability that the next outcome of the random variable X is equal to X_k." For a discrete random variable, the component $p(X)$ of its probability distribution can be a list of the probability values corresponding to each possible outcome of X. Hence

$$p(X_k) = \Pr(X = X_k) \tag{9.1}$$

For now we can proceed with a commonsense notion of what "probability" means, and we will use the term interchangeably with "chance." In Section 9.2 we will try to arrive at an understanding of "probability" in a more formal way.

An example will help make all this clearer and will serve to introduce some new ideas. Consider a game based on putting 10 numbered balls into a hat. The hat is shaken well, a single ball is pulled out and its number read aloud, and the ball is replaced. This is repeated over and over, without end. This activity can be described mathematically as a discrete random variable.

TABLE 9.1 Probability Distribution of X

k	X_k	$p(X_k)$
1	5	.1
2	6	.2
3	7	.3
4	8	.4

Suppose that among the 10 balls one is labeled "5," two "6," three "7," and four "8." Since it is equally likely that any one of the 10 balls will be picked on any draw, simple logic and common sense lead us to see that the probability of getting a 5 is 1/10, of getting a 6 is 2/10, of getting a 7 is 3/10, and of getting a 8 is 4/10.

Letting X denote the discrete random variable describing this game, Table 9.1 shows a compact way of writing down all that we know about its probability distribution. The column headed X_k gives the list $\{X\}$ and the column labeled $p(X_k)$ gives the values of $p(X)$. For a discrete random variable this $p(X_k)$ is simply Pr $(X = X_k)$. Alternatively, all the information about the probability distribution could be graphed, as in Figure 9.1. Finally, it is possible to write down the probability distribution in an equation format:

$$p(X_k) = \frac{X_k - 4}{10} \qquad X = 5, 6, 7, 8 \tag{9.2}$$

In this example the equation format is not an especially useful way of writing down the distribution, but in other cases it will be.

The general notion that outcomes of a random variable are independent sometimes leads to surprises for people who are not accustomed to games of chance. For example, suppose that a 7 has just occurred twice in a row. We might

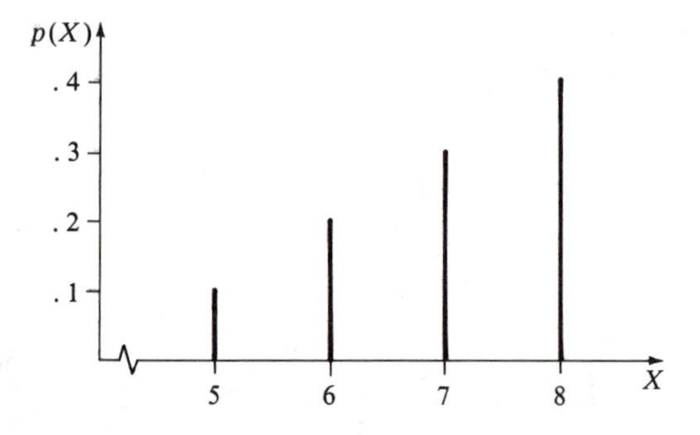

FIGURE 9.1 This graph represents the probability distribution of X, a discrete random variable that is also represented by Equation (9.2) and Table 9.1. For each specific value X_k, the value $p(X_k)$ is the probability that the outcome of X will be equal to X_k; that is, $p(X_k)$ is Pr $(X = X_k)$.

wonder now what the probability is that the next outcome of the game will be a 7 again [i.e., what is Pr $(X = 7)$?]. The implication of independence is that this probability is still 3/10, because the previous history of outcomes has no effect on the probabilities of the different values in $\{X\}$ occurring this time. In the ball-in-hat game, it is essential that the balls be replaced after being drawn and that they be shaken up very well; otherwise, successive drawings would not be independent.

Two simple aspects of discrete probabilities should be noted. First, if X_k is a possibly occurring value for X, we must mean that its probability is positive. That is,

$$p(X_k) > 0 \qquad \text{for } X_k \text{ in } \{X\} \tag{9.3}$$

Second, the sum of all the probabilities associated with the X values is exactly equal to 1:

$$\sum p(X_k) = 1 \qquad \text{including all } X_k \text{ in } \{X\} \tag{9.4}$$

These two aspects are in accord with our commonsense understanding of what "probability" means.

As another example, consider a random variable Y that has a probability distribution given by Table 9.2 and by Figure 9.2. One should verify that $\Sigma\, p(Y_k) = 1$. Without thinking about any underlying physical activity, it should be understandable that Y is some chance process such that the probability that the outcome will be zero is 1/64, that the outcome will be 4 is 15/64, and so on.

It is straightforward to use a probability distribution to determine the probability or chance that the outcome of the random variable will be in some range or interval of numbers. In general, an ***event*** is a set of possible outcomes of a random variable, and for a discrete random variable the probability of an event

TABLE 9.2 Probability Distribution of Y and Computation of μ_Y and σ_Y

(1) k	(2) Y_k	(3) $p(Y_k)$	(4) $Y_k p(Y_k)$	(5) $(Y_k - \mu_Y)^2 p(Y_k)$
1	0	1/64	0	9/64
2	1	6/64	6/64	24/64
3	2	15/64	30/64	15/64
4	3	20/64	60/64	0
5	4	15/64	60/64	15/64
6	5	6/64	30/64	24/64
7	6	1/64	6/64	9/64
			192/64	96/64
			$\mu_Y = 3$	$\sigma_Y^2 = 1.5$
				$\sigma_Y = 1.22$

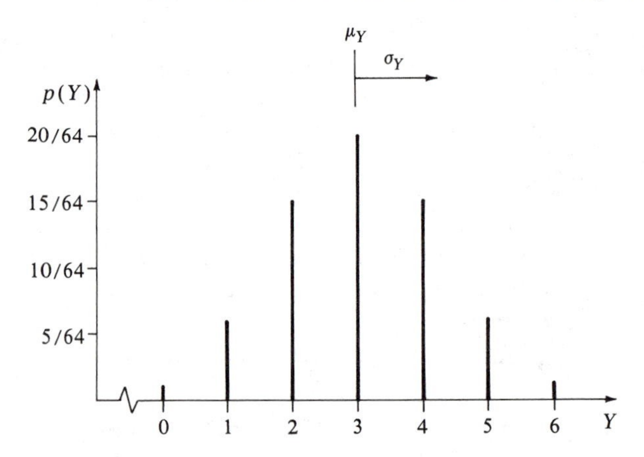

FIGURE 9.2 This graph represents the probability distribution of Y, a discrete random variable that is also represented by Table 9.2. The values of the mean and standard deviation of Y, which are calculated in Table 9.2, are also indicated here ($\mu_Y = 3$, $\sigma_Y = 1.22$). One possible description of Y is that it is the number of heads that might result from tossing a fair coin six times. In Section 9.3 we see that Y has a binomial distribution with $n = 6$ and $P = 1/2$.

is the sum of the probabilities associated with each of the outcomes that make up the event. For example, in the balls-in-hat game

$$\begin{aligned} \Pr(X \le 7) &= p(5) + p(6) + p(7) \\ &= 1/10 + 2/10 + 3/10 = 6/10 = .6 \end{aligned} \tag{9.5}$$

and in the second example above it is easy to see that

$$\Pr(Y \le 3) = 42/64 \tag{9.6}$$

Similarly, it is possible to consider evaluating probabilities such as $\Pr(X_1 \le X \le X_2)$—the probability that X takes on a value between two specified values X_1 and X_2, including these endpoint values. It should be clear that

$$\Pr(6 \le X \le 8) = .9 \tag{9.7}$$

and

$$\Pr(2 \le Y \le 4) = 50/64 \tag{9.8}$$

Often we work with "greater than" inequalities such as

$$\Pr(X \ge 7) = p(7) + p(8) = .7 \tag{9.9}$$

Mean and Standard Deviation

The final concepts we need to develop for our treatment of discrete random variables are the summary features of their probability distributions. Recall that in working with data variables, we found that the mean and standard deviation could summarize a relative frequency distribution. The same ideas apply here.

The ***mean*** of a random variable X is denoted by μ_X (μ is lowercase "mu," the Greek "m") or by $E[X]$, and its definition is

$$\mu_X = E[X] = \sum X_k p(X_k) \tag{9.10}$$

As with data variables, the mean is a typical value, but because of the probabilistic nature of random variables the mean sometimes is referred to as the ***expected value.*** Most often, we use the notation μ_X when we would like to say "mean" and the notation $E[X]$ when we would like to say "expected value." Comparing (9.10) with (4.9), we note how similar the definition of μ_X is to that of $\overline{X}$, the mean of a data variable, with a probability now appearing instead of a relative frequency.

The ***variance*** of X is denoted by σ_X^2 or $\sigma^2(X)$ (σ is lowercase "sigma," the Greek "s"), and its definition is

$$\sigma_X^2 = \sigma^2(X) = \sum (X_k - \mu_X)^2 p(X_k) \tag{9.11}$$

We will use two alternative notations for the variance, σ_X^2 and $\sigma^2(X)$, because sometimes the symbolic name of the random variable will be too large to write as a subscript.

The ***standard deviation,*** denoted alternatively by σ_X or $\sigma(X)$, is simply the square root of the variance:

$$\sigma_X = \sigma(X) = \sqrt{\sigma_X^2} = \sqrt{\sum (X_k - \mu_X)^2 p(X_k)} \tag{9.12}$$

and it is a measure of the typical deviation of X from its mean. The definition of σ_X is analogous to that of the root mean squared deviation of a data variable [see (4.10)], except that a probability appears instead of a relative frequency.

A worked-through set of calculations for the mean, variance, and standard deviation of the probability distribution of a random variable is given in Table 9.2 for the random variable Y defined above. It is helpful to do the calculations in a neat table like this rather than to try to plug numbers into formulas. The values of μ_Y and σ_Y are graphed in Figure 9.2, which shows the probability distribution of Y.

9.2 Random Variables and Samples of Data

We now examine more carefully what the relation is between a random variable and the data it generates, and we reconsider what is meant by "probability."

We have thought of a random variable as a number-generating process that can produce outcomes (numbers) forever. The probability distribution consisting of $\{X\}$ and $p(X)$ is said to completely characterize the uncertain nature of the random variable: for a discrete random variable this means that the value $p(X_k)$ is the probability that the random variable takes on the specific value X_k, as stated in (9.1).

TABLE 9.3 One Relative Frequency Distribution

X	f_k
5	
6	.33
7	
8	.67

If we collect a series of outcomes from a random variable, we have a set of numbers that are a sample from the random variable. This sample also is a data variable: each outcome is an observation from the given process. The data variable can be analyzed with the methods of descriptive statistics, and we can construct the relative frequency distribution. For a discrete data variable, each relative frequency value, f_k, gives the proportion of the observations that take on the specific value X_k.

Consider again the balls-in-hat game and its description as a discrete random variable X whose probability distribution is given in Table 9.1 and Figure 9.1. If we take a sample of just three observations ($n = 3$), we might get two 8s and a 6. The complete relative frequency distribution for this sample is given in Table 9.3. (Note that the distribution in Table 9.4 cannot possibly describe this sample, because three observations cannot yield nonzero relative frequencies for all four X_k values.) Although the relative frequencies in Table 9.3 differ greatly in magnitude from the probabilities in Table 9.1, we are not greatly surprised that this is the situation in such a small sample.

Now, if we take a sample of 100 observations from the random variable described by Table 9.1, the relative frequency distribution that we calculate might be that in Table 9.4. Here the f_k values are quite similar to the $p(X_k)$ values of the probability distribution, and this is in accord with commonsense expectations. By contrast, if the frequency distribution turned out to be the one in Table 9.3, we would be greatly surprised; this would mean that we got thirty-three 6s and sixty-seven 8s. Although this situation is surely unlikely, we should realize that our understanding of the balls-in-hat game does not preclude it from occurring.

Three general ideas emerge from this thinking. First, we realize that the relative frequency values, f_k, describing a sample of data taken from a random

TABLE 9.4 Another Relative Frequency Distribution

X	f_k
5	.11
6	.18
7	.33
8	.38

variable cannot be expected to be identical to the corresponding probability values, $p(X_k)$.

Second, we expect that the similarity between the relative frequency values and the probability values will tend to be greater in large samples than in small samples.

Third, if we take two samples of the same size from a given random variable, we expect each of their relative frequency distributions to be similar to the probability distribution of the random variable but probably not identical to it. Thus we expect the two samples' frequency distributions to be similar to each other, but probably somewhat different. This difference illustrates the phenomenon of ***sampling variability,*** which is very important.

It can be shown mathematically that as we consider letting the size of the sample approach infinity, the relative frequency f_k associated with each distinct outcome X_k approaches some limiting value. In the limit, each f_k will be equal to the corresponding $p(X_k)$; that is, the relative frequencies will be equal to the probabilities.

Turning this idea around, we now have a way of constructively defining what we mean by ''probability'': the probability of a distinct outcome X_k occurring is given by the proportion of the observations in a sample that would take on the value X_k, if the sample size were infinite. In other words, an infinite historical record of the various outcomes of X informs us about what to expect on the next outcome. Although this is not a practical way to determine the exact probabilities characterizing some process, it helps us understand what ''probability'' means. Also, we might be able to take a very large sample and be willing to accept the calculated relative frequencies as estimates of the underlying exact probabilities.

An alternative approach to conceptualizing how we determine probabilities is based on making an a priori analysis of the process. For the balls-in-hat game a thorough understanding of the laws of physics, coupled with careful examination of the materials, could allow us to make statements of the probability values. In the preceding section we used logic and common sense to approximate this scientific approach.

In our study we will not be concerned with how the probabilities describing a given process are determined. In this chapter and the next, our attention is focused on learning about the nature of random variables and their probability distributions. Later in the book we are concerned with the relation between random variables and samples of data.

9.3 The Binomial Distribution

If two or more chance processes are similar or related to each other in a certain way, the probability distributions describing these processes will also be similar or related to each other in a certain way. Under appropriate circumstances, a group of similar distributions that share a common mathematical formulation can be viewed as being members of a particular ***family.***

One of the most important families of discrete distributions is that of the ***binomial distribution.*** This family is built up from combinations of a basic chance process known as a ***Bernoulli trial*** (named after the mathematician James Bernoulli). Each Bernoulli trial is a process that has only two possible outcomes, either a "success" or a "failure." In any trial the probability of a success occurring is denoted by P, and therefore the probability of a failure is $1 - P$. In a series of trials the probability of a success remains fixed at P, so the outcomes of the trials are independent: the previous history of successes or failures does not affect the probability of success in any given trial. For example, flipping a coin can be described as a Bernoulli trial, with the probability of getting "heads" (a success, say) being P. If the coin is physically symmetrical, the laws of physics tell us that $P = .5$; if the coin is specially weighted, it might be that $P = .6$.

Now suppose that we are going to examine the results of a series of n Bernoulli trials. Let X be the number of successes that occur over this whole series. Clearly, the minimum number of successes is 0 and the maximum is n. Also, X can take on any integer value between 0 and n, so X is discrete. A statement of the probability distribution of X consists of a list of the possibly occurring values of X and the probabilities that X will take on each of these values. It can be shown mathematically that this probability distribution is given by

$$p(X) = \frac{n!}{X!\,(n - X)!}\,(P)^X(1 - P)^{n-X} \qquad X = 0, 1, 2, \ldots, n \qquad (9.13)$$

In this notation, the exclamation mark stands for the factorial operation, which specifies that the given integer is to be multiplied by all the lower positive integers. Thus, $5! = (5)(4)(3)(2)(1)$, for example. By special definition, $0! = 1$. [In (9.13), the ratio of factorials gives the number of ways of getting exactly X successes in n independent trials, and the other terms give the probability of one of those ways occurring.]

For example, suppose that we think of tossing a fair coin six times and that we are interested in the number of heads that might occur. Each toss is a Bernoulli trial with $P = 1/2$. Let Y denote the number of heads. Clearly, Y is a random variable having a binomial distribution with $n = 6$ and $P = 1/2$. Using (9.13), we can compute the probability of each specific number of heads occurring; for example, the probability of getting exactly 2 heads in the 6 tosses is

$$p(2) = \frac{6!}{2!\,4!}\left(\frac{1}{2}\right)^2\left(\frac{1}{2}\right)^4 = \frac{(6)(5)(4)(3)(2)(1)}{(2)(1)(4)(3)(2)(1)}\left(\frac{1}{2}\right)^2\left(\frac{1}{2}\right)^4 = 15/64 \qquad (9.14)$$

The complete probability distribution of Y is given in Table 9.2 and Figure 9.2, which we considered earlier without knowing its origin.

In general, it can be shown that if $P = .5$, the graphed probability distribution is symmetrical around the midpoint of the range of X values. If $P < .5$, there is a tendency for relatively low values of X to occur more frequently so that the distribution is skewed with a tail to the right. If $P > .5$, the distribution is skewed to the left. When n is large, the distribution appears to be fairly symmetrical even if $P \neq .5$, as long as P is not extremely small or large; however, if $P \neq .5$, this fairly symmetrical distribution is not centered at the middle of the range of X.

A particular binomial distribution can be summarized by its mean and standard deviation. These can be calculated according to (9.10) and (9.12), as in Table 9.2. However, for a binomially distributed random variable, it turns out that the exact mean and standard deviation also are given by

$$\mu_X = nP \tag{9.15}$$

and

$$\sigma_X = \sqrt{nP(1 - P)} \tag{9.16}$$

For example, in the series of 6 tosses of a fair coin, the mean of Y (i.e., the expected number of heads) is $(6)(1/2) = 3$, and the standard deviation is $\sqrt{(6)(1/2)(1/2)} = 1.22$. It should be noted that these values are the same as those calculated in Table 9.2.

The binomial distribution is not used much in econometric applications, although it is quite important in some other statistical techniques. However, the binomial distribution does play an important role in some types of economic analysis. For example, consider the situation of an unemployed worker who plans to look for a job each week for 5 weeks and who keeps looking even if he is offered a job. Suppose that the probability of his getting a job offer in any given week is .3, that this probability is not affected by the history of his search activity, and that he cannot get more than one offer in a week. Thus a week's search is a Bernoulli trial. Let X be the number of offers that the searcher finds in the five-week period. Clearly, X is a random variable having a binomial distribution with $n = 5$ and $P = .3$. For the worker, the expected number of job offers is 1.5. However, he may get no job offers at all, and the probability of this occurring is computed from (9.13):

$$p(0) = \frac{5!}{0!\ 5!}(.3)^0(.7)^5 = .1681 \tag{9.17}$$

The complete probability distribution, with each value similarly computed from (9.13), is given in Table 9.5 and Figure 9.3. The probability that the worker does get a job offer (i.e., that the worker gets one or more) could be determined as $p(1) + p(2) + p(3) + p(4) + p(5)$, but we can more quickly calculate this as $1 - p(0) = .8319$.

If we assume that a large group of searchers face the same probability of getting a job offer each week, the search process of the previous example stands

TABLE 9.5 Distribution of the Number of Job Offers

X	$p(X)$
0	.1681
1	.3602
2	.3087
3	.1323
4	.0284
5	.0024

as a partial model of search activity in the labor market. The experience of the group hypothetically constitutes a sample from the random variable. Based on our previous results, we predict that about 17 percent of workers searching for a job for five weeks would not get even one job offer.

9.4 Continuous Random Variables

A ***continuous random variable*** is one in which the set of possible outcomes is some range or interval of real numbers. We maintain the notion that the random variable represents a chance process, which can produce an unlimited number

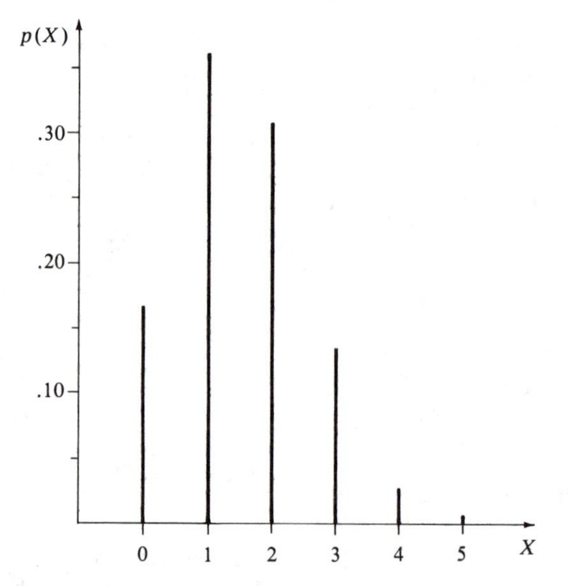

FIGURE 9.3 The discrete random variable X is the number of job offers received by a searcher in a five-week period. As discussed, X has a binomial distribution with $n = 5$ and $P = .3$; in this distribution, $\mu_X = 1.5$ and $\sigma_X = 1.02$. Since $P < .5$, the distribution is skewed with a tail to the right.

of independent outcomes one after another. However, because there are an infinite number of possible outcomes, our basic ideas regarding probability differ importantly from the discrete case.

For a continuous random variable X, consider how we might determine the probability that X will take on the specific value X_k. In the special case when each of the infinite number of outcomes has an equally likely chance of occurring, we would think that the probability of X_k occurring is $1/\infty$, which is zero. Although this may seem wrong, it makes sense. With a continuous random variable the probability of any particular value occurring is essentially zero, so we no longer focus on that kind of probability calculation.

Instead, the basic probability concept relates to the occurrence of an event defined as the outcome of X being in some specific interval. With X_1 being less than X_2, such a probability is denoted by Pr $(X_1 \leq X \leq X_2)$. Note that since the probability that X equals X_k equals zero, it follows that Pr $(X \leq X_k)$ = Pr $(X < X_k)$, so either inequality notation can be used.

As with the case of a discrete random variable, all the probability information about a continuous random variable X is contained in a probability distribution, which has two important components. The first is a statement $\{X\}$ of the possibly occurring values; this could be a closed interval such as $-1 \leq X \leq 1$, or an open interval such as $0 \leq X \leq \infty$. The second component is a function $p(X)$ that gives information about probabilities. The probability distribution is summarized by its mean and standard deviation; these have the same interpretation as in the discrete case, but they are defined in calculus terms.

The function $p(X)$, which is known as a ***probability density function,*** is quite different from the discrete case. For a continuous random variable, the value of $p(X)$ does not give Pr $(X = X_k)$, because we recognize that this probability amount is zero. Instead, the function $p(X)$ is constructed in such a way as to allow us to determine a basic probability like Pr $(X_1 \leq X \leq X_2)$. In graphical terms, the method of determination is to find the area under the curve $p(X)$ from X_1 to X_2. [In calculus terms, the probability is found by calculating the definite integral of $p(X)$ from X_1 to X_2.] Although this seems like a cumbersome way to organize information about probabilities, it works out quite well.

Two important aspects of the probability density function $p(X)$ should be noted. First, $p(X) > 0$ for any value in $\{X\}$. Second, the area under the graph of $p(X)$ within the whole range $\{X\}$ is exactly equal to 1. This aspect corresponds to $\Sigma\, p(X) = 1$ in the discrete case.

These ideas are illustrated in Figure 9.4a, which shows a graph of the probability density function $p(X)$ for a continuous random variable whose values occur only in the interval $\{X\}$ from $X_{\min}$ to $X_{\max}$. The function $p(X)$ is made up in such a way that the total area under the curve from $X_{\min}$ to $X_{\max}$ is equal to 1. The probability amount Pr $(X_1 \leq X \leq X_2)$ is given by the shaded area, which is necessarily less than 1 in value. Calculation of such a value is not an easy task. Mathematically, calculus is needed to find areas, except in some cases where simple geometry is sufficient.

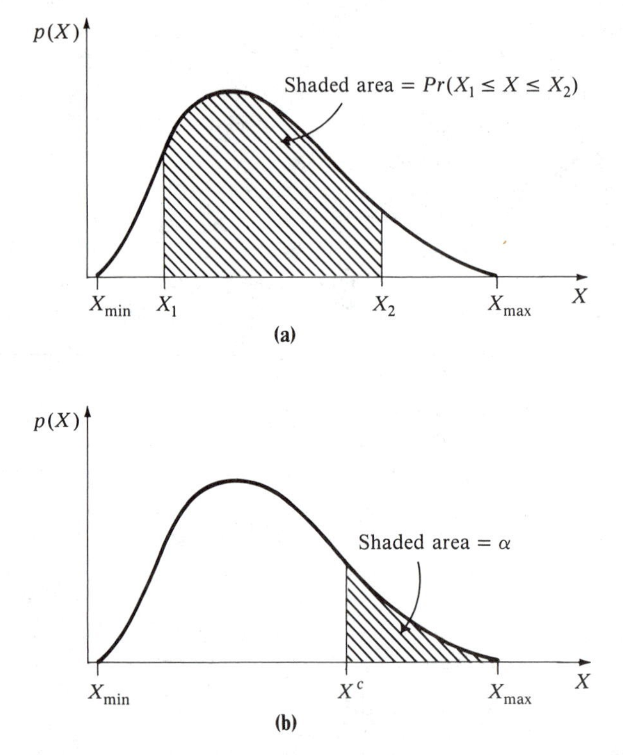

FIGURE 9.4 For a continuous random variable, probabilistic information regarding the random variable is contained in a probability density function (or probability distribution). The same distribution is graphed in (a) and (b). In both, the total area under $p(X)$, from X_{min} to X_{max}, is equal to 1, as must be true for any density function. In (a), the shaded area gives Pr $(X_1 \leq X \leq X_2)$. In (b), the shaded area is specified to be an amount α, and this determines the value X^c such that Pr $(X \geq X^c) = \alpha$.

That an area under a curve should give probability amounts seems very odd at first, but it becomes intuitive when we think about the relation between a description of the random variable and a description of a sample drawn from it. Following the logic of Section 9.2, consider taking a very large sample from the random variable and organizing the data in a frequency table. Using the methods of Section 4.2 for continuous data variables, a relative frequency histogram could be constructed. With a great number of observations, the class widths could be made very small and the steplike pattern of the bar tops (see Figure 4.4a) would blend into a smooth curve or line. Among all the observations in the data of the sample, the proportion occurring between two specific values X_1 and X_2 is given in the histogram by the relevant area divided by the total area.

If the sample size were infinite, the proportion Prop $(X_1 \leq X \leq X_2)$ would be a certain amount and that amount would be equal to Pr $(X_1 \leq X \leq X_2)$ for the random variable. Turning this around, one way to interpret what we mean by Pr $(X_1 \leq X \leq X_2)$ for a continuous random variable is the proportion Prop $(X_1 \leq X \leq X_2)$ that would occur in an infinite sample from the random variable.

The shape of the probability density function $p(X)$ can be thought of as being the same as the shape of the smoothed histogram in the infinite sample. The height values for $p(X)$ are redefined from the histogram so that the area under the curve equals 1; with the total area being equal to 1, we no longer need to be concerned with dividing a specified area by the total area. The mean μ_X and standard deviation σ_X of the probability distribution can be thought of as being equal to the mean $\overline{X}$ and standard deviation S_X of the histogram for the data in the infinite sample.

For this continuous random variable X, suppose that we want to determine the value X^c such that

$$\Pr (X \geq X^c) = \alpha \tag{9.18}$$

where α is some specified probability amount, $0 \leq \alpha \leq 1$ (α is lowercase "alpha," the Greek "a"). That is, we want to find the specific value X^c such that the probability that the outcome of the random variable X will be greater than or equal to X^c is α. The X^c value is determined from the probability density function, $p(X)$, as illustrated in Figure 9.4b: the X^c we seek is the value that bounds exactly α area to the right of itself under the curve.

For example, consider target shooting as an activity subject to chance and uncertainty. Let X be the distance from the center of the target to the hit, with positive values measuring distance to the right and negative values measuring distance to the left. Suppose that no hit is farther from the center than 1 meter (left or right). Further, suppose that we are dealing with a good shooter who is more likely to hit close to the center than far away.

We can formalize this description by considering X to be a continuous random variable whose probability distribution is graphed in Figure 9.5a. As described, the range of possible values is $-1 \leq X \leq 1$. The shape of $p(X)$ is a particular specification of the fact that the shooter is more likely to hit near the center than far away. Since the total area under $p(X)$ must be equal to 1, simple geometry requires that the peak height, $p(0)$, is also equal to 1 in this case. The mean of X is obviously 0, and the standard deviation looks as though it might be between 1/3 and 1/2. (Calculus is required for the exact determination; it turns out that $\sigma \approx 0.4$.)

The probability that a hit will occur farther than 1/2 meter to the left of the target's center, Pr $(-1 \leq X \leq -1/2)$, is given by the shaded area in Figure 9.5a. Simple geometry is all that is needed to determine that this equals 1/8. Similarly, from the figure we see that Pr $(-1/2 \leq X \leq 1/2) = 3/4$, so we can say that there is a 75 percent chance that a shot will land within 1/2 meter of the center. Clearly, the value X^c such that Pr $(X \geq X^c) = 1/8$ is 1/2.

For a continuous random variable, the values of $p(X)$ may turn out to be greater than 1, which never occurs in the discrete case. For example, if a triangular density function like that in Figure 9.5a covered only the interval

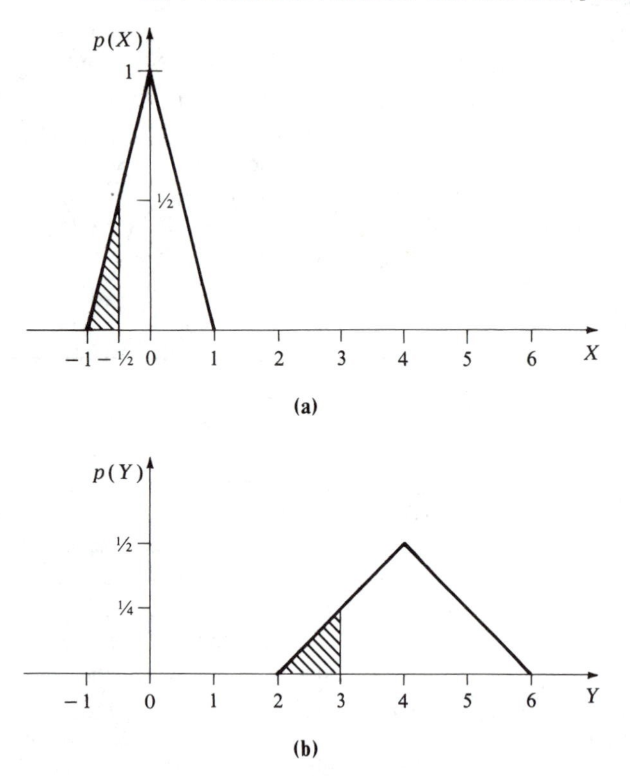

FIGURE 9.5 Part (a) shows the probability density function of a continuous random variable, X, that measures the distance from the center for outcomes of a specific target-shooting activity. In this distribution, $\mu_X = 0$ and $\sigma_X = 0.4$, roughly. The shaded area gives Pr $(X \leq -1/2)$. Part (b) shows the probability distribution (i.e., density function) of Y, which is defined as a linear transformation of X: $Y = 4 + 2X$. The graph of $p(Y)$ is obtained from $p(X)$ by expanding $p(X)$ by a factor of 2 and then moving the result four units to the right. The shaded area gives Pr $(Y \leq 3)$, which is equal to Pr $(X \leq -1/2)$.

$-1/2 \leq X \leq 1/2$, the height $p(0)$ would have to equal 2 for the total area to be 1. The absolute magnitudes of the density values have no direct meaning, and we can think of them being adjusted as necessary to make the total area equal to 1.

By contrast, the relative magnitudes of the different $p(X)$ values for a particular density function do have a meaning and interpretation: they give the relative ***likelihood*** of different values occurring. For example, in Figure 9.4a, if $p(X_1)$ is twice as great as $p(X_2)$, it is twice as likely that the outcome of X will be near X_1 as it is that the outcome will be near X_2.

Finally, recall that every probability distribution has two components: a statement $\{X\}$ of the values that X might ever take on and a function $p(X)$ giving probability information. To economize on terminology and notation, we sometimes will refer to the function $p(X)$ alone simply as "the probability distribution,'' both in the continuous and discrete cases. As long as we remember that $\{X\}$ is being specified implicitly, this should cause no confusion. Also, as long as the distinction between continuous and discrete random variables is understood, it is not necessary to refer to $p(X)$ specially as ''the density function'' in the continuous case.

9.5 Transformations

If we start with knowledge of a random variable and its probability distribution, we can think of creating a new random variable by transforming the existing one according to some fixed rule, as we did with data variables. A useful class of transformations is that of linear transformations, whose general form is

$$Y_k = a + bX_k \tag{9.19}$$

Except if $b = 0$, in which case all $Y_k = a$, there is a one-to-one correspondence between Y_k and X_k values. In addition, there is an important equivalence between the probability distributions of the original and transformed random variables. This concept applies somewhat differently in discrete and continuous cases.

For discrete random variables related by (9.19), the probability associated with a particular X_k is also associated with the corresponding Y_k. In graphical terms, the transformation changes the distance between the plotted vertical lines—each distance is multiplied by $|b|$—and shifts them all to the left or right. (If $b < 0$, there are also changes of order involved.) For example, letting X be the random variable described by (9.2), the transformation $Y = -1 + (1/2)X$ yields a random variable whose probability distribution $p(Y)$ is given in Table 9.6 and Figure 9.6; these should be compared with the probability distribution of X.

For continuous random variables related by the linear transformation (9.19), $p(Y)$ is obtained from $p(X)$ by compressing (for $|b| < 1$) or expanding (for $|b| > 1$) the distribution and then shifting it to the left or right. (If $b < 0$, there are also changes of order involved.) Since the area under $p(Y)$ must equal unity, compression [i.e., squeezing the original $p(X)$ shape] will cause all the $p(\cdot)$ values to increase proportionately and expansion will cause all of them to decrease proportionately.

TABLE 9.6 Probability Distribution of Y $(Y = -1 + (1/2)X)$

k	Y_k	$p(Y_k)$
1	1.5	.1
2	2.0	.2
3	2.5	.3
4	3.0	.4

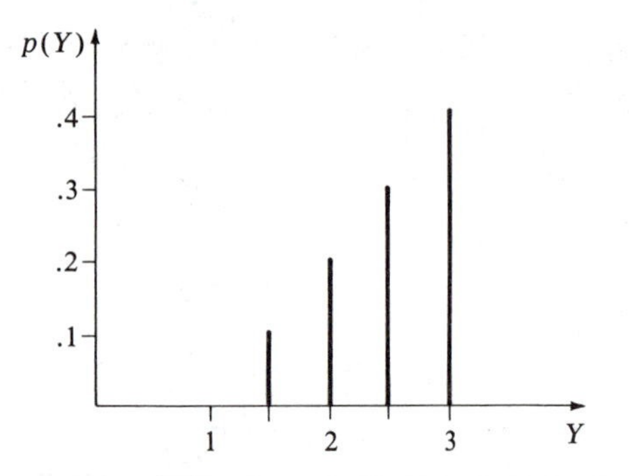

FIGURE 9.6 The probability distribution of Y graphed here is also represented in Table 9.6. The discrete random variable Y is related to the random variable X, which is described by Table 9.1 and Figure 9.1, through the linear transformation $Y = -1 + (1/2)X$. To obtain $p(Y)$ graphically, $p(X)$ is compressed by a factor of 2 and then shifted to the left. With discrete random variables related through a linear transformation, the probability associated with a particular X_k is also associated with the corresponding Y_k. Thus the heights of $p(\cdot)$ are preserved.

For continuous random variables, the probability equivalence is that the probability associated with Y being in a certain interval of Y values is the same as the probability of X being in the corresponding interval of X values. For example, letting X be the random variable whose probability distribution is given in Figure 9.5a, the transformation $Y = 4 + 2X$ yields a random variable whose probability distribution is given in Figure 9.5b. We can determine Pr $(Y \leq 3)$ directly from the shaded area: by geometry, this equals 1/8. Alternatively, we can see that the X_k value that corresponds to any Y_k is found by solving (9.19) for X_k:

$$X_k = \frac{Y_k - a}{b} \tag{9.20}$$

In this case, $X_k = -1/2$ corresponds to $Y = 3$, and the probability-equivalence concept of linear transformations assures us that Pr $(Y \leq 3) =$ Pr $(X \leq -1/2)$, which was determined to be 1/8 in Section 9.4.

Given the original X distribution and the parameters a and b, we can construct the complete probability distribution of Y and go on to calculate its mean, variance, and standard deviation. If we are interested only in these summary characteristics of Y and already know the corresponding characteristics of X, our task is simplified by knowing that

$$\mu_Y = a + b\mu_X \tag{9.21}$$

$$\sigma_Y^2 = b^2\sigma_X^2 \tag{9.22}$$

$$\sigma_Y = |b|\sigma_X \tag{9.23}$$

These relations, whose proofs follow along lines similar to the corresponding results for data variables, are often useful to us. For example, the mean and standard deviation of the discrete random variable Y described in Table 9.6

can be determined directly from its probability distribution. However, since we know how Y is related to X, we know that $\mu_Y = -1 + (1/2)\mu_X$ and that $\sigma_Y = (1/2)\sigma_X$.

Problems

Section 9.1

9.1 For the random variable X described in Equation (9.2), determine
(a) $\Pr(X = 6)$
(b) $\Pr(X \leq 6)$
(c) $\Pr(X \geq 7)$
(d) $\Pr(6 \leq X \leq 8)$

9.2 For the random variable Y described in Table 9.2, determine
(a) $\Pr(Y = 2)$
(b) $\Pr(Y \leq 2)$
(c) $\Pr(Y \geq 3)$
(d) $\Pr(2 \leq Y \leq 4)$

⋆ **9.3** Consider "$p(Y_k) = Y_k/20$, for $Y = 1, 2, 3, 4, 5, 6$." Could this be the probability distribution of a random variable? If so, graph the distribution.

9.4 Consider "$p(X_k) = (X_k)^2/10$, for $X = -2, -1, 0, 1, 2$." Could this be the probability distribution of a random variable? If so, graph the distribution.

⋆ **9.5** Calculate the mean and standard deviation of X, the random variable defined in Equation (9.2).

⋆ **9.6** Let X be the outcome on the toss of a fair die. Construct the probability distribution of X and determine its mean and standard deviation.

9.7 In the balls-in-hat game, if the first two balls drawn are 7s and if they are not replaced after being drawn, what is the probability of getting a 7 on the third draw?

9.8 Suppose that X takes on the value 1 with probability P and the value 0 with probability $(1 - P)$. Determine the mean and variance of X.

Section 9.2

9.9 Consider the process of flipping a single coin. Let $X = 1$ if a head occurs, and let $X = 0$ if a tail occurs. For a fair coin, $\Pr(X = 0) = 1/2$ and $\Pr(X = 1) = 1/2$. Based on this information, construct the probability distribution of X. Now, flip a real coin 10 times, thereby taking a sample from X. Construct a relative frequency distribution table for this sample and compare it with the probability distribution of X.

9.10 Extending Problem 9.9, flip a fair coin 10 more times and construct a relative frequency table for this sample. Compare it with the relative frequency distribution from Problem 9.9.

★ **9.11** Continuing with coin flipping from Problem 9.9, let $X = 2$ if the coin lands on its edge. Suppose that Pr $(X = 2) = 1/1000$. Is the frequency distribution determined for a particular sample likely to be identical to the probability distribution of X?

9.12 Construct the relative frequency distributions that might reasonably describe two different samples of size 5 taken from the random variable X defined in Equation (9.2).

9.13 Determine the means of X for the samples in Tables 9.3 and 9.4. What would be the mean of X in an infinite sample drawn from the random variable described in Table 9.1?

Section 9.3

9.14 Suppose that X has a binomial distribution with $n = 3$ and $P = 1/3$. Construct the complete probability distribution of X and graph it. Determine the mean and standard deviation.

★ **9.15** Suppose that a worker plans to look for work for five weeks and that the probability of getting an offer in any week is 1/5. What is his expected number of job offers? What is the probability that he will get fewer than the expected number?

9.16 Among members of the binomial family all having the same value of n, which has the largest variance? (Try the set of values $P = .1, .3, .5, .7, .9$, or use calculus.)

Section 9.4

9.17 Based on Figure 9.4a, which is greater: Pr $(X \leq X_2)$ or Pr $(X \geq X_2)$?

9.18 Based on $p(X)$ shown in Figure 9.5a, determine
(a) Pr $(X \geq 0)$
(b) Pr $(X \geq 1/2)$
(c) Pr $(0.2 \leq X \leq 0.8)$
(d) X^c such that Pr $(X \geq X^c) = .32$

★ **9.19** Based on $p(X)$ shown in Figure 9.5a, explain why the chance that X will be about 0.2 is greater than the chance that X will be about 0.8.

9.20 Let X be a continuous random variable that takes on values only between 0 and 1 and for which $p(X) = 2X$ over this interval.
(a) Graph the density function and verify that the area under $p(X)$ equals 1.
(b) Determine Pr $(X \geq 0.5)$.
(c) Determine X^c such that Pr $(X \geq X^c) = .5$.

9.21 Suppose that a sample of size 10 is taken from the random variable described by Figure 9.5a. Using four class intervals (-1 to -0.5,

-0.5 to 0, etc.), construct a frequency table and relative frequency histogram that might reasonably describe the sample.

Section 9.5

9.22 Comparing Y and X in Figure 9.5,

(a) What Y values correspond to the following X values: 0, 1/4, 1/2, 1?

(b) What X values correspond to the following Y values: 2, 3, 4, 5, 6?

9.23 For Y given in Figure 9.5b,

(a) Determine Pr $(Y \geq 4.5)$.

(b) Given that $\mu_X = 0$ and $\sigma_X \approx 0.4$, determine the mean and standard deviation of Y.

⋆ **9.24** From its probability distribution, calculate the mean and standard deviation of Y, the random variable defined in Table 9.6. Compare the results with those from Problem 9.5.

Appendix

9.25 Based on the discussion in the text, draw (on the same set of axes) the probability distribution of earnings for a person with a small amount of wealth and the probability distribution of earnings for a person with a large amount of wealth.

⋆ **9.26** Let W be the sum of the outcomes of a pair of dice. Determine the mean and standard deviation of W.

9.27 Based on Figure 9.7, construct a small table showing

(a) $p(Y|X = 5)$

(b) $p(X|Y = 4)$

APPENDIX*

This appendix extends the discussion of random variables and probability distributions to cases involving two or more random variables.

Joint Probability Distributions

Our concept of a single random variable likens it to a number-generating process that produces outcomes whose values are governed by the probability distribution of the random variable. We now can think of a more complex process that produces two numbers simultaneously. A gambling activity that serves as an example is throwing a pair of dice, one red and the other green. A second example is the economic process that determines a person's earnings

*This appendix is relatively difficult and can be skipped without loss of continuity.

(labor income) and his wealth at a particular time. The key idea in both cases is that two numbers are produced, but that they are separately identifiable. In the first example, one number is red and the other green; in the second, one is earnings and the other wealth. Although in each example the two numbers are separately identifiable, they are jointly determined: the number-generating process produces the two numbers simultaneously, and together they constitute an outcome. Were we to draw a sample from this process, each pair of simultaneously produced numbers would constitute an observation.

To generalize, we can think of a number-generating process that determines jointly two separately identifiable numbers, which are outcomes of the random variables Y and X. The one overall statement that describes the likelihood that different Y values will be produced, the likelihood that different X values will be produced, and the extent (if any) to which Y and X values are related, is called the ***joint probability distribution*** of Y and X. This joint distribution can be presented as a mathematical equation, a three-dimensional figure, or (in the case of discrete Y and X) a table. The actual representation is not important to us, but we can denote it by $p(Y, X)$. For discrete Y and X, the joint distribution $p(Y, X)$ gives the probability of occurrence for each possible combination of Y and X values.

Although Y and X are jointly determined, it is possible to focus on one of the variables, such as Y. Since Y is a random variable, it has a regular (univariate) probability distribution, $p(Y)$, that describes the probability or likelihood that Y will take on different values for the next outcome. This distribution has a mean μ_Y and variance σ_Y^2. It turns out that the exact specification of $p(Y)$ can be determined from the joint distribution $p(Y, X)$ if that is given. Similarly, from the joint distribution $p(Y, X)$ it is possible to determine the probability distribution of X alone, $p(X)$, along with its mean μ_X and variance σ_X^2. In the context of thinking that Y and X are jointly determined in a process described by $p(Y, X)$, the regular (univariate) distributions $p(Y)$ and $p(X)$ are known as ***marginal probability distributions*** because they can be displayed easily in the margins of a table that gives the joint distribution (for a discrete random variable).

A numerical example will help make these fundamental ideas clearer. Let Y and X be two jointly determined discrete random variables; we will not consider what physical or economic process they represent. The possible outcomes $\{X\}$ are 5 and 10, and the possible outcomes $\{Y\}$ are 4, 5, and 6. The joint distribution $p(Y, X)$ is shown in Figure 9.7, along with the marginal distributions. Each entry in the joint distribution is the probability that a specific pair of Y and X values will occur. For example, the probability that the pair $(Y = 5, X = 10)$ occurs is 30 percent. Notice that the marginal probabilities for Y are simply the row sums in the joint distribution. For example, the outcome that Y equals 5 corresponds both to the pair $(Y = 5, X = 5)$ and the pair $(Y = 5, X = 10)$; hence $p(Y)$ for $Y = 5$ is the sum of the two probabilities in the joint distribution. Similarly, the marginal probabilities for X are simply the column sums in the joint distribution.

Now it is possible also to focus on what would be the probability distribution

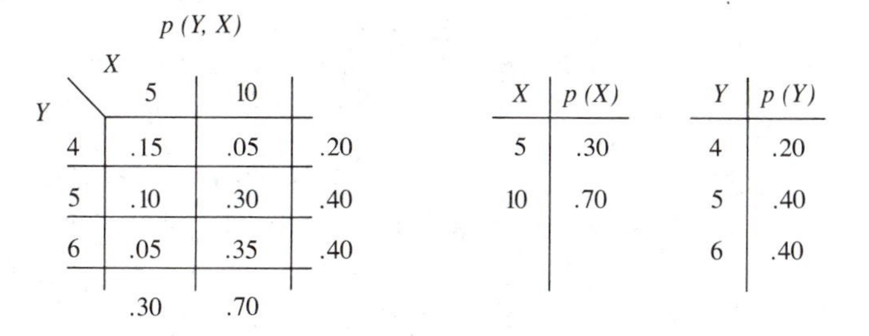

$p(Y, X)$

Y \ X	5	10	
4	.15	.05	.20
5	.10	.30	.40
6	.05	.35	.40
	.30	.70	

X	p (X)
5	.30
10	.70

Y	p (Y)
4	.20
5	.40
6	.40

FIGURE 9.7 The left side of this figure displays the joint probability distribution $p(Y, X)$ for two discrete random variables. On the right side are the marginal probability distributions for X and for Y, which are shown also as the column and row sums, respectively, in the joint distribution.

of Y if X takes on a specific value X_k. This distribution, which is denoted by $p(Y|X_k)$, is known as the ***conditional probability distribution*** of Y (i.e., the probability distribution of Y conditional on a specific value for X). In the case of dice tossing, the conditional distributions of Y (the red outcome) for the different possible values of X (the green outcome) are all exactly the same as the unconditional (marginal) probability distribution, $p(Y)$. This is because the physics of tossing fair dice tells us that whatever happens to the green die has no effect on what happens to the red die. Hence, the likelihood of different values occurring on the red die is not affected by the actual outcome on the green die. In our new terminology, this means that the conditional probability distributions for Y, $p(Y|X_k)$, are all identical to the unconditional (marginal) probability distribution for Y, $p(Y)$. By contrast, the likelihood of a person's receiving various possible levels of earnings (Y) is probably related to how much wealth (X) he has; the economic process is probably such that persons with considerable amounts of wealth have a greater chance also to have high levels of earnings. Hence, at least some of the conditional probability distributions of earnings, $p(Y|X_k)$, for different levels of wealth (X_k) are different from each other and different from $p(Y)$, the unconditional probability distribution of earnings.

Suppose that Y and X are two random variables having a joint probability distribution, $p(Y, X)$. If all the conditional distributions $p(Y|X_k)$ are identical to each other, so that the likelihood of different values occurring for Y is not at all affected by the X value that is simultaneously determined, the Y and X are ***independent.*** In the case of tossing a pair of dice, Y (the red outcome) and X (the green outcome) are independent. On the other hand, if not all the conditional distributions $p(Y|X_k)$ are identical to each other, so that the likelihood of different values occurring for Y may be affected by the X value that is simultaneously determined, then Y and X are not independent. In the joint determination of earnings and wealth, Y (earnings) and X (wealth) are not independent.

In the example of Figure 9.7, the conditional probability distributions for Y are denoted by $p(Y|X = 5)$ and $p(Y|X = 10)$. When $X = 5$ the most likely value for Y is 4; when $X = 10$ the most likely value for Y is 6. Since the value that would occur for Y is affected by the value of X with which it is jointly determined, Y and X are not independent. [The conditional distribution $p(Y|X = 5)$ is obtained from the first column of the joint distribution by dividing each of the three entries by $p(X)$ for $X = 5$. Since that $p(X)$ value is obtained marginally by adding up the probabilities in the first column, the effect of dividing each of the entries in the first column by the $p(X)$ value for that column is to make the sum of the "new entries" equal to 1.]

A measure of the relation, or dependence, between two random variables having the joint distribution $p(Y, X)$ is given by the ***covariance*** between them; for the discrete case,

$$\sigma_{YX} = \sigma(Y, X) = \sum (Y - \mu_Y)(X - \mu_X)p(Y, X) \tag{9.24}$$

This is a somewhat cumbersome calculation and we do not dwell on it. Suffice it to say that the covariance σ_{YX} between two random variables is the theoretical analog of the covariance S_{YX} between two data variables. As might be expected, one can define the ***correlation*** between two random variables:

$$\rho_{YX} = \frac{\sigma_{YX}}{\sigma_Y \sigma_X} \tag{9.25}$$

where ρ is the theoretical analog of the correlation coefficient, r, between two data variables (ρ is lowercase "rho," the Greek "r"). If two random variables are independent, the covariance between them is zero and therefore so is the correlation. However, the converse of this statement in not always true: if the covariance and correlation between two jointly determined random variables is zero, it is not necessarily true that they are independent.

All the basic ideas presented here are symmetrical. There are conditional distributions $p(X|Y_k)$ of the same nature as the $p(Y|X_k)$ discussed above, $\sigma_{XY} = \sigma_{YX}$, and independence is a mutual relation—Y cannot be independent of X without X being independent of Y.

These concepts regarding jointly determined random variables can be extended to any number of separately identifiable numbers occurring simultaneously as a single outcome. We can denote the different random variables as $X_1, X_2, \ldots, X_n$ and consider that their outcomes are governed by a joint probability distribution, $p(X_1, X_2, \ldots, X_n)$. This distribution is characterized in part by n means, n variances, and $(n^2 - n)/2$ different covariances. All covariances are pairwise relations, and there is no similar concept that encompasses more than two variables together. If all the n variables are independent, the value occurring for any one has no effect on the likelihood of different values occurring for any of the others and all the covariances are zero.

Linear Combinations

Sometimes we might want to define a new random variable, say W, in terms of two random variables, Y and X, that have a joint distribution $p(Y, X)$. One of the possible ways in which W might be defined in terms of X and Y is as a linear combination, of the general form

$$W = a_1Y + a_2X \tag{9.26}$$

For example, if Y and X are the outcomes of red and green dice, $W = Y + X$ (i.e., $a_1 = 1$, $a_2 = 1$) is a variable that gives the sum of the outcomes. If Y and X are earnings and wealth, and if 0.07 is the annual rate of return to wealth, then $W = Y + 0.07X$ (i.e., $a_1 = 1$, $a_2 = 0.07$) is a variable giving total annual income.

Since W is a random variable, it has a (univariate) probability distribution $p(W)$ that describes the probability or likelihood of its taking on various values. The exact form of $p(W)$ depends on the joint distribution $p(Y, X)$ and the linear combination defining W. Determining $p(W)$ can be very complex. However, $p(W)$ is a genuine distribution with mean μ_W and variance σ_W^2, and these two summary characteristics are nicely related to the characteristics of $p(Y, X)$. It turns out that

$$\mu_W = a_1\mu_Y + a_2\mu_X \tag{9.27}$$

$$\sigma_W^2 = (a_1)^2\sigma_Y^2 + (a_2)^2\sigma_X^2 + 2(a_1)(a_2)\sigma_{YX} \tag{9.28}$$

If Y and X are independent, the covariance between Y and X is zero, and therefore the last term in (9.28) is equal to zero. In the example of the dice, $\mu_W = \mu_Y + \mu_X$ and $\sigma_W^2 = \sigma_Y^2 + \sigma_X^2$ because Y and X are independent. In the earnings and wealth example, $\mu_W = \mu_Y + 0.07\mu_X$ and $\sigma_W^2 > \sigma_Y^2 + 0.0049\sigma_X^2$, assuming it is true that labor income and wealth are positively related (i.e., $\sigma_{YX} > 0$).

Extending these ideas to the special case of n independent random variables having a joint distribution $p(X_1, X_2, \ldots, X_n)$ with all covariances equal to zero, we can consider a linear combination:

$$W = a_1X_1 + a_2X_2 + \cdots + a_nX_n = \sum_{j=1}^{n} a_jX_j \tag{9.29}$$

It turns out that the means are related as

$$\mu_W = a_1\mu_1 + a_2\mu_2 + \cdots + a_n\mu_n = \sum_{j=1}^{n} a_j\mu_j \tag{9.30}$$

and (in the case of independence) the variances as

$$\sigma_W^2 = (a_1)^2\sigma_1^2 + (a_2)^2\sigma_2^2 + \cdots + (a_n)^2\sigma_n^2 = \sum_{j=1}^{n} (a_j)^2\sigma_j^2 \tag{9.31}$$

where μ_j is the mean of X_j and σ_j^2 is the variance of X_j.

This analysis of linear combinations will be an important element in the derivation of certain results later in the book. Outside the field of statistics, it is also used in economics and finance to determine features of the probability distribution of the return to a mutual fund when the fund is made up of stocks whose return is uncertain.

The Normal and *t* Distributions

In this chapter we examine some families of continuous probability distributions that are of importance in regression analysis. The two main sections of the chapter discuss the normal and t distributions, which are basic to the study of econometrics. The appendix discusses the chi-square and F distributions, which are somewhat more advanced.

10.1 The Normal Distribution

In Section 9.3 we examined the family of binomial distributions. Each member of that family is characterized by specific values for two parameters, n and P, and each has a probability distribution whose equation form is the same except for these values. Also, we noted that a discrete random variable having a specific binomial distribution can be thought of as arising from a sequence of independent Bernoulli trials, such as flipping a coin or searching for a job.

We now consider a family of continuous probability distributions that is of central importance in the study of statistics and econometrics. In Chapter 11 we will adopt the assumption that the disturbance in a regression model is a random variable having a probability distribution in this family. Like the discrete binomial family, each of these distributions is characterized by specific values for two parameters in a general equation form. We will not try to explain how any realistic chance process leads to a distribution in this family, but such an explanation is possible with advanced mathematics. In this section we first consider the family in its most general form and then go on to consider specific cases.

A random variable, X, whose possibly occurring values run all the way from $-\infty$ to ∞ is said to have a ***normal distribution*** if its probability distribution (i.e., density function) is of the form

$$p(X) = \frac{1}{b\sqrt{2\pi}} \exp\left[-\frac{1}{2}\left(\frac{X-a}{b}\right)^2\right] \qquad -\infty < X < \infty \qquad (10.1)$$

where exp[·] indicates that the terms in brackets are an exponent to which e, the base of the natural logarithms, is to be raised. Also in this expression, π is the number 3.14 . . . , and a and b are fixed parameters. The distribution is graphed in Figure 10.1. It is perhaps surprising that any expression that is so complicated could be called normal or would turn out to be useful, but such is the case. Many magnitudes that occur in nature or that result from measurement errors seem to be samples drawn from normally distributed random variables.

The normal distribution is defined so that $p(X) > 0$ for all values of X. It may be somewhat surprising, again, that the area under the $p(X)$ curve is not infinite, but it is not. The area under $p(X)$ over the range $-\infty < X < \infty$ is equal to 1, as is required for $p(X)$ to be a genuine density function. It turns out that $\mu_X = a$ and $\sigma_X = b$, so it is common to rewrite the general formulation as

$$p(X) = \frac{1}{\sigma_X\sqrt{2\pi}} \exp\left[-\frac{1}{2}\left(\frac{X-\mu_X}{\sigma_X}\right)^2\right] \qquad -\infty < X < \infty \qquad (10.2)$$

This distribution is the archetypical bell-shaped curve. It is symmetrical around the value $X = \mu_X$, as can be determined by inspecting (10.2). Also, it may be

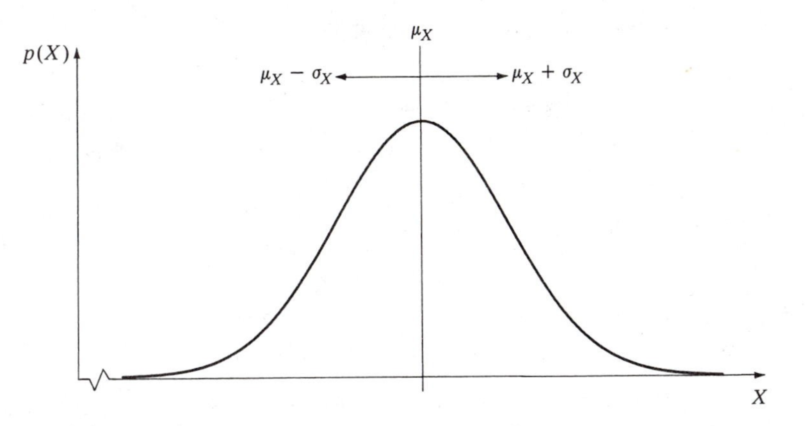

FIGURE 10.1 The density function for a continuous random variable having a normal distribution is given by Equation (10.1) or (10.2). Members of the family of normal distributions differ only with regard to two parameters: the mean, μ_X, and the standard deviation, σ_X. For any member of the family, 68.2 percent of the probability is concentrated within one standard deviation of the mean. That is, the probability that the random variable will take on a value between $\mu_X - \sigma_X$ and $\mu_X + \sigma_X$ is .682.

noted that the highest value for $p(X)$ occurs when $X = \mu_X$, and that $p(X)$ decreases as X moves away from the mean in either direction.

The Standard Normal

The most important member of the family is the normal distribution with zero mean ($\mu = 0$) and unit standard deviation ($\sigma = 1$). It is called the ***standard normal,*** or ***unit normal,*** and a random variable with this distribution is commonly denoted by Z:

$$p(Z) = \frac{1}{\sqrt{2\pi}} \exp\left[-\frac{1}{2}Z^2\right] \qquad -\infty < Z < \infty \tag{10.3}$$

It is depicted in Figure 10.2. Note that the $p(Z)$ values get very small for $Z < -3$ and $Z > 3$, but they are positive.

It would require burdensome calculations to determine probabilities directly from the density function, and our effort is spared by the use of a table that gives us a record of the values. In an appendix at the end of the book, Table A.1 gives

$$\alpha = \Pr(Z \geq Z^*) \tag{10.4}$$

which is the probability that a standard normal random variable exceeds a specified value Z^*, for various positive values of Z^*. In Figure 10.2a, the shaded area represents this probability, $\Pr(Z \geq Z^*)$, for the Z^* indicated. By taking advantage of the symmetry of the distribution, it is possible to use this information to solve a wide variety of probability problems involving the standard normal.

The first class of probability problems we commonly confront are of the form:

$$\text{Find } \alpha \text{ such that } \Pr(Z \geq Z^*) = \alpha \tag{10.5}$$

where Z^* is a specified value of Z. For example, if $Z^* = 1.5$, then α is equal to the area in the tail of the distribution to the right of $Z = 1.5$. A quick glance at Table A.1 gives the answer, $\alpha = .067$. In other words, the probability that a random variable with a standard normal distribution will take on a value greater than 1.5 is 6.7 percent. Because of the symmetry of the standard normal distribution around its zero mean, if Z^* is a particular positive value, then

$$\Pr(Z \leq -Z^*) = \Pr(Z \geq Z^*) \tag{10.6}$$

Hence, $\Pr(Z \leq -1.5) = .067$, for example. In this way, Table A.1 can be used for negative values of Z as well as positive values.

We often are interested in determining the probability amounts in two symmetrical tails of the distribution. It is helpful in this connection to use absolute value notation:

$$|Z| \geq Z^* \quad \text{means} \quad Z \leq -Z^* \quad \text{and} \quad Z \geq Z^* \quad \text{together} \tag{10.7}$$

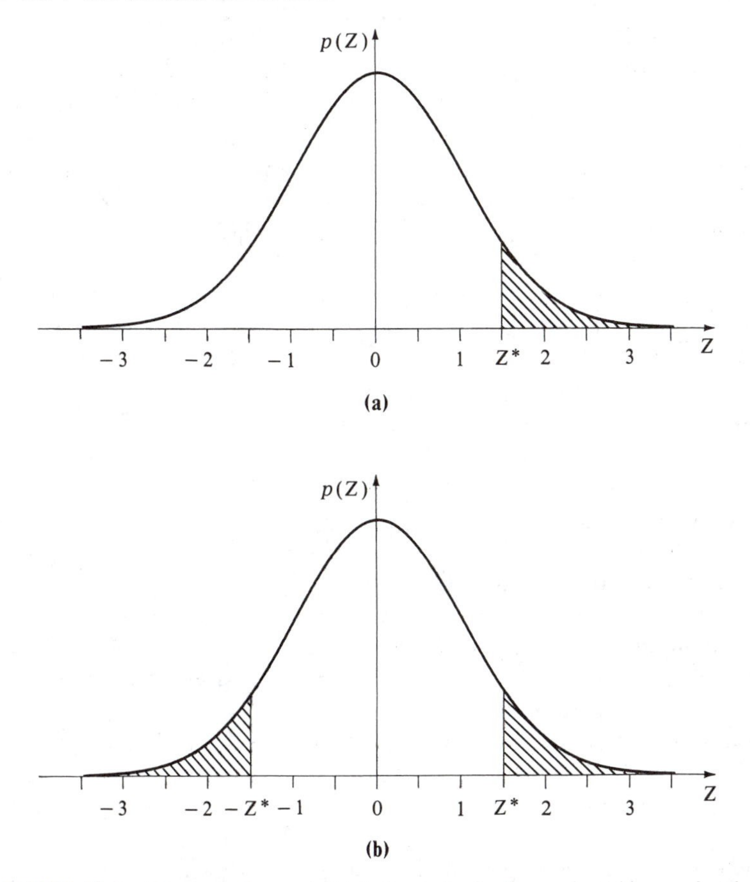

FIGURE 10.2 The standard normal distribution is the family member with $\mu = 0$ and $\sigma = 1$. A random variable having a standard normal distribution is usually denoted by Z. In (a), the shaded area represents Pr $(Z \geq Z^*)$. In (b), the shaded area represents Pr $(|Z| \geq Z^*)$, and half the probability is located in each tail. In practice, probability calculations such as these are made using Table A.1.

That is, the range of Z values for which the absolute value of Z is greater than or equal to Z^* consists of the Z values less than or equal to $-Z^*$ and the Z values greater than or equal to Z^*. Thus it should be clear that the probability of being in one tail or the other is

$$\begin{aligned} \Pr(|Z| \geq Z^*) &= \Pr(Z \leq -Z^*) + \Pr(Z \geq Z^*) \\ &= 2 \Pr(Z \geq Z^*) \end{aligned} \tag{10.8}$$

In Figure 10.2b the shaded areas together represent this probability. The probability of not being in either tail, Pr $(|Z| \leq Z^*)$, is represented by the unshaded area. Since an outcome is either in the tails or not,

$$\Pr(|Z| \leq Z^*) = 1 - \Pr(|Z| \geq Z^*) \tag{10.9}$$

For example, since $\Pr(Z \geq 1.5) = .067$, then $\Pr(|Z| \geq 1.5) = .134$ and $\Pr(|Z| \leq 1.5) = .866$.

More complex probability calculations can be carried out using simple tail areas and basic knowledge of probabilities. In complex problems, as well as simple ones, it is always useful to draw a diagram to represent the probabilities as areas. For example, if we wish to determine $\Pr(-0.31 \leq Z \leq 1.5)$, we realize that we need to determine the shaded area in Figure 10.3. The shaded area equals the total area minus the unshaded areas; these unshaded areas represent simple tail probabilities, whose amounts are given in Table A.1. Mathematically, the desired probability equals $1 - \Pr(Z \leq -0.31) - \Pr(Z \geq 1.50)$, which equals $1 - .378 - .067$ or $.555$.

A second class of probability problems that are important for us are of the form:

$$\text{Find } Z^c \text{ such that } \Pr(Z \geq Z^c) = \alpha \tag{10.10}$$

where α is a specified amount of probability and Z^c is the ***critical value*** of Z that bounds exactly α probability in the right-hand tail. To solve these problems we use Table A.1 in the opposite way: for the given probability, we search for the corresponding Z value. For example, to find Z^c such that $\Pr(Z \geq Z^c) = .25$, we use the table to see that Z^c is between 0.67 and 0.68. Taking the closer value, we say that Z^c is about 0.67; if more precision is needed, interpolation can be used. Because of the symmetry of the standard normal distribution, we see that the critical value $-Z^c$ such that $\Pr(Z \leq -Z^c) = .25$ is about -0.67. (For consistency throughout this book, critical value notation like "Z^c" always refers to a positive value. A negative critical value is clearly signed: here, $-Z^c \approx -0.67$.)

We often deal with a two-tailed probability of the general form:

$$\text{Find } Z^c \text{ such that } \Pr(|Z| \geq Z^c) = \alpha \tag{10.11}$$

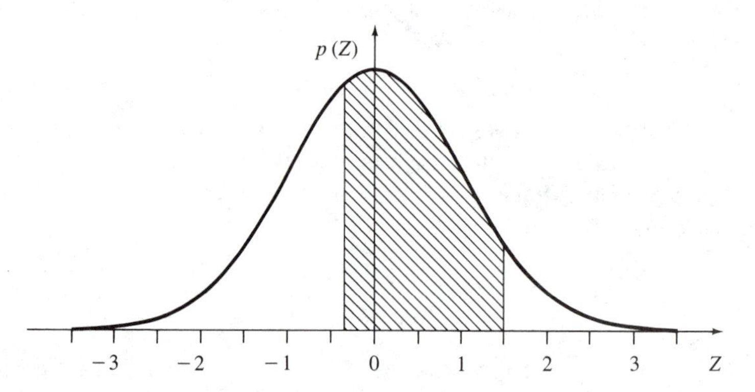

FIGURE 10.3 In this graph of the standard normal distribution, the shaded area gives $\Pr(-0.31 \leq Z \leq 1.5)$. This may be determined by first using Table A-1 to find the probabilities in the unshaded tail areas.

Referring to Figure 10.2b, if we let α be the probability measure of all the shaded area, then the area in each tail is $\alpha/2$. Thus we solve (10.11) by determining Z^c such that

$$\Pr(Z \geq Z^c) = \frac{\alpha}{2} \tag{10.12}$$

For example, a common need is to determine the symmetrical values of Z that bound a total of 5 percent of the probability in both tails. We do this by finding Z^c such that $\Pr(Z \geq Z^c) = .025$; the foot of the table shows that $Z^c = 1.960$ precisely to three decimals because this is such an important reference.

As before, we may be interested in the probability of not being in the tails. For example, an interesting question is: what symmetrical values of Z contain between them 50 percent of the total probability? This can be rephrased as: find Z^c and $-Z^c$ such that $\Pr(|Z| \leq Z^c) = .5$. These are found by determining Z^c such that $\Pr(Z \geq Z^c) = .25$. As before, we choose the closest Z value in Table A.1 and say that Z^c is about 0.67. Thus, the answer to our question is that the Z values -0.67 and 0.67 contain between them (about) 50 percent of the probability. (A more precise determination shows that Z^c is about 0.6745.)

Other Normal Distributions

We turn now to other members of the family of normal probability distributions. Each family member is characterized by its mean and standard deviation, μ_X and σ_X, which are the only parameters in (10.2). Since there are an unlimited number of combinations of these two parameter values, there are an unlimited number of members of the family. The general shape of all normal distributions is the same, and if we look at the density function graphed without labels on the axes, we cannot tell which distribution it is. Figure 10.4 shows three normal distributions on the same set of axes. The probability distributions of X_1 and X_2 have the same standard deviation but different means, and the density functions look alike except that $p(X_2)$ is shifted over to the right relative to $p(X_1)$. The probability distributions of X_2 and X_3 have the same mean but different standard deviations; they are centered at the same value but $p(X_3)$ has a greater standard deviation and is shorter at the mean value. When drawn on the same axes, all normal distributions have the same amount of graphical area under them.

In solving probability problems involving a member of the normal family, we might hope to have a table similar to Table A.1 that is determined for the specific distribution under consideration. Clearly, this approach would necessitate the construction of many tables. A more practical approach is to take advantage of the fact that all members of the normal distribution family are linear transformations of each other. Loosely speaking, one distribution can be obtained from any other by compressing or expanding it and then shifting it to the left or right.

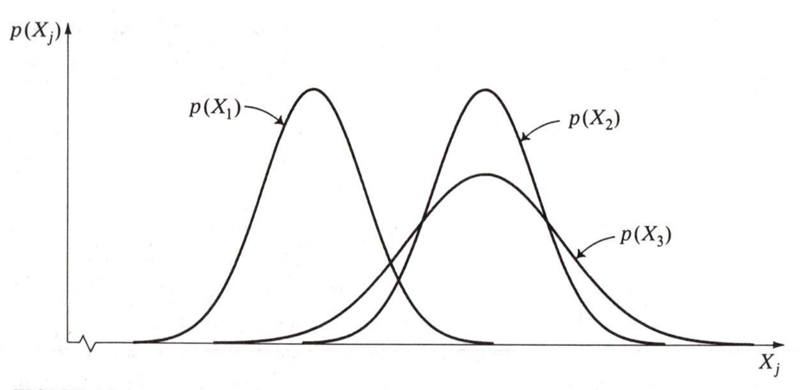

FIGURE 10.4 In this graph, three members of the normal distribution family are plotted together. The probability distributions of X_1 and X_2 have different means but the same standard deviation. The probability distributions of X_2 and X_3 have the same mean, but $\sigma(X_3) > \sigma(X_2)$. Any member of the normal family can be obtained as a linear transformation of some other member. Thus, the graph of one distribution can be obtained from that of another by compressing or expanding it and then shifting it to the left or right.

In particular, any member can be thought of as a transformation of the standard normal distribution. Starting with a standard normal random variable Z and two parameters a and b, we can create a new random variable X by letting each possible value Z_k be transformed into a value X_k according to

$$X_k = a + bZ_k \tag{10.13}$$

It can be shown that the probability distribution of X is given by (10.1); that is, it is normally distributed with $\mu_X = a$ and $\sigma_X = b$. Hence it is common to think of the correspondence as

$$X_k = \mu_X + \sigma_X Z_k \tag{10.14}$$

Putting all this in reverse, if we start with a normal random variable X having mean μ_X and standard deviation σ_X, it is possible to think of any particular value X_k as corresponding to a particular Z_k value found by

$$Z_k = \frac{X_k - \mu_X}{\sigma_X} \tag{10.15}$$

which is just (10.14) solved for Z_k.

Now, recall the fundamental probability equivalence between two continuous random variables, such as X and Z, that are related by a linear transformation: the probability associated with X being in a certain interval of X values is the same as that of Z being in the corresponding interval of Z values. Thus Pr $(X \geq X_k)$ is exactly the same as Pr $(Z \geq Z_k)$ when Z_k and X_k correspond to each other through (10.14) or (10.15). Hence, any probability problem involving a normal variable X can be solved by seeing it as equivalent to a probability problem involving Z, a standard normal variable. For example, suppose that X has a normal probability distribution with $\mu_X = 5$ and $\sigma_X = 2$, and that we want to determine Pr $(X \geq 6)$. The value $X_k = 6$ corresponds to a particular Z value, determined by $Z_k = (X_k - \mu_X)/\sigma_X = (6 - 5)/2 = 0.5$. We know, from Table A.1, that Pr $(Z \geq 0.5) = .309$; because of the probability equivalence between X and Z, this probability amount is the same as Pr $(X \geq 6)$ and it is the answer to our problem. This is illustrated in Figure 10.5.

For future reference we may rework and generalize this example more formally:

$$
\begin{aligned}
\alpha &= \Pr (X \geq X_k) \\
&= \Pr (Z \geq Z_k), \quad \text{where} \quad Z_k = \frac{X_k - \mu_X}{\sigma_X} \qquad (10.16) \\
&= \Pr (Z \geq 0.5) = .309
\end{aligned}
$$

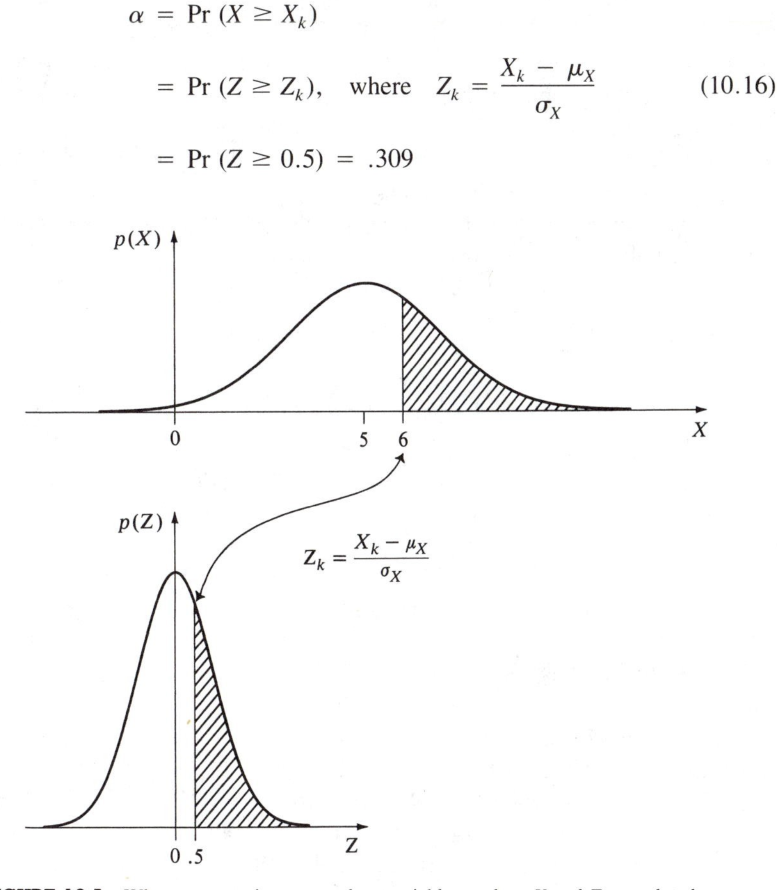

FIGURE 10.5 When two continuous random variables such as X and Z are related through a linear transformation, the probability associated with X being in a certain interval of X values is the same as that of Z being in the corresponding interval of Z values. Hence, a probability problem involving any member of the normal family can be solved by seeing it as equivalent to a problem involving the standard normal distribution. In turn, this problem is solved using Table A.1. In the case illustrated here $\mu_X = 5$ and $\sigma_X = 2$, so the Z value corresponding to $X_k = 6$ is $Z_k = 0.5$. The shaded area in both distributions is the same probability amount, $\alpha = .309$.

The first line restates the problem in a general form. The second line formulates the equivalent problem in terms of Z, which is a linear transformation of X, preserving the probability amount α. Finally, the third line computes Z_k for this example and shows the probability value determined from Table A.1.

A useful interpretation of the Z value corresponding to any X value can be seen from (10.15). The numerator, $X_k - \mu_X$, measures how far X_k is from its mean. Dividing this difference by σ_X expresses this in terms of how many standard deviations, σ_X, it constitutes. For example, in the previous example $X_k = 6$ corresponds to $Z_k = 0.5$; the X_k value is one-half of a standard deviation (of X) above the mean (of X). Similarly, for the same random variable $X_k = 1$ corresponds to $Z_k = -2$; the X_k value is two standard deviations below its mean. The amount of probability associated with any interval of X values for one normal random variable is exactly the same as the amount of probability associated with a corresponding interval of values for another normal random variable when the endpoints of these intervals are the same number of standard deviations from the respective means.

We can now entertain two familiar questions. First, what is the probability that a normal random variable Z will take on a value within one standard deviation of its mean? This question is formally posed and solved as:

$$\begin{aligned}\alpha &= \Pr(\mu_X - \sigma_X \leq X \leq \mu_X + \sigma_X) \\ &= \Pr(Z_1 \leq Z \leq Z_2), \quad \text{where} \quad Z_1 = \frac{(\mu_X - \sigma_X) - \mu_X}{\sigma_X} = -1 \\ &\qquad\qquad\qquad\qquad\qquad\qquad\quad Z_2 = \frac{(\mu_X + \sigma_X) - \mu_X}{\sigma_X} = +1 \\ &= \Pr(-1 \leq Z \leq 1) = .682 \end{aligned} \tag{10.17}$$

Similarly, we can show that the probability that any normal random variable X will take on a value within two standard deviations of its mean is .954 [i.e., $\Pr(\mu_X - 2\sigma_X \leq X \leq \mu_X + 2\sigma_X) = .954$]. These two findings, combined with the fact that in many cases observed relative frequency distributions are similar to those taken from normal random variables, are the source of the empirical approximation that about 68 percent of the observations on a data variable lie within one standard deviation of the mean and about 95 percent lie within two.

More complex probability calculations for normally distributed random variables can be carried out by transforming them into equivalent problems stated in terms of the standard normal. For example, continuing with the random variable X having a normal distribution with $\mu_X = 5$ and $\sigma_X = 2$, we might seek to determine $\Pr(4.38 \leq X \leq 8.00)$. Using (10.15) and the fundamental fact of probability equivalences, this probability is equal to $\Pr(-0.31 \leq Z \leq 1.5)$; this was calculated earlier to be .555 and that is the answer to our current problem also.

10.2 The *t* Distribution

Another family of continuous probability distributions is that of the ***t distribution,*** which is sometimes referred to as ***Student's t distribution*** after the pseudonym of its originator, W. S. Gossett. This is perhaps the most frequently used distribution in the everyday work of applied econometricians.

The members of this family have probability distributions that have the same general equation form but differ with regard to the value of one parameter; the equation itself is quite complex, so it is not given here. The parameter that distinguishes one family member from another is known as the ***number of degrees of freedom,*** which we denote by df. This takes on only positive integer values: $1, 2, 3, \ldots, \infty$.

The shape of the probability distribution for all members of the t family is a symmetrical, bell-shaped curve, very much like that of the standard normal distribution. In all cases, the mean is equal to zero. Comparing different members of the family, as the number of degrees of freedom increases, the standard deviation decreases toward a lower limit of one. Figure 10.6a compares t distributions with 5 and 50 degrees of freedom. The overall difference between the two distributions appears to be small, but their standard deviations are 1.29 and 1.02, respectively. Also, the differences in tail areas [e.g., $\Pr(t \geq 2)$] are substantial, and these are of special importance.

In the limit, as the number of degrees of freedom approaches infinity, the t distribution becomes identical with the standard normal distribution. Figure 10.6b shows a t distribution with 5 degrees of freedom and also a standard normal distribution. The overall difference between these two distributions is barely greater than the difference between those in Figure 10.6a, and we realize that a t distribution having 50 or more degrees of freedom must be nearly indistinguishable from a standard normal distribution.

In contrast to the normal family, the different t distributions are not linear transformations of each other. Thus, it is necessary to carry out separate mathematical computations to determine probabilities for each distribution. Some of this work is already done, and the results are given in Tables A.2 and A.3. Since each distribution is different, these tables can display only a limited amount of probability information for each of a limited number of members of the t distribution family. This turns out to be satisfactory in practice, and computers can be used to determine specific probability amounts that are not given in the tables.

One type of probability problem that we sometimes encounter is of the form:

$$\text{Find } \alpha \text{ such that} \quad \Pr(t \geq t^*) = \alpha \tag{10.18}$$

where t is a random variable having a t distribution with some particular number of degrees of freedom and t^* is a specified value of t. Table A.2 is arranged to

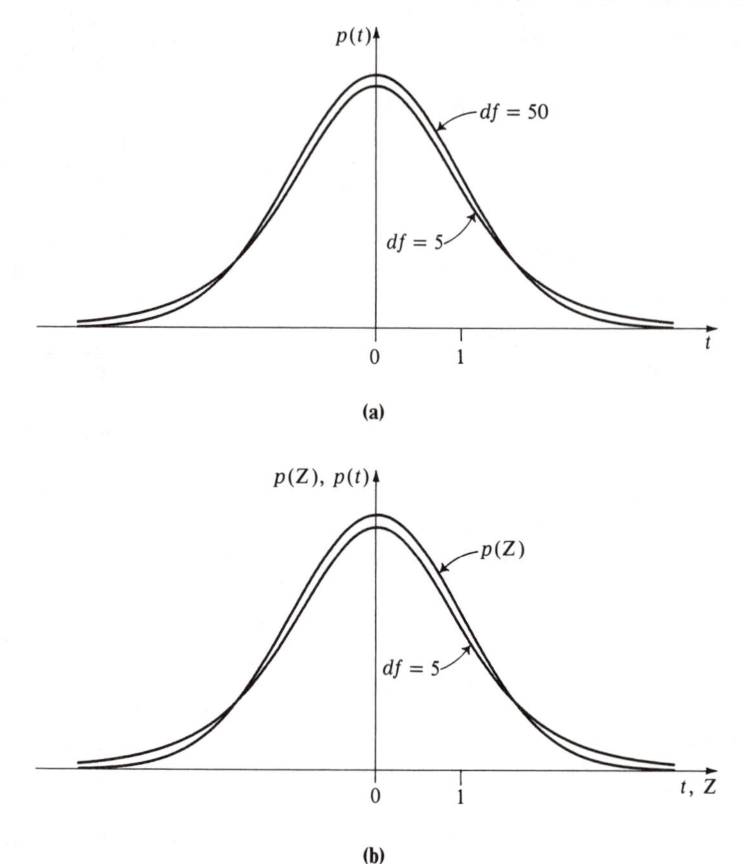

FIGURE 10.6 Part (a) compares two t distributions, those with 5 and 50 degrees of freedom. The overall difference between them is small, but the differences in the tails are more substantial. Part (b) compares the t distribution with 5 degrees of freedom with the standard normal. As the number of degrees of freedom approaches infinity, the limiting form of the t distribution is the standard normal. Comparing the two parts of this figure, we realize that when df = 50, the t distribution is already quite close to its limiting form.

provide the answer to a problem of this form. In the table, each row provides probability information about a t distribution with a particular df. The table entries in the row give the probabilities that a random variable having this particular distribution will exceed the corresponding t^* values specified in the column headings. In other words, each entry gives the probability α that we are seeking in (10.18).

For example, we see that for df = 5, Pr $(t \geq 1.5)$ = .097 and Pr $(t \geq 2.5)$ is .027. Looking down any column, we see that the probability of exceeding the value t^* decreases as the number of degrees of freedom increases. This reflects the fact that the t distribution becomes more compressed as df increases. For example, Pr $(t \geq 2.0)$ = .051 when df = 5, and Pr $(t \geq 2.0)$ = .025

when df = 50. [Note that for the standard normal distribution, Pr $(Z \geq 2.0)$ = .023.]

Our most frequent use of the *t* distribution is to determine what value, t^c, leads to a certain desired probability (α) being bounded in one or both tails. In the case of two tails, half the probability ($\alpha/2$) is to be in each tail. For the random variable *t* having a *t* distribution with df degrees of freedom, we encounter problems of the forms:

$$\text{Find } t^c \text{ such that Pr } (t \geq t^c) = \alpha \tag{10.19}$$

$$\text{Find } t^c \text{ such that Pr } (t \leq -t^c) = \alpha \tag{10.20}$$

$$\text{Find } t^c \text{ such that Pr } (|t| \geq t^c) = \alpha \tag{10.21}$$

Problem forms (10.19) and (10.21) are illustrated in Figure 10.7a and b, respectively.

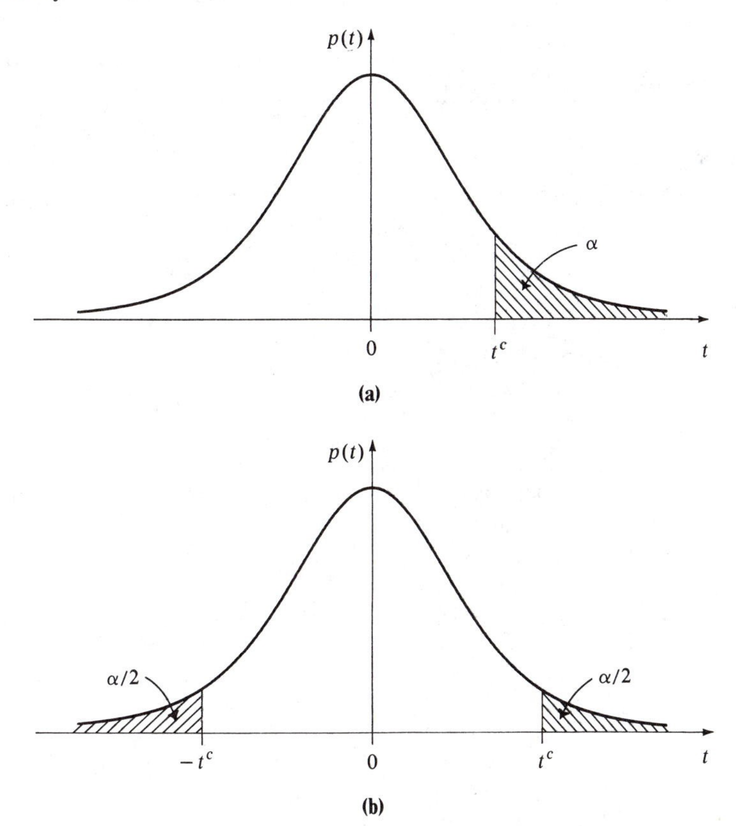

FIGURE 10.7 Suppose that a certain random variable is known to have a *t* distribution. We often need to determine what value, t^c, leads to a certain probability amount (α) being bounded in one or both tails. In (a), the shaded area represents Pr $(t \geq t^c) = \alpha$; in (b), the shaded area represents Pr $(|t| \geq t^c) = \alpha$, with $\alpha/2$ probability being bounded in each tail. In both cases, when α is specified, t^c can be determined from Table A.3.

Table A.3 is arranged to provide the answers to these probability problems. Each row provides information about a t distribution with a particular number of degrees of freedom. For the double row of probability values (α) given in the column headings, the table entries are the t^c values such that (10.19) or (10.21) is satisfied. For example, if df = 20 and α = .05, then the t^c such that Pr ($t \geq t^c$) = .05 is 1.725; in other words, for a random variable that has a t distribution with 20 degrees of freedom, the value that is exceeded with a probability of 5 percent is 1.725. By symmetry, Pr ($t \leq -1.725$) = .05 also. For α = .10, which is twice as big, the t^c value satisfying the two-tailed probability statement (10.21) is the same: Pr ($|t| \geq 1.725$) = .10, because $\alpha/2$ = .05 probability is in each tail.

More complex probability problems can be handled using these basic elements. To avoid errors and to see the simple elements of a complex problem, it is a good idea to draw a diagram to represent the probabilities as areas.

Problems

Section 10.1

10.1 For the random variable Z having a standard normal distribution, determine the following probabilities and draw figures showing the areas representing these probabilities:

(a) Pr ($Z \geq 1.1$)
(b) Pr ($Z \leq -0.5$)
(c) Pr ($0.5 \leq Z \leq 1.5$)
(d) Pr ($-0.5 \leq Z \leq 1.5$)
(e) Pr ($|Z| \geq 0.5$)
(f) Pr ($|Z| \leq 1.5$)

★ **10.2** For the random variable Z having a standard normal distribution, determine the value of Z^c such that

(a) Pr ($Z \geq Z^c$) = .02
(b) Pr ($Z \leq -Z^c$) = .10
(c) Pr ($Z \leq Z^c$) = .85
(d) Pr ($|Z| \geq Z^c$) = .02
(e) Pr ($|Z| \geq Z^c$) = .10
(f) Pr ($|Z| \leq Z^c$) = .85

10.3 What symmetrical values of Z contain between them 20 percent of the total probability?

10.4 Consider a variable X having a normal distribution with mean 100 and standard deviation 15. Determine

(a) Pr ($X \geq 100$)
(b) Pr ($X \geq 110$)
(c) Pr ($X \leq 90$)

(d) Pr $(90 \leq X \leq 110)$
(e) Pr $(-10 \leq X - \mu_X \leq 10)$

★ **10.5** For X defined in Problem 10.4:
(a) What symmetrical values of X contain between them 60 percent of the total probability?
(b) What is the probability of X occurring within one-half standard deviation of its mean?

10.6 For the standard normal distribution, compute the value of $p(Z)$ for $Z = 0, 1, 2,$ and 3.

10.7 For normal distributions, compute the value of $p(X)$ at the mean (i.e., $X = \mu_X$) in the following cases: $\sigma_X = 0.2, 0.5, 1.0, 2.0,$ and 10.0.

10.8 For a normally distributed random variable u having a mean of zero, determine
(a) Pr $(u \geq 7)$ if $\sigma_u = 3$
(b) Pr $(u \geq 7)$ if $\sigma_u = 13$
(c) Pr $(|u| \leq 5)$ if $\sigma_u = 4$
(d) u^c if Pr $(u \geq u^c) = .05$, with $\sigma_u = 5$
(e) u^c if Pr $(|u| \leq u^c) = .25$, with $\sigma_u = 5$

Section 10.2

10.9 For a variable t having a t distribution with 21 degrees of freedom, determine
(a) Pr $(t \geq 1.00)$
(b) Pr $(|t| \geq 1.50)$
(c) Pr $(t \leq -2.00)$
(d) Pr $(t \leq 2.00)$

★ **10.10** Compare Pr $(Z \geq 1.00)$ with Pr $(t \geq 1.00)$, for t distributions with various degrees of freedom. What does this show about the shape of the probability distribution of t?

★ **10.11** For a variable t having a t distribution with 21 degrees of freedom, determine t^c such that
(a) Pr $(t \geq t^c) = .01$
(b) Pr $(t \leq -t^c) = .05$
(c) Pr $(t \leq t^c) = .90$
(d) Pr $(|t| \geq t^c) = .01$
(e) Pr $(|t| \geq t^c) = .05$
(f) Pr $(|t| \leq t^c) = .90$

10.12 Use a diagram to compare the results of Problem 10.11 parts (b) and (e).

10.13 For a random variable t having a t distribution, under what conditions is
(a) Pr $(|t| \geq 2) \approx .05$?

(b) Pr ($|t| \geq 2$) $\leq .05$?
(c) Pr ($|t| \geq 1.7$) $\leq .10$?
(d) Pr ($|t| \geq 1.25$) $\leq .30$?

10.14 For a random variable t having a t distribution with 1000 degrees of freedom, determine t^c such that
(a) Pr ($t \geq t^c$) = .05
(b) Pr ($|t| \geq t^c$) = .05
(c) Pr ($t \geq t^c$) = .15
(d) Pr ($|t| \geq t^c$) = .25

Appendix

10.15 For a random variable χ^2 having a chi-square distribution with 20 degrees of freedom, determine $(\chi^2)^c$ such that
(a) Pr ($\chi^2 \geq (\chi^2)^c$) = .025
(b) Pr ($\chi^2 \geq (\chi^2)^c$) = .100
(c) Pr ($\chi^2 \geq (\chi^2)^c$) = .900
(d) Pr ($\chi^2 \leq (\chi^2)^c$) = .025
(e) Pr ($\chi^2 \leq (\chi^2)^c$) = .010

⋆ **10.16** In each column of Table A.4, the values of $(\chi^2)^c$ increase with the number of degrees of freedom. What does this illustrate about the probability distributions in the chi-square family?

⋆ **10.17** For a random variable $F_{5,20}$, find F^c such that Pr ($F_{5,20} \geq F^c$) = .05.

10.18 Except for cases in which the number of degrees of freedom in the denominator is very small, use the F^c values in Table A.5 to describe the differences among the probability distributions in the F family.

APPENDIX*

In this appendix we discuss two families of probability distributions that are related to the normal distribution.

The Chi-Square Distribution

Suppose that we have d independent random variables, $Z_1, Z_2, \ldots, Z_d$, each having a standard normal distribution. We can define a new random variable, χ^2, as

$$\chi^2 = \sum_{j=1}^{d} (Z_j)^2 \qquad (10.22)$$

*This appendix is relatively difficult and can be skipped without loss of continuity.

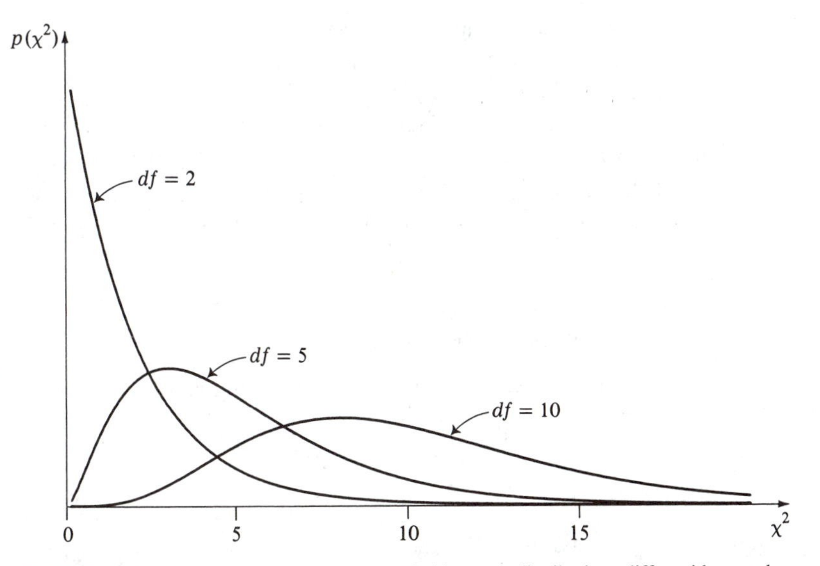

FIGURE 10.8 The members of the family of chi-square distributions differ with regard to only one parameter, the number of degrees of freedom (df). The distributions are skewed to the right, but they become more symmetric as df increases. Table A.4 is used to find the critical values of χ^2 that bound specified probability amounts in the right-hand tail.

where χ^2 should be thought of as the name of the variable, rather than as the square of a variable (χ is lowercase "chi," the Greek "ch"). The possibly occurring values of χ^2 are all nonnegative, because χ^2 is defined as a sum of squares.

A random variable that can be thought of as arising through (10.22) has a ***chi-square distribution*** with d degrees of freedom. The number of degrees of freedom, df $= d$ here, is the characteristic that makes one member of the family different from another. As illustrated in Figure 10.8, the graph of the probability distribution is skewed to the right with a tail that extends out to infinity, but the distribution becomes more symmetric as the number of degrees of freedom increases.

The chi-square distribution is of practical importance to us in Chapters 18 and 19, where we encounter problems of the forms:

$$\text{Find } (\chi^2)^c \text{ such that } \Pr(\chi^2 \geq (\chi^2)^c) = \alpha \tag{10.23}$$

$$\text{Find } (\chi^2)^c \text{ such that } \Pr(\chi^2 \leq (\chi^2)^c) = \alpha \tag{10.24}$$

Table A.4 is arranged to provide directly the answer to a problem posed like (10.23). For example, if a random variable has a chi-square distribution with 10 degrees of freedom and if we are concerned with $\alpha = .10$, then the random variable will take on a value greater than or equal to $(\chi^2)^c = 15.99$ with the stated probability. A problem like (10.24) seeks the critical value that bounds

α probability in the left-hand side. If $\alpha = .05$, for example, then this problem is equivalent to one posed like (10.23) with $\alpha = 1 - .05 = .950$. For df = 10, again the critical value is $(\chi^2)^c = 3.94$.

The chi-square distribution is a logical bridge between the standard normal distribution and two other families that are of interest to us. Mathematically, a *t* disribution arises in the following way. Suppose that Z has a standard normal distribution and χ_d^2 has a chi-square distribution with d degrees of freedom. If Z and χ_d^2 are independent,

$$t = \frac{Z}{\sqrt{\chi_d^2/d}} \tag{10.25}$$

defines a new random variable, t, having a t distribution with d degrees of freedom.

The *F* Distribution

Suppose that we have two independent random variables, χ_n^2 and χ_d^2, having chi-square distributions with n and d degrees of freedom, respectively. We can define a new random variable, F, as the ratio of these independent chi-squares:

$$F = \frac{(\chi_n^2)/n}{(\chi_d^2)/d} \tag{10.26}$$

A random variable that can be thought of as arising this way is said to have an ***F* distribution** with n and d degrees of freedom, and we use the notation $F_{n,d}$

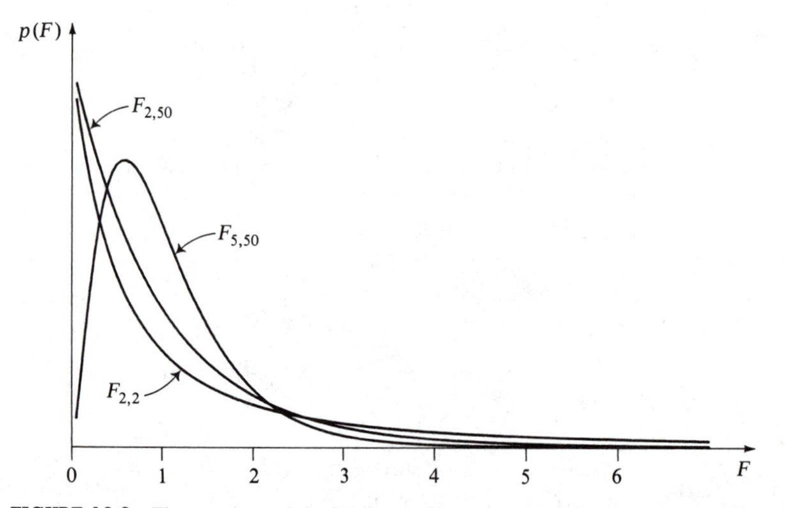

FIGURE 10.9 The members of the family of F distributions differ with regard to two parameters, which are known as the numbers of degrees of freedom in the numerator (n) and denominator (d), respectively. Table A.5 is used to find the critical value of F that bounds 5 percent probability in the right-hand tail, and Table A.6 is used for 1 percent.

to indicate this sometimes. Note that n and d are the parameters in the numerator and denominator, respectively, of (10.26).

The F distribution is defined for the range $0 \leq F < \infty$. The members of the family differ from each other only by the values of the parameters n and d. As illustrated in Figure 10.9, the shape of the distribution is skewed to the right (i.e., there is a long tail coming out to the right). However, as n and d become large, the F distribution looks much like a member of the normal family.

In Chapters 15 and 19 we encounter problems of the form:

$$\text{Find } F^c \text{ such that} \quad \Pr(F_{n,d} \geq F^c) = \alpha \tag{10.27}$$

Table A.5 gives the F^c values for various combinations of n and d, when $\alpha = .05$, and Table A.6 gives F^c when $\alpha = .01$. For example, for an F distribution with 5 and 10 degrees of freedom, the critical value that bounds 5 percent probability in the right-hand tail is 3.33. For an F distribution with 10 and 1500 degrees of freedom, the critical 5 percent probability value is $F^c = 1.83$, using ∞ to approximate the high number of degrees of freedom in the denominator.

In econometric applications, n is often relatively small whereas d may be small or large because it is closely related to the number of observations.

Inference in Regression

Sampling Theory in Regression

While discussing the estimation of regression coefficients in Chapter 5, we noted that the actual $\hat{\beta}_0^*$ and $\hat{\beta}_1^*$ values that we calculate will not be equal to the true β_0 and β_1 values of the economic process that generated the data, except by coincidence. Our interest now is in going beyond the estimates themselves and making some probabilistic statements about the true values of the parameters. The key to doing this is based on the theories of ***sampling*** and ***statistical inference,*** which we develop and apply to regression in this and the next two chapters.

Some of the material in these chapters is fairly mathematical and contains some subtle reasoning. These difficult topics are worth mastering, because only then can the results be applied properly and with understanding.

11.1 The Normal Regression Model

The basic model of simple regression developed in Chapter 5 states that for a given set of values for the variable X, the corresponding values of Y are determined by

$$Y_i = \beta_0 + \beta_1 X_i + u_i \tag{11.1}$$

for $i = 1, 2, 3, \ldots, n$. As defined previously, the expression $\beta_0 + \beta_1 X_i$ is the systematic portion of the determination of Y_i, and the disturbance u_i is the random portion.

The disturbance is meant to represent a host of factors that help determine Y, including the impacts of unconsidered explanatory variables and possible error in the measurement of Y. These determinants are considered to be secondary in importance to the explanatory variable X, and their impacts on Y are not analyzed directly. Instead, these factors collectively are considered equivalent to pure chance, and the disturbance is taken to be a random variable. The random nature of the disturbance is extremely important in the development of sampling theory.

To begin, we focus on the determination of Y for a typical observation, the ith. The value of the explanatory variable X_i is fixed by some other economic process, and the value of the systematic portion $\beta_0 + \beta_1 X_i$ is thereby determined. With regard to the disturbance, we assume that u_i is a random variable having a probability distribution that is normal with

$$E[u_i] = 0 \qquad (11.2)$$

$$\sigma(u_i) = \sigma_u \qquad (11.3)$$

By (11.1), the value of Y_i depends on the outcome of this random variable u_i; therefore, Y_i itself is a random variable.

It is essential to understand that the result of assuming the disturbance to be a random variable is that the value of Y_i we observe in a data set is simply one possible outcome that might have occurred for that individual; pure chance might have led to a different outcome, even though X_i is fixed. For economy of notation we do not distinguish between Y_i used as the name of the random variable (e.g., what the earnings of the ith person might be) and as its outcome (e.g., the actual earnings of the ith person in our data).

In this view of things, the determination of Y_i is illustrated in Figure 11.1, which shows $p(Y_i)$, the probability distribution of Y_i. A particular outcome is indicated by the asterisk (*) value, and it is this value that we obeserve and denote by Y_i in our data. Since the random variable Y_i is a linear transformation of u_i (with $a = \beta_0 + \beta_1 X_i$ and $b = 1$), $p(Y_i)$ can be thought of as being the probability distribution of u_i (normal, $\mu = 0$, $\sigma = \sigma_u$) simply moved over to the right or left by the amount $\beta_0 + \beta_1 X_i$. Thus Y_i has a probability distribution that is normal. Its mean is

$$E[Y_i] = \beta_0 + \beta_1 X_i \qquad (11.4)$$

and its standard deviation is

$$\sigma(Y_i) = \sigma_u \qquad (11.5)$$

Equation (11.4) provides the technically correct basis for the nontechnical use of this notation in Chapter 5.

With Y_i being viewed as a random variable, we are in a position to make probabilistic statements about the outcome that we will observe in data. For example, suppose that somehow we know that $E[Y_i] = 70$ for the ith observation

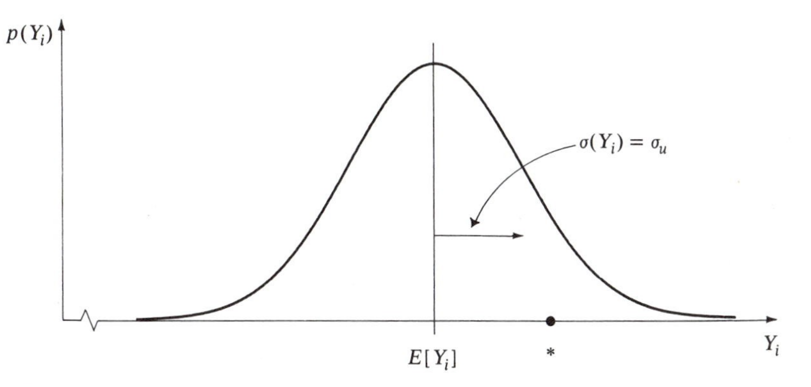

FIGURE 11.1 In the normal regression model, X_i is fixed and Y_i is thought of as a random variable because it is determined, in part, by a random disturbance. The probability distribution of Y_i is normal, with a mean equal to $\beta_0 + \beta_1 X_i$ and a standard deviation equal to σ_u. The value we denote by Y_i in a data set now is thought of as one outcome from this probability distribution, and such a value is indicated by the asterisk in this figure.

and that the disturbance has a normal distribution with $E[u_i] = 0$ and $\sigma(u_i) = 5$. Our new view of Y_i leads us to see that Y_i is normally distributed with $\mu = 70$ and $\sigma = 5$. Based on our familiarity with the normal family, we can say that the probability is about 68 percent that the observed value of Y_i (i.e., its outcome) will be between 65 and 75; there is a 95 percent chance that it will be between 60 and 80. Similarly, the probability that the observed value of Y will be less than or equal to 65 is about 16 percent.

Now we consider the situation for all the observations. Each observation is characterized by a particular value for X, although more than one observation can have the same value of course. For each observation the value of Y is determined according to the process just discussed. We make two explicit assumptions about the relations among the disturbances. First, we assume that all the u_i are independent: the outcome of the disturbance for one observation in no way affects the outcome for another. Second, as already embodied in (11.3), we assume that $\sigma(u_i)$ is the same value, σ_u, for all observations. Figure 11.2 illustrates the situation for a set of three observations in which the X values are given as X_1, X_2, and X_3. This figure represents a three-dimensional diagram in which the familiar Y and X axes are drawn in a horizontal plane and the probability density values are drawn vertically. For each observation the probability distribution of Y_i is shown, as though it were taken from Figure 11.1 and placed upright on the Y, X plane. A set of particular outcomes of the Y_i are shown, traced over to the Y axis, and labeled as Y_1, Y_2, and Y_3.

The assumptions we have made about the disturbances, together with the basic tenet that the X values are determined outside the process described by the regression model, imply that the disturbance values are not related to the X

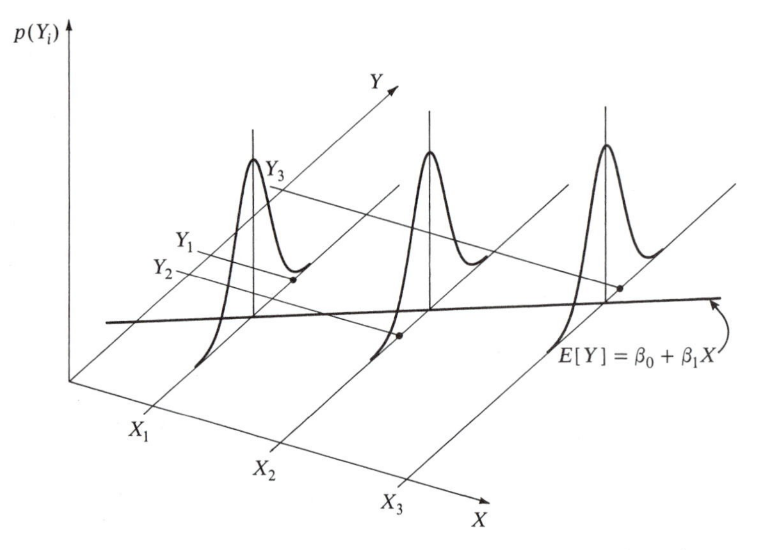

FIGURE 11.2 In the normal regression model, the value of Y for each observation is considered to be an outcome of a separate normal probability distribution. Each distribution has a mean of $\beta_0 + \beta_1 X_i$ and a standard deviation of σ_u, as in Figure 11.1. In this three-dimensional diagram, probability density values are drawn vertically. The set of outcomes labeled here as Y_1, Y_2, and Y_3 are thought of as just one possible set of outcomes corresponding to the three given values of X.

values. Thus, for instance, the disturbance is no more likely to be positive when X_i is large than when it is small.

The determination of Y values as discussed in this section is known as the ***normal regression model.*** The ideas extend to multiple regression without substantial change, except that a three-dimensional diagram no longer suffices.

11.2 The Sampling Distributions of Coefficient Estimators

Suppose we have a set of data on Y and X that we believe was generated by a process described by the normal regression model. In Chapter 5 we developed the OLS estimator for β_1:

$$\hat{\beta}_1 = \frac{\sum (X_i - \overline{X})Y_i}{\sum (X_i - \overline{X})^2} \tag{11.6}$$

For a given set of X values, the value of $\hat{\beta}_1$ depends partly on the outcomes of the disturbances because the Y values depend on the disturbances. The same can be said for the value of $\hat{\beta}_0$. This notion leads to a theoretical analysis that views the actual estimates in our data, $\hat{\beta}_0^*$ and $\hat{\beta}_1^*$, as being outcomes of certain probability distributions.

To understand this theory we conduct a thought experiment, which consists of a hypothetical set of statistical calculations. Following the discussion of the preceding section, we consider taking a sample of size $n = 3$ for which the X values are given as X_1, X_2, and X_3. Based on (11.1), the Y values are determined in part by drawing values for the disturbance terms. We then have a set of data with three observations: (X_1, Y_1), (X_2, Y_2), and (X_3, Y_3). These are plotted in the top panel of Figure 11.3, where also is plotted the true regression line with height $\beta_0 + \beta_1 X$. For each observation, whether the actual Y_i value is more or less than $\beta_0 + \beta_1 X_i$ depends on pure chance; that is, it depends on the outcome of the disturbance u_i. The OLS fitted line, which is based on the parameter estimates $\hat{\beta}_0^*$ and $\hat{\beta}_1^*$, is also drawn in.

In the next step of the experiment we keep the same three X_i values and draw three new values for the disturbances, yielding a new set of Y_i values. This new set is plotted in the second panel of Figure 11.3, where the true regression line is repeated and the new sample's fitted regression line is also drawn. Except by coincidence, the Y_i values in this second sample will be different from the corresponding Y_i values in the first sample because of the random nature of the disturbances. Hence, except by coincidence, the estimated values $\hat{\beta}_0^*$ and $\hat{\beta}_1^*$ will also be different from those in the first sample.

In the same fashion we can draw a total of N samples, all having the same three X values. In each sample we calculate the values $\hat{\beta}_0^*$ and $\hat{\beta}_1^*$.

Finally, we collect the N different $\hat{\beta}_0^*$ values and display them in a relative frequency histogram, and we collect the N different $\hat{\beta}_1^*$ values and display them similarly. This is done near the bottom of Figure 11.3. That different values for $\hat{\beta}_0^*$ and $\hat{\beta}_1^*$ occur in different samples drawn from the same economic process illustrates the phenomenon of ***sampling variability.*** Looked at another way, sampling theory makes us realize that when we have a set of real data our situation might be like that of any one of the samples illustrated or hinted at in Figure 11.3—but we cannot know which one.

In an experiment like this, mathematical analysis shows that if N is increased toward infinity, the relative frequency distributions become smoother and each approaches a limiting form. These limiting forms can be interpreted as probability distributions, as illustrated at the bottom of Figure 11.3. Because of the situation in which they arise, these two probability distributions are given special names: they are the ***sampling distributions*** of $\hat{\beta}_0$ and $\hat{\beta}_1$, respectively.

The actual specification of these sampling distributions can be developed theoretically. Based on the assumptions of the normal regression model, it can be shown that the sampling distribution of $\hat{\beta}_0$ is normal with a mean and standard deviation given by

$$E[\hat{\beta}_0] = \beta_0 \tag{11.7}$$

$$\sigma(\hat{\beta}_0) = \sigma_u \sqrt{\frac{\sum X_i^2}{n \sum (X_i - \bar{X})^2}} \tag{11.8}$$

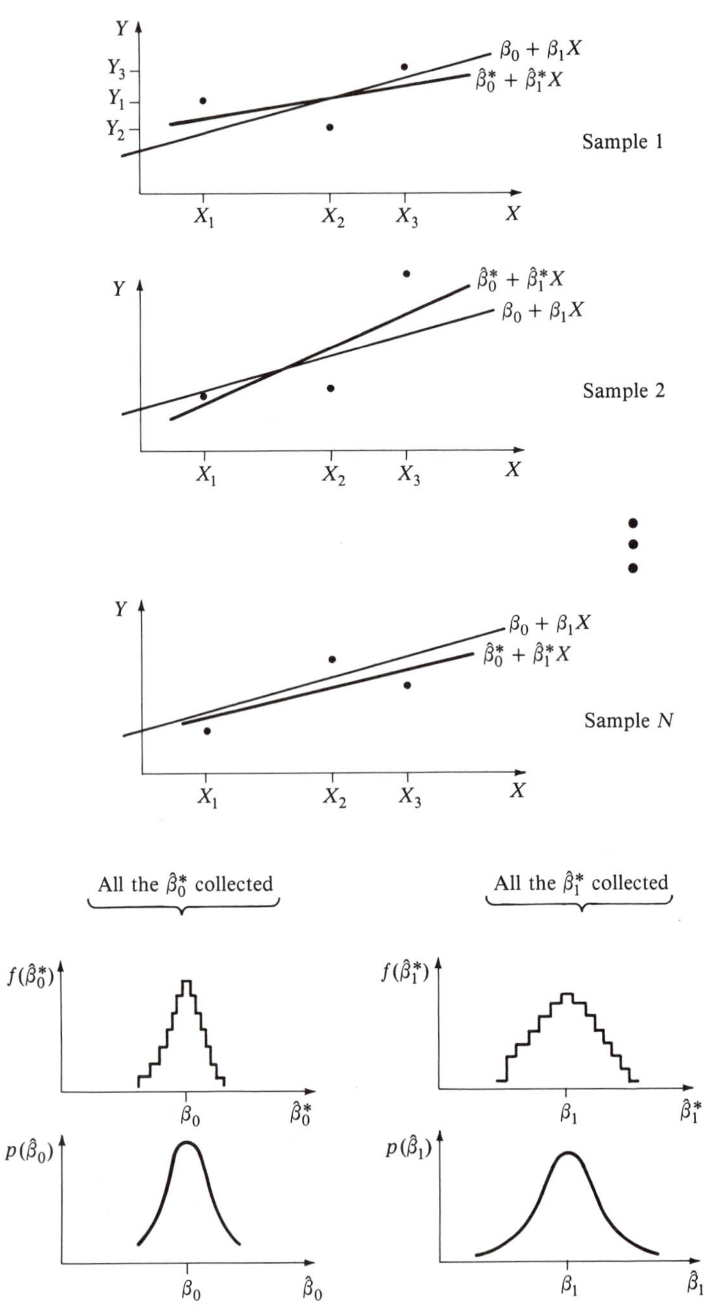

FIGURE 11.3 A thought experiment helps explain the nature of the sampling distributions of the OLS estimators of β_0 and β_1. The same three X values are used in each sample, but different sets of Y values are produced because different values for the disturbances occur in each sample. The $\hat{\beta}_0^*$ and $\hat{\beta}_1^*$ values computed in each sample are collected, and their frequency distributions are constructed. The sampling distribution of an estimator is the limiting form of the frequency distribution, as N approaches infinity. The actual $\hat{\beta}_0^*$ and $\hat{\beta}_1^*$ that we calculate from a set of data are thought of as just one pair of outcomes from these sampling distributions.

Similarly, in the appendix to this chapter we show that the sampling distribution of $\hat{\beta}_1$ is normal with

$$E[\hat{\beta}_1] = \beta_1 \tag{11.9}$$

$$\sigma(\hat{\beta}_1) = \sigma_u \sqrt{\frac{1}{\sum (X_i - \bar{X})^2}} \tag{11.10}$$

For reasons elaborated below, the standard deviation of a sampling distribution is often called the ***standard error*** of the estimator. Thus $\sigma(\hat{\beta}_0)$ is the standard error of $\hat{\beta}_0$, and $\sigma(\hat{\beta}_1)$ is the standard error of $\hat{\beta}_1$.

A sampling distribution gives a complete description of the likelihood of various possible values occurring as the actual estimate of a coefficient when a set of data is drawn. When we work with a set of actual data, the $\hat{\beta}_0^*$ and $\hat{\beta}_1^*$ that we calculate are thought of as just one pair of outcomes from their respective sampling distributions. Thus we realize that chance might have led to other values.

In practical applications σ_u is unknown, and it is estimated by the standard error of the regression, *SER*. (The *SER* is the typical residual, and hence it reasonably serves as an estimate of the typical disturbance.) When *SER* replaces σ_u in (11.8) and (11.10), the resulting estimators of the standard errors $\sigma(\hat{\beta}_0)$ and $\sigma(\hat{\beta}_1)$ are denoted by $s(\hat{\beta}_0)$ and $s(\hat{\beta}_1)$:

$$s(\hat{\beta}_0) = SER \sqrt{\frac{\sum X_i^2}{n \sum (X_i - \bar{X})^2}} \tag{11.11}$$

$$s(\hat{\beta}_1) = SER \sqrt{\frac{1}{\sum (X_i - \bar{X})^2}} \tag{11.12}$$

For example, the simplest earnings function reported in Chapter 6 is based on the presumption that the true process determining earnings is

$$EARNS_i = \beta_0 + \beta_1 ED_i + u_i \tag{11.13}$$

The estimated form of the model is

$$\begin{aligned} \widehat{EARNS}_i &= -\underset{(1.540)}{1.315} + \underset{(0.128)}{0.797}ED_i \\ R^2 &= .285 \qquad SER = 4.361 \end{aligned} \tag{11.14}$$

The calculated standard errors of the coefficient estimators are reported here in parentheses below the corresponding coefficients; for β_1, $\hat{\beta}_1^* = 0.797$ and $s^*(\hat{\beta}_1) = 0.128$.

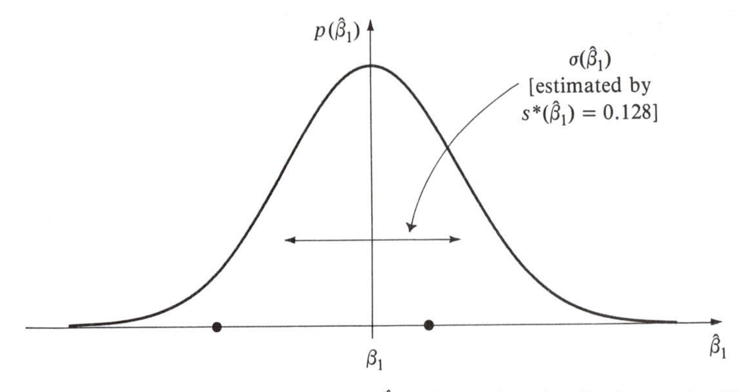

FIGURE 11.4 The sampling distribution of $\hat{\beta}_1$ in the earnings function is normal, with a mean equal to β_1 (unknown) and a standard deviation (i.e., standard error) estimated to be 0.128. The estimate of β_1 is $\hat{\beta}_1^* = 0.797$; this might correspond to either of the unlabeled points on the horizontal axis.

The information we have about β_1 is illustrated in Figure 11.4: the sampling distribution of $\hat{\beta}_1$ is normal, with a mean equal to the unknown true β_1 and a standard deviation estimated to be 0.128. It might be that the calculated $\hat{\beta}_1^*$ lies at the unlabeled point to the right, in which case the true β_1 would be less than 0.797; or it might be that $\hat{\beta}_1^*$ lies at the unlabeled point to the left, in which case the true β_1 would be greater than 0.797. In any practical situation we never know where the calculated $\hat{\beta}_1^*$ lies relative to the true β_1. Although it is tempting, we never think of a sampling distribution as being centered on the $\hat{\beta}_1^*$ we happen to calculate.

In multiple regression also, each of the estimated coefficients can be thought of as being the outcome of the sampling distribution for the estimator of that coefficient. Each of the sampling distributions is normal, with

$$E[\hat{\beta}_j] = \beta_j \tag{11.15}$$

and a standard deviation that depends on σ_u and the values of all the regressors.

Estimation Errors

Consider any coefficient estimator $\hat{\beta}_j$, where β_j stands for either β_0 or β_1 in simple regression or for the jth coefficient in multiple regression. Sampling theory tells us that the actual estimate $\hat{\beta}_j^*$ will not be equal to the value of the true coefficient β_j, except by coincidence. For a particular set of data, the difference $\hat{\beta}_j^* - \beta_j$ might be positive or negative.

From the point of view of the task of estimation, such a difference is an error. Considering possible values for the outcome of an estimator $\hat{\beta}_j$, we define the associated ***estimation error*** by

$$\text{estimation error for } \hat{\beta}_j = \hat{\beta}_j - \beta_j \tag{11.16}$$

For example, in Figure 11.4 the estimation error of any outcome for $\hat{\beta}_1$ is simply the distance from β_1 to that value.

Since the mean of $\hat{\beta}_j$ is β_j, the estimation error for $\hat{\beta}_j$ is the same as the deviation of $\hat{\beta}_j$ from its mean: both are equal to $\hat{\beta}_j - \beta_j$. As for any random variable the expected deviation of $\hat{\beta}_1$ from its mean is equal to zero, and therefore so is the expected estimation error. Because of this, the OLS estimator is said to be ***unbiased.*** This means that on average (i.e., in expectation) the estimate is right on target. However, we realize that in any particular case the actual estimation error will not be zero, except by coincidence.

Applying similar logic, we can see that the typical size (neglecting sign) of the estimation error for $\hat{\beta}_1$ is given by the standard deviation of $\hat{\beta}_1$. It is for this reason that $\sigma(\hat{\beta}_1)$ is also called the "standard error" of $\hat{\beta}_1$: this name makes sense when we think of it as being short for "standard estimation error."

11.3 Probability Calculations

The sampling theory we have developed permits us to pose and answer questions about our estimates and the associated estimation errors. For example, suppose that we are omniscient and know that the true regression specification is

$$Y_i = 7 + 12X_i + u_i \tag{11.17}$$

and that the u_i satisfy all the assumptions of the normal regression model with $\sigma_u = 5$. We do not face an estimation problem at all, because we know that $\beta_0 = 7$ and $\beta_1 = 12$. For the fun of it, however, we can do some experiments with data.

Suppose first that we select a set of data for which the total variation of X [i.e., $\sum (X_i - \overline{X})^2$] is equal to 9. Given these particular X values, the corresponding Y values that we find are thought of as having been produced by the normal regression model (11.17). Hence focusing on $\hat{\beta}_1$, we realize that the $\hat{\beta}_1^*$ we calculate by applying (11.6) is not likely to be equal to 12, the true value. Rather, we realize that the calculation of $\hat{\beta}_1$ will result in an estimation error whose typical size is equal to

$$\sigma(\hat{\beta}_1) = \sigma_u \sqrt{\frac{1}{\sum (X_i - \overline{X})^2}} = (5)\sqrt{1/9} \approx 1.67 \tag{11.18}$$

Further, knowing that $\hat{\beta}_1$ has a sampling distribution that is normal with a mean of 12 and a standard deviation of 1.67, we might ask ourselves what the chance would be that $\hat{\beta}_1$ in this sample would turn out to be between 11 and

13. Applying the method of calculating probabilities in a normal distribution, this works out to about 45 percent:

$$\alpha = \Pr(11 \le \hat{\beta}_1 \le 13) = 1 - 2\Pr(\hat{\beta}_1 \ge 13)$$

$$= 1 - 2\Pr(Z \ge Z_k),$$

$$\text{where } Z_k = \frac{(\hat{\beta}_1)_k - E[\hat{\beta}_1]}{\sigma(\hat{\beta}_1)}$$

$$= \frac{13 - 12}{1.67} = 0.6$$

$$= 1 - 2\Pr(Z \ge 0.6) = 1 - (2)(.274)$$

$$= .452 \tag{11.19}$$

As a second experiment, suppose that we now select a set of data for which the total variation of X is equal to 25. In this case the standard error computed through (11.18) is $\sigma(\hat{\beta}_1) = 1.0$, rather than 1.67 as before. The probability that $\hat{\beta}_1$ in this sample would turn out to be between 11 and 13 is about 68 percent, rather than about 45 percent as before.

Comparing these two experiments, we see that when the standard error is smaller there is a greater probability that $\hat{\beta}_1$ will take on a value in some interval centered on the true β_1 value (e.g., in the interval 12.0 ± 1.0). If we think of some interval around β_1 that we would want to call "close" to β_1, we see that the smaller is the standard error of $\hat{\beta}_1$ the greater will be the probability that the outcome of $\hat{\beta}_1$ will be "close" to the true β_1 value. Because of this, we say that the smaller is the standard error, the more ***precise*** is $\hat{\beta}_1$ as an estimator of β_1.

Generally speaking, it is good to have a lot of variation among the X values in a data set used for regression, because this makes the standard errors small. We can see this in (11.18), where the "total variation of X" is in the denominator under the square-root sign.

11.4 *Z* and *t* Statistics

Suppose someone conjectures that the true value of β_1 is 0.7—that an additional year of schooling increases annual earnings by exactly seven hundred dollars. Are the results of the regression (11.14) consistent with this? If $\beta_1 = 0.700$, this implies that the actual estimation error for $\hat{\beta}_1$ is $0.797 - 0.700 = 0.097$. This is about three-fourths as large as the calculated standard error for $\hat{\beta}_1$, and therefore $\hat{\beta}_1$ would correspond to the unlabeled point to the right of β_1 in Figure 11.4. Its distance from the mean is not so extreme as to demand an explanation as a "statistical fluke" in order for us to accept the original conjecture that

$\beta_1 = 0.7$. Since the outcome is not much of a surprise, we might say that the data are consistent with the conjecture; at least, they are not too inconsistent.

By contrast, given the results in (11.14) we would be very surprised by the conjecture that the true $\beta_1 = 3.5$. This would imply an actual estimation error equal to $0.797 - 3.500 = -2.703$, which is more than 21 times as large as the calculated standard error—that is, 21 times as large as the typical estimation error expected for this regression. In Figure 11.4, $\hat{\beta}_1^*$ would be way off the page, to the left. Although sampling theory tells us that there is no theoretical limit on the size of the estimation error, we would be very surprised to witness such a "statistical fluke" in our lifetime. Hence because we react to surprise with disbelief, we would tend to say that the data are inconsistent with the conjecture that $\beta_1 = 3.5$.

This example suggests that whether a given estimation error should be judged large or small—whether it leads us to disbelieve a conjecture about the value of β_1, or not—involves a comparison of the estimation error with the standard error. Following up on this idea, we first return to the hypothetical situation in which we know the true value of σ_u. Consider the coefficient β_j, which stands for either β_0 or β_1 in simple regression, or for the jth coefficient in multiple regression. The ratio of the estimation error to the standard error defines a new statistic

$$Z = \frac{\text{estimation error for } \hat{\beta}_j}{\text{standard error of } \hat{\beta}_j} = \frac{\hat{\beta}_j - \beta_j}{\sigma(\hat{\beta}_j)} \tag{11.20}$$

This random variable Z measures the size of the estimation error for any outcome relative to the standard error, and we interpret it as a measure of the surprise we feel when considering that the conjectured value for β_1 might be the true value.

Since $\hat{\beta}_j$ has a sampling distribution that is normal with mean β_j and standard deviation $\sigma(\hat{\beta}_j)$, the Z defined by (11.20) has a standard normal distribution. Hence probability calculations regarding the chance that a certain surprise level would be achieved or exceeded can be easily calculated.

In the more realistic situation in which σ_u is not known, we can still entertain and evaluate conjectures that β_j is some specific value. In this case the ratio of the estimation error to the (calculated) standard error defines a new statistic

$$t = \frac{\text{estimation error for } \hat{\beta}_j}{\text{(calculated) standard error of } \hat{\beta}_j} = \frac{\hat{\beta}_j - \beta_j}{s(\hat{\beta}_j)} \tag{11.21}$$

This newly defined random variable, t, has the same interpretation as Z above: it serves as a measure of the surprise we feel when considering that the conjectured value for β_j might be the true value.

Since $s(\hat{\beta}_j)$ is itself an estimate of the true standard error $\sigma(\hat{\beta}_j)$, we might anticipate that the probability distribution of this t would be very similar to the standard normal distribution. Indeed, this is so: it can be shown that this newly defined random variable has a t distribution with $n - k - 1$ degrees of freedom (k is the number of regressors, so df $= n - 2$ for simple regression).

For example, returning to the finding reported in (11.14), we consider again the conjecture that $\beta_1 = 0.7$. If this were true, the actual estimation error would be $0.797 - 0.700 = 0.097$. In relation to the (calculated) standard error, this is

$$t^* = \frac{\hat{\beta}_1^* - \beta_1}{s^*(\hat{\beta}_1)} = \frac{0.797 - 0.7}{0.128} = \frac{0.097}{0.128} = 0.76 \qquad (11.22)$$

One way to evaluate this is to determine the chance that t^* would turn out to be 0.76 or greater. Since t has a t distribution with 98 degrees of freedom, we can see from Table A.2 that Pr $(t \geq 0.76)$ is between .309 and .160; a computer calculation yields .225 as the exact value. Since the chance that t^* would turn out to be 0.76 or greater is more than 22 percent, we are not greatly surprised by the conjecture that $\beta_1 = 0.7$ and we would tend to say that the data are consistent with it.

In Chapter 12 we formalize and extend this analysis about surprise and conjectures to develop procedures for hypothesis testing. The term "measure of surprise" as used here is not conventional, but it is a natural interpretation of an important statistic.

11.5 Sample Selection

In Chapter 2 we introduced the idea that the essential criterion for selecting data to be used in econometric regression analysis is that all the observations are generated by the same economic process. We maintain this idea. In addition, two further considerations regarding the selection of a sample of observations are suggested by sampling theory.

The first consideration is that some care must be taken so that the disturbances for the observations we use can reasonably be thought of as a random sample from independent, normally distributed random variables. We usually cannot address this consideration fully, because we know so little about the disturbances.

Having a random sample is important in order that we may apply the results of sampling theory for hypothesis testing and related techniques of inference. Loosely speaking, a ***random sample*** is one in which the outcomes of a random variable are accepted, or selected, without regard to their values. For example, if we plan to accept the next 10 outcomes from a random variable, those outcomes would constitute a random sample; all the samples dealt with in Chapter 9 are of this type. By contrast, if we plan to accept the next 10 *positive* outcomes from a normal random variable, those outcomes would not constitute a random sample.

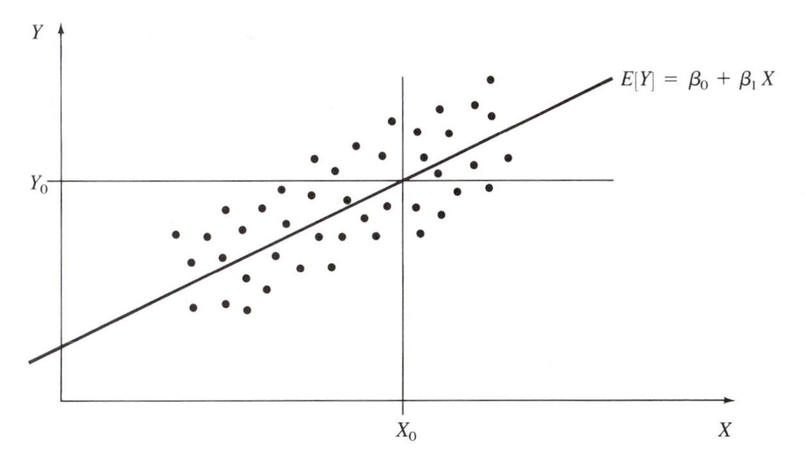

FIGURE 11.5 For the given set of X values, the Y values generated around the true regression line constitute a random sample. For these data, the OLS slope estimator is unbiased: $E[\hat{\beta}_1] = \beta_1$. If those observations for which $Y > Y_0$ are excluded, the slope estimator is biased: the estimated slope in the reduced sample will tend to be smaller than the estimated slope in the full random sample. However, if only those observations for which $X > X_0$ are excluded, the slope estimator remains unbiased.

In this regard we can see that care should be taken to avoid selecting observations on the basis of the value of the dependent variable, because this is tantamount to selecting them on the basis of the value of the disturbances. Were we to select observations on the basis of the outcomes of their disturbances, the outcomes in our data would not constitute a random sample.

This situation and its consequences are illustrated in Figure 11.5. Following regression theory, the underlying true regression line $E[Y] = \beta_0 + \beta_1 X$ is plotted first. Based on a given set of values for X, a set of observations resulting from this process is shown. If all these observations were used to make estimates of the coefficients, the sampling theory of this chapter could be applied appropriately.

Suppose, however, that for some reason the observations with Y values greater than Y_0 are excluded from the sample. Looking at the figure, this would tend to exclude observations having relatively high values for both X and Y, but would tend to let observations with relatively high values of X but low values of Y remain in the sample. In this case a regression line estimated with the reduced sample would have a flatter slope than a regression line estimated with the full sample. Since the expected value of the estimated slope equals β_1 with the full sample, the expected value of the estimated slope in the reduced sample would be less than β_1. In other words, the slope estimator is biased in the reduced sample.

The exclusion of observations having $Y_i > Y_0$ effectively means that observations are being excluded, in part, on the basis of the value of their disturbances. Among the observations with $X_i \geq X_0$, all those having $u_i \geq 0$ will be excluded; in addition, those whose negative disturbance is sufficiently small in absolute value will be excluded. Similarly, among the observations with $X_i < X_0$, those with sufficiently large positive disturbances will be excluded. Thus the observations in the reduced sample will tend to have disturbances that are negative, and this tendency will be accentuated the greater is the value of X.

Interestingly, these ideas do not carry over to possible selection of the sample on the basis of the explanatory variables. For example, referring to Figure 11.5 again, if we were to exclude those observations having $X_i > X_0$, none of the previous difficulties would appear. The outcomes of the disturbances for the observations remaining in the sample have not influenced the sample selection, and they constitute a random sample.

The second consideration we are concerned with here recognizes that it is appropriate to select observations on the basis of the values of the explanatory variables. Given this, is there any advantage to selecting one type of sample rather than another? Assuming that all the assumptions of the normal regression model hold appropriately, the only advantage that one sample can have compared with another is smaller standard errors. Let us focus on the estimating the slope in a simple regression. For a given economic process the standard error of $\hat{\beta}_1$ depends inversely on the total variation of X in the data, as can be seen in (11.10). In general, the larger is the sample, the greater this will be, so it is advantageous to have as large a sample as possible.

More interestingly, among samples of the same size the total variation of X depends on how widely dispersed the values of X are in the data. So a sample can be selected with this in mind. For example, in Section 11.3 we looked at two experiments. The total variation in X was 9 in the first case and 25 in the second, and the corresponding standard errors for $\hat{\beta}_1$ were 1.67 and 1.0, respectively. These samples could be the same size, and if we were interested in making the estimate of β_1 as precise as possible, the second sample would be preferable.

11.6 A Compendium of Reported Regressions

Tables 11.1 through 11.6 bring together nearly all the regressions reported in Chapters 6 and 7. This tabular style of presentation is a common way of reporting alternative specifications of regressions having the same dependent variable. Each column displays a separate estimated regression, and a blank entry indicates that the variable listed in the left-hand stub is not included in the specification. When only one or two regressions are reported, the style of (11.14) is satisfactory.

The number in parentheses below each estimated coefficient is the corresponding standard error. Often researchers report results following another convention: instead of giving the standard error, the number that appears in paren-

TABLE 11.1 Earnings Functions
(Dependent Variable: *EARNS*, $n = 100$)

	(6.7)	(7.15)	(7.46)	(7.33)	(7.41)
Constant	−1.315	−6.179	−9.791	−0.778	−0.803
	(1.540)	(2.780)	(3.780)	(1.615)	(1.778)
ED	0.797	0.978	0.995	0.762	0.794
	(0.128)	(0.153)	(0.152)	(0.131)	(0.130)
EXP		0.124	0.471		
		(0.060)	(0.254)		
EXPSQ			−0.00751		
			(0.00536)		
DRACE				−1.926	
				(1.759)	
DNCENT					0.288
					(1.166)
DSOUTH					−0.828
					(1.184)
DWEST					−1.992
					(1.329)
R^2	.285	.315	.329	.293	.310
SER	4.361	4.288	4.267	4.356	4.351

TABLE 11.2 Semilog Earnings Functions
(Dependent Variable: *LNEARNS*, $n = 100$)

	(6.40)	(7.52)
Constant	0.673	−2.031
	(0.158)	(0.974)
ED	0.107	0.106
	(0.013)	(0.016)
EXP		0.0501
		(0.0253)
EXPSQ		−0.000930
		(0.000533)
LNMONTHS		0.908
		(0.375)
DRACE		−0.239
		(0.177)
DNCENT		−0.00469
		(0.116)
DSOUTH		−0.193
		(0.119)
DWEST		−0.162
		(0.130)
R^2	.405	.511
SER	0.446	0.420

TABLE 11.3 Consumption Functions (Dependent Variable: *CON*)

	(6.10)	(6.22)	(7.29)
Constant	0.568	10.913	−2.370
	(6.973)	(12.167)	(10.024)
DPI	0.907		0.910
	(0.010)		(0.034)
DPILAG		0.923	
		(0.017)	
RAAA			0.500
			(3.902)
RINF2			−0.562
			(1.444)
R^2	.997	.993	.997
SER	8.935	14.953	9.309
n	25	24	25

TABLE 11.4 Trend Analyses (Dependent Variable: As Indicated)

	(6.14) *APC*	(6.45) *LNGNP*
Constant	0.911	6.456
	(0.0046)	(0.012)
T	−0.000213	0.0354
	(0.000309)	(0.00082)
R^2	.021	.988
SER	0.011	0.0297
n	25	25

TABLE 11.5 Phillips Curves (Dependent Variable: *RINF1*, $n = 15$

	(6.19)	(7.36)
Constant	−1.984	−0.803
	(1.219)	(1.524)
UINV	22.234	15.030
	(5.594)	(7.965)
D (1965–70)		0.894
		(0.717)
R^2	.549	.600
SER	0.956	0.936

TABLE 11.6 Money Demand Functions (Dependent Variable: *LNM*, n = 25)

	(6.33)	**(7.51)**
Constant	3.948	3.759
	(0.165)	(0.325)
LNGNP	0.215	0.246
	(0.024)	(0.053)
LNRTB		−0.0205
		(0.0304)
R^2	.780	.785
SER	0.0305	0.0309

theses is the ratio of the estimated coefficient to its standard error: $\hat{\beta}_j^*/s^*(\hat{\beta}_j)$. This ratio is the same as the outcome of the sampling statistic t defined in (11.21) *if and only if* $\beta_j = 0$. (In the next chapter we see why this is interesting to consider.) Both styles of reporting give essentially the same information: given the estimated coefficient, the t ratio can be determined from the standard error, and vice versa.

Problems

Section 11.1

11.1 Consider a regression model in which $\sigma_u = 2.5$. If the expected value of Y is 10 for a given observation, what is the probability that its actual value will be between 9 and 11?

⋆ **11.2** Suppose that three observations are observed sequentially. If the disturbance is positive for the first two, what is the probability that the disturbance will be positive for the third observation?

⋆ **11.3** Suppose that the mean of the disturbance is not zero, but a constant for all observations. How could the regression model (11.1) be rewritten to conform to the assumption in Equation (11.2)?

11.4 Show graphically what the generated data might look like if the disturbance is negative for low values of X and positive for high values of X.

11.5 Suppose that the true regression is $E[Y] = 18 - 2X$, with $\sigma_u = 3$. If $X_i = 7$, determine the probability that Y_i will be negative.

Section 11.2

11.6 Consider the relative frequency histograms in Figure 11.3.
(**a**) How are they affected by increases in N?
(**b**) How are they affected by increases in n?

Section 11.3

11.7 Suppose we know that $\beta_1 = 10$ and that the standard deviation of the sampling distribution of its estimator is 1.2. What is the probability that the actual estimate of β_1 will be between 8.5 and 11.5?

⋆ **11.8** Suppose we know that $\beta_1 = 10$ and $\sigma(\hat{\beta}_1) = 8$. What is the probability that the actual estimate of β_1 will be negative?

11.9 In a statistical experiment, suppose that the true regression line is $E[Y] = 19 - 2X$. If the standard error for $\hat{\beta}_0$ equals 2, what is the probability that $\hat{\beta}_0$ will be greater than 20?

11.10 Suppose that we are doing a statistical experiment in which we know that $E[Y] = 5 + 3X$, $\sigma_u = 2$, and the total variation of X is 1. Determine the probability that an estimate of β_1 will be less than 2. If the total variation of X were 9, what would this probability be?

11.11 In simple regression, if the total variation of X is quadrupled, what happens to the standard error of $\hat{\beta}_1$?

Section 11.4

11.12 Suppose that the exact standard errors for coefficient estimators are known in a statistical experiment. What is the chance that the absolute value of an estimation error will be more than twice as large as the standard error?

11.13 Suppose that a simple regression is estimated with 22 observations. What is the chance that the absolute value of an estimation error will be more than twice as large as the calculated standard error?

11.14 In the consumption function reported as Equation (6.10), the estimated intercept is 0.568 and the calculated standard error for $\hat{\beta}_0$ is 6.973. Would these results be surprising if we knew that the true β_0 were equal to zero?

11.15 Suppose it is somehow known that the true value of the impact of *ED* on *EARNS* is 1.0.

(a) Determine the value of the t statistic for the regression reported in Equation (11.14).

(b) Use Table A.2 to determine, approximately, the probability that a t statistic would turn out to be this large or larger in absolute value.

⋆ **11.16** Suppose it is conjectured that the true value of the impact of *ED* on *EARNS* is zero. What is the value of the t statistic for the regression reported in Equation (11.14)? Is it likely that such an outcome would often occur?

Section 11.5

11.17 Consider a random variable X having a standard normal distribution. If all outcomes having a value greater than 1 are thrown away, would you anticipate that the mean of X in the remaining sample would equal zero? Explain.

11.18 Suppose that we considered estimating a simple earnings function using as observations only workers with annual earnings between 15 and 25 thousand dollars. What effect would this selection have on our estimation?

Appendix

11.19 Show that $\Sigma\, w_i = 0$.

11.20 Show that $\Sigma\, w_i X_i = 1$.

APPENDIX: Derivation of a Sampling Distribution*

In this appendix we derive the sampling distribution of $\hat{\beta}_1$ in the simple regression model and show some related results.

We start by accepting the specification of the simple regression model including all the assumptions about the disturbances, as presented in Section 11.1. Based on (5.12), the ordinary least squares estimator of β_1 is given by

$$\begin{aligned}\hat{\beta}_1 &= \frac{\sum (X_i - \bar{X})Y_i}{\sum (X_i - \bar{X})^2} = \sum \left[\frac{(X_i - \bar{X})}{\sum (X_i - \bar{X})^2}\right] Y_i \\ &= \sum w_i Y_i, \quad \text{where} \quad w_i = \frac{(X_i - \bar{X})}{\sum (X_i - \bar{X})^2}\end{aligned} \tag{11.23}$$

Since the X values are taken as fixed rather than as random variables, the w_i values are constants. Substituting for Y_i yields

$$\begin{aligned}\hat{\beta}_1 &= \sum w_i(\beta_0 + \beta_1 X_i + u_i) \\ &= \beta_0 \sum w_i + \beta_1 \sum w_i X_i + \sum w_i u_i\end{aligned} \tag{11.24}$$

From the definition of w_i, it can be shown that $\Sigma\, w_i = 0$ and $\Sigma\, w_i X_i = 1$. Hence

$$\hat{\beta}_1 = (\beta_0)(0) + (\beta_1)(1) + \sum w_i u_i = \beta_1 + \sum w_i u_i \tag{11.25}$$

Let us define

$$V = \sum w_i u_i \tag{11.26}$$

Recall that the u_i are independent identical random variables each having a normal distribution with mean 0 and variance σ_u^2 and that each w_i is a constant. Since V is a linear combination of normally distributed random variables, it can

*This appendix is relatively difficult and can be skipped without loss of continuity.

be shown that V itself is normally distributed. Based on (9.30) and (9.31), the mean and variance of V are

$$E[V] = \sum w_i \mu_u = \sum w_i 0 = 0 \tag{11.27}$$

$$\sigma^2(V) = \sum (w_i)^2 \sigma_u^2 = \sigma_u^2 \sum (w_i)^2 \tag{11.28}$$

Noting that the denominator in the ratio called w_i is the same for all i, we have

$$\begin{aligned}\sum (w_i)^2 &= \sum \left[\frac{(X_i - \bar{X})}{\sum (X_i - \bar{X})^2}\right]^2 = \frac{\sum (X_i - \bar{X})^2}{\left[\sum (X_i - \bar{X})^2\right]^2} \\ &= \frac{1}{\sum (X_i - \bar{X})^2}\end{aligned} \tag{11.29}$$

so that

$$\sigma^2(V) = \sigma_u^2 \frac{1}{\sum (X_i - \bar{X})^2} \tag{11.30}$$

Now, since $\hat{\beta}_1 = \beta_1 + V$, $\hat{\beta}_1$ is a random variable equal to the random variable V plus a constant. Thus

$$E[\hat{\beta}_1] = \beta_1 + E[V] = \beta_1 + 0 = \beta_1 \tag{11.31}$$

$$\sigma^2(\hat{\beta}_1) = \sigma^2(V) = \sigma_u^2 \frac{1}{\sum (X_i - \bar{X})^2} \tag{11.32}$$

Since V has a normal distribution, so does $\hat{\beta}_1$. Equation (11.31) underlies (11.9), and taking the square root of (11.32) gives (11.10).

A similar derivation shows that the sampling distribution of $\hat{\beta}_0$ in a simple regression is normal with mean and variance given by

$$E[\hat{\beta}_0] = \beta_0 \tag{11.33}$$

$$\sigma^2(\hat{\beta}_0) = \sigma_u^2 \frac{\sum X_i^2}{n \sum (X_i - \bar{X})^2} \tag{11.34}$$

Finally, we note that both $\hat{\beta}_0$ and $\hat{\beta}_1$ depend on the same X values and the same disturbance terms. Their individual sampling distributions can be thought of as being derived from a joint sampling distribution $p(\hat{\beta}_0, \hat{\beta}_1)$, and it should not be surprising to find a nonzero covariance between $\hat{\beta}_0$ and $\hat{\beta}_1$. The form of this joint distribution turns out to be "bivariate normal," and the covariance between $\hat{\beta}_0$ and $\hat{\beta}_1$ is

$$\sigma(\hat{\beta}_0, \hat{\beta}_1) = \frac{-\bar{X}\sigma_u^2}{\sum (X_i - \bar{X})^2} \tag{11.35}$$

Since σ_u^2 and $\Sigma (X_i - \bar{X})^2$ are always positive, the sign of the covariance is opposite to that of $\bar{X}$. Suppose that $\bar{X}$ is positive. If $\hat{\beta}_1^*$ turns out to be greater than β_1 in a particular sample, we would expect that $\hat{\beta}_0^*$ in that sample would be less than β_0; in other words, too high an estimate of β_1 tends to be compensated for by too low an estimate of β_0.

Hypothesis Testing

In this chapter we develop methods for testing hypotheses about individual coefficients in simple and multiple regression models. Our attention is limited to a procedure known as a ***significance test.*** Tests of hypotheses involving more than one coefficient are discussed in Chapter 15.

12.1 Specification of Hypotheses

In regression analysis, a hypothesis is a statement or conjecture about the value of a particular coefficient, β_j, that is made before the empirical analysis is undertaken. (In simple regression j is either 0 or 1, representing the intercept or the slope; in multiple regression β_j represents any of the coefficients.) The statement may be deduced from theoretical principles or induced from previous empirical findings.

Formal hypothesis testing involves a choice between two contradictory hypotheses, one called the ***null hypothesis*** (denoted by H_0) and the other called the ***alternative hypothesis*** (H_1). The null hypothesis must state that the true value of the regression coefficient β_j is equal to a specific value, which is denoted by β_j^0. The alternative hypothesis may be one of two vague (or, composite) forms: (1) that β_j is simply not equal to the specified null value, or (2) that β_j is greater than β_j^0 (or, instead, that β_j is less than β_j^0).

For example, the null hypothesis might be that $\beta_1 = 25$. One alternative hypothesis is that β_1 is not equal to 25. In this case, the hypotheses are compactly written as

$$H_0: \beta_1 = 25 \qquad H_1: \beta_1 \neq 25 \tag{12.1}$$

A second situation is that the researcher thinks that if β_1 is not equal to 25, it must be greater than 25; in this case the alternative is of the second form, and the hypotheses are compactly written as

$$H_0: \beta_1 = 25 \qquad H_1: \beta_1 > 25 \tag{12.2}$$

(Instead, the alternative might be H_1: $\beta_1 < 25$.) In both of these sets of hypotheses, the null is specific and the alternative is appropriately vague.

Whether it is the null hypothesis or the alternative that the researcher really believes in is not a factor that is built into the testing procedure. Sometimes it is the null hypothesis that is scientifically interesting; sometimes it is the alternative that is so. For example, the hypotheses in (12.1) are used both to test a theory that predicts that $\beta_1 = 25$ and to test a theory that $\beta_1 \neq 25$. In the first case the interesting proposition is formulated as the null hypothesis, whereas in the second the interesting proposition is formulated as the alternative.

In the procedures of hypothesis testing that we develop, the null and the alternative hypotheses play very different roles. Attention is focused on the null hypothesis, which is a specific conjecture about the value of a coefficient. As in Chapter 11, sampling theory is used to describe (probabilistically) the outcomes of $\hat{\beta}_j$ that might occur in data and to assess the degree of surprise we feel about the outcome that actually occurs. The role of the alternative hypothesis is to help shape a decision rule that leads us either to reject the null hypothesis in favor of the alternative or to not reject it.

12.2 The Basic Significance Test

The most common test in regression analysis involves the following hypotheses:

$$H_0: \beta_j = 0 \qquad H_1: \beta_j \neq 0 \tag{12.3}$$

This test, which we call the ***basic significance test,*** serves to answer the question: Does X_j affect Y? If the null hypothesis H_0 is true, the product term $\beta_j X_j$ is equal to zero no matter what the value of X_j is and therefore the explanatory variable X_j has no effect on the dependent variable Y. Alternatively, if H_1 is true, $\beta_j X_j$ depends on the value of X_j and therefore X_j affects Y. (If β_j is the intercept, the hypotheses simply state whether or not it is equal to zero.)

It may seem odd to raise the question of whether an explanatory variable affects the dependent variable, because in the specification of regression models we have always presumed such an effect. However, the question is apt because the validity of our decision to include a certain explanatory variable is always open to doubt. Also, it can be very interesting to find that some variable has no effect on the dependent variable.

The test is based on a comparison of the actual estimated coefficient $\hat{\beta}_j^*$ with the value for β_j that is conjectured by the null hypothesis—which is zero in this test. (As before, an asterisk indicates some particular value calculated from data.) Loosely speaking, if $\hat{\beta}_j^*$ is very different from zero, we tend to doubt that the null hypothesis is true. By contrast, if $\hat{\beta}_j^*$ is not very different from zero, there is not much doubt. The judgment is based on *how* different $\hat{\beta}_j^*$ is from zero, and we rely on sampling theory to guide us in this assessment.

The formal logic of the hypothesis test is to consider what would follow if the null hypotheses were true (i.e., if $\beta_j = 0$). Sampling theory tells us that $\hat{\beta}_j$ would have a normal distribution with a mean of zero (because $E[\hat{\beta}_j] = \beta_j$) and some positive standard deviation $\sigma(\hat{\beta}_j)$. In theory, no value for $\hat{\beta}_j^*$ that we ever find in an estimated regression could stand as definite evidence that the null hypothesis is false, because the normal distribution can yield any outcome from $-\infty$ to ∞. Thus we must give up hope of ever determining that the null hypothesis is totally, definitely false. Instead, we move toward a different approach, one based on probabilistic evidence.

If the null hypothesis were true, the implied estimation error associated with any value of the estimator $\hat{\beta}_j$ would simply be that value: the estimation error equals $\hat{\beta}_j - \beta_j$, which equals $\hat{\beta}_j$ when $\beta_j = 0$. Even though the estimation error can take on any value from $-\infty$ to ∞, we are surprised when it is much larger than the standard error (i.e., much larger than the typical estimation error, neglecting sign). As in Chapter 11, we interpret the statistic

$$t = \frac{\text{estimation error}}{\text{standard error}} = \frac{\hat{\beta}_j - 0}{s(\hat{\beta}_j)} = \frac{\hat{\beta}_j}{s(\hat{\beta}_j)} \tag{12.4}$$

as a measure of surprise, because it tells us how large the estimation error is in relation to the standard error. We are surprised when the value of t is large, and this leads us to doubt that the null hypothesis is true. This statistic has a t distribution with $n - k - 1$ degrees of freedom ($n - 2$ for simple regression), which is illustrated in Figure 12.1.

Our approach will be to use an occurrence of t that is considered very unlikely (and therefore is a surprise) as probabilistic evidence that the null hypothesis is false. For example, if

$$t^* = \frac{\hat{\beta}_j^*}{s^*(\hat{\beta}_j)} \tag{12.5}$$

equals 2.5, the estimation error is two and one-half times as large as the standard error. When there are more than 20 degrees of freedom, we see from Table A.2

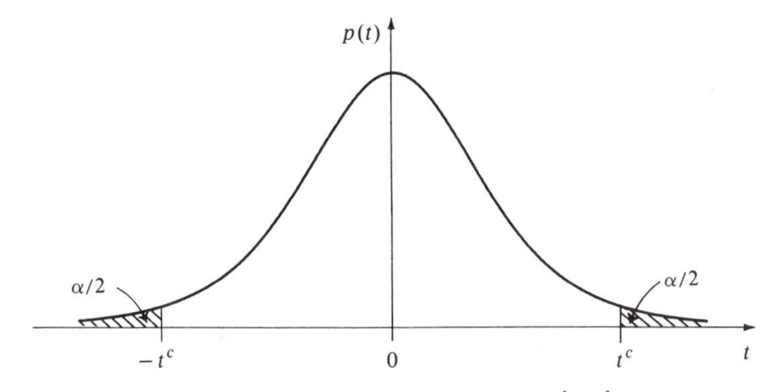

FIGURE 12.1 If the null hypothesis that $\beta_j = 0$ is true, $t = \hat{\beta}_j/s(\hat{\beta}_j)$ has a t distribution with $n - k - 1$ degrees of freedom. The possible values of t range from $-\infty$ to ∞, and thus no observable value t^* is totally inconsistent with the null hypothesis. In the basic significance test, the decision rule is to reject the null hypothesis if $|t^*| \geq t^c$. This specifies a two-tailed critical region, in which $\alpha/2$ probability is bounded in each tail. Therefore, if the null hypothesis is true, the probability of deciding to reject it (i.e., of making a Type I error) is α. The critical region corresponds to the shaded area in the figure, for which $\alpha = .05$.

that there is only a 1 percent chance that an estimation error would be this large or larger relative to $s(\hat{\beta}_j)$ on the positive side, and another 1 percent on the negative side. Thus we consider that $t^* = 2.5$ is rather unlikely to occur, and we are surprised to see it. Therefore, we view its occurrence as evidence against H_0.

Based on this logic, the method of hypothesis testing leads us to set up a ***decision rule*** to reject the null hypothesis if it is sufficiently unlikely that the t^* from the regression result would have occurred with $\beta_j = 0$. We will determine two ***critical values*** of t, denoted by t^c and $-t^c$, and we will reject the null hypothesis if $t^* \geq t^c$ or if $t^* \leq -t^c$. Put formally, the decision rule is of the form:

$$\text{Reject } H_0 \quad \text{if} \quad |t^*| \geq t^c \tag{12.6}$$

Referring to Figure 12.1, the values of t that (if taken on by t^*) lead to rejection are called the ***critical region.*** The basic significance test is called a ***two-tailed test,*** because the critical region consists of two open intervals, $t \leq t^c$ and $t \geq t^c$. Note that the decision rule can be reformulated or read simply as: reject H_0 if t^* is in the critical region.

The boundaries of the critical region are determined by the t^c value such that

$$\Pr\,(|t| \geq t^c) = \alpha \tag{12.7}$$

where α is a probability amount known as the ***level of significance,*** which we choose in the test procedure. Table A.3 is set up to give the values for t^c corresponding to five values of α. The level of significance is the probability amount that corresponds to occurrences of t that we consider "sufficiently unlikely" enough to make us reject the null hypothesis. In other words, the level

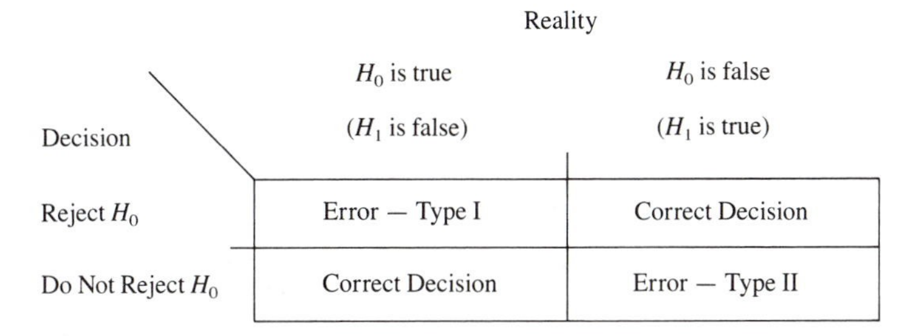

Decision \ Reality	H_0 is true (H_1 is false)	H_0 is false (H_1 is true)
Reject H_0	Error — Type I	Correct Decision
Do Not Reject H_0	Correct Decision	Error — Type II

FIGURE 12.2 The possible errors and correct decisions that can be made in hypothesis testing are classified in this array. The probability of making a Type I error is determined by the choice of the level of significance. This choice affects the probability of making a Type II error, but does not alone determine it.

of significance is a measure of how much surprise or doubt is necessary to cause us to reject the null hypothesis; it is an inverse measure because greater surprise is associated with a smaller probability.

In applying the decision rule (12.6), there are two types of mistakes that can be made: (1) we may decide to reject the null hypothesis when in fact it is true, or (2) we may decide to not reject the null hypothesis when in fact it is false. These mistakes are known formally as ***Type I*** and ***Type II error,*** respectively. A classification of the possible errors and correct decisions that can be made in hypothesis testing is illustrated in Figure 12.2.

When the reality is that H_0 is true, the probability of making a Type I error is exactly equal to the level of significance. This follows directly from our determination of t^c in (12.7). Thus our choice of α *is* the choice of the probability of making a Type I error in the test. When the reality is that H_0 is false, there is a certain probability of making a Type II error. Although we cannot determine this probability, it increases if α decreases. In other words, in a particular significance test, the smaller is the probability of making a Type I error, the greater is the probability of making a Type II error. These probabilities are affected by the choice of the level of significance; although we cannot eliminate making mistakes in applying the decision rule, we can control the balance of the two types.

In principle, the appropriate level of significance is chosen after careful consideration of the relative undesirability of committing Type I and Type II errors in a particular empirical setting. In practice, researchers choose a conventional level: 5 percent ($\alpha = .05$) is the most common choice, but 10 percent or 1 percent levels are also used frequently. The impact of this choice is quite important, as we will see in Section 12.3.

Only after the decision rule is designed do we look at the regression results. If the value of the t statistic, t^*, is in the critical region, the null hypothesis is rejected; if t^* is not in the critical region, H_0 is not rejected (sometimes it is said that H_0 is "accepted"). If the null hypothesis is rejected, it is common to say that the coefficient β_j is ***significant*** or "significantly different from zero," or that the explanatory variable X_j is "significant." If X_j is significant, we say that X_j does affect Y.

It should be noted carefully that whether or not a variable is "significant" is different from whether or not it is "important." Often the two findings go hand in hand, but they need not do so. The question of significance has to do with whether or not a hypothesis test concludes that an explanatory variable has *some* effect on the dependent variable. By contrast, the question of importance has to do with whether or not an economic assessment concludes that an explanatory variable has a *substantial* effect on the dependent variable. To emphasize the distinction, sometimes the term "statistical significance" is used instead of just "significance."

For completeness, it is a good idea to carry out the following five steps in conducting the test:

1. State the hypotheses clearly.
2. Choose the level of significance, α.
3. Construct the decision rule.
4. Determine the value of the test statistic, t^*.
5. State the conclusion of the test.

For example, the simplest earnings function reported in Chapter 6 is

$$\begin{aligned} \widehat{EARNS}_i &= -\underset{(1.540)}{1.315} + \underset{(0.128)}{0.797}ED_i \\ R^2 &= .285 \qquad SER = 4.361 \end{aligned} \tag{12.8}$$

with standard errors reported in parentheses beneath the corresponding coefficients. If we are interested in the question "Does education affect earnings?" then we are interested in the basic significance test. First, we clearly state the hypotheses, as in (12.3). Second, we choose $\alpha = .05$, by convention. Third, given this α and noting that the number of degrees of freedom (df) is 98, we see from Table A.3 that $t^c = 1.984$, approximately. Thus our decision rule is to reject H_0 if $|t^*| \geq 1.984$. Fourth, we compute the value $t^* = \hat{\beta}_1^*/s^*(\hat{\beta}_1) = 0.797/0.128 = 6.23$. Fifth, seeing that t^* is in the critical region, we decide to reject the null hypothesis and we conclude that education is significant in the determination of earnings.

Note that it is not possible to look at the magnitude of the estimated coefficient, here $\hat{\beta}_1^* = 0.797$, and judge whether it is significantly different from zero. A comparison must be made to the standard error, which leads naturally to the t statistic. Indeed, if the units of measurement of earnings were changed to dollars (instead of thousands of dollars), the estimated coefficient would be 797.0, which is a much larger magnitude. However, the standard error would also increase by a factor of 1000, so t^* would be 6.23 as before.

After the procedure is understood fully, it is possible to reach the conclusion of the test very quickly in many cases. For example, regarding the estimated intercept in (12.8) it is quickly seen that t^* is less than 1 in absolute value, so an experienced econometrician would immediately say that the intercept is not statistically significant. An experienced listener would understand that this is the conclusion of a basic significance test carried out at some conventional level of significance. However, there is plenty of room for ambiguity and error, and carrying out a complete test is advisable.

Practical researchers sometimes use a rule-of-thumb procedure in which a coefficient is judged to be significant if its estimate is at least twice as large as its standard error in absolute value (i.e., if $|t^*| \geq 2$). In formal terms, this amounts to setting $t^c = 2$ and letting the level of significance be whatever is consistent with this. Looking at Table A.2 or A.3 we see that the value of α would be about 5 percent so long as df is not small. Putting all this in reverse, we see that the critical value of the basic significance test is roughly $t^c = 2$ when the level of significance is 5 percent.

The testing procedure in multiple regression is the same as in simple regression, except that the relevant number of degrees of freedom is $n - k - 1$. For example, consider the consumption function (7.29), which is reported fully in Table 11.3. In this multiple regression model, consumption is theorized to depend on income (*DPI*), an interest rate (*RAAA*), and the rate of inflation (*RINF2*). In Section 7.2 we noted that economic theory does not clearly specify whether the effects of *RAAA* and *RINF2*, individually, would be positive, negative, or zero. Thus the basic significance test is quite appropriate for these two variables. For each of these tests, the hypotheses are as given in (12.3). Choosing a 5 percent level of significance, the critical value is $t^c = 2.080$ because there are 21 degrees of freedom. The decision rule is: reject H_0 if $|t^*| \geq 2.080$. Looking quickly at Table 11.3, we see that $|t^*| \leq 1$ in both cases, because the estimated coefficients are smaller in absolute value than their standard errors. It is hardly necessary to calculate the t^* values with any greater precision because it is clear that they are not in the critical region. The conclusions of the two separate tests are that both the interest rate and the rate of inflation are ***insignificant:*** we cannot reject the null hypotheses that they have no effect at all on aggregate consumption. The tests do not prove that *RAAA* and *RINF2* have no effects on *CON*, but we conclude that we cannot reject those propositions.

Because the basic significance test is so commonly used, many computer programs print out a value labeled "t statistic" or "t ratio." This is simply $\hat{\beta}_j^*/s^*(\hat{\beta}_j)$, and it is the appropriate t^* for the basic significance test. Many researchers routinely report these values rather than the standard errors, because this simplifies making judgments in the basic test. However, for some of the tests discussed later in this chapter, the computer's "t statistic" is irrelevant and definitely not the appropriate t^* value.

12.3 The Test for Sign

Sometimes the substantive hypothesis that a researcher is interested in is simply that β_j is positive. Although the complete logical opposite to this statement is that β_j is negative or zero, the formal procedure of hypothesis testing leads us to set up the following:

$$H_0: \beta_j = 0$$
$$H_1: \beta_j > 0 \tag{12.9}$$

The substantive hypothesis is set up as the alternative, and the null is that $\beta_j = 0$. This test serves to answer the question: Does X_j have a positive effect on Y?

As before, the procedure focuses on whether or not to reject the null hypothesis. It is important to understand that neither "$\beta_j \leq 0$" nor "$\beta_j > 0$" can be used as the null hypothesis. The null hypothesis must be specific so that we can examine the precise sampling distribution of $\hat{\beta}_j$ under the supposition that the null is true.

The alternative hypothesis plays an important role in the design of the decision rule, a role that was hidden in our discussion of the basic significance test. If the null hypothesis is true, so that $\beta_j = 0$, then the t statistic (12.4) has a t distribution with $n - k - 1$ degrees of freedom, as before. Any outcome from $-\infty$ to ∞ is possible, but with an eye on H_1 we use the unlikely occurrences of t^* being positive and large as evidence against the null hypothesis and in favor of the alternative. Note that t^* being positive and large reflects the situation that $\hat{\beta}_j^*$ is positive and large relative to its standard error.

The decision rule we construct is of the form:

$$\text{Reject } H_0 \text{ if } t^* \geq t^c \tag{12.10}$$

where the value of t^c is determined from

$$\Pr(t \geq t^c) = \alpha \tag{12.11}$$

As illustrated in Figure 12.3, the critical region consists only of the values $t \geq t^c$, which are in the right-hand tail of the distribution. For this reason, the test

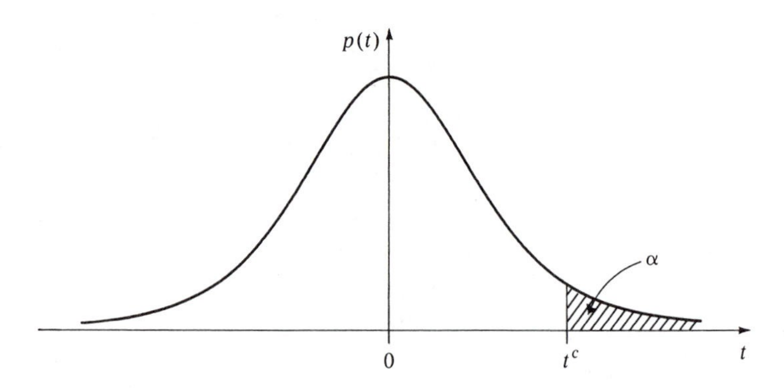

FIGURE 12.3 In the test for positive sign, the decision rule is to reject the null hypothesis if $t^* \geq t^c$. This specifies a one-tailed critical region, which bounds α probability in the right-hand tail. The critical region corresponds to the shaded area in the figure, for which $\alpha = .05$.

for sign is a ***one-tailed test.*** For a given level of significance, the critical value t^c is found in Table A.3. For example, at a 5 percent level of significance, $t^c = 1.708$ for df $= 25$. Note that for a given α, t^c is smaller in a one-tailed test than in a two-tailed test because in a two-tailed test t^c is chosen to bound $\alpha/2$ probability in each tail.

The test just described is a test for positive sign. If the substantive hypothesis of interest is that $\beta_j < 0$, we are led to a test for negative sign. The hypotheses are:

$$\begin{aligned} H_0&: \beta_j = 0 \\ H_1&: \beta_j < 0 \end{aligned} \tag{12.12}$$

and the decision rule is:

$$\text{Reject } H_0 \text{ if } t^* \leq -t^c \tag{12.13}$$

where t^c is determined as in (12.11).

When the null hypothesis is rejected in a test for positive sign, it is common to conclude that the coefficient is "significantly positive," or that X_j has a "significantly positive" effect on Y. In a test for negative sign the conclusions are changed accordingly. Again, one should bear in mind that significance is different from importance: a finding that X_j has a significantly positive effect on Y does not mean that it has a *big* effect, but just that we reject the hypothesis that it has *no* effect.

In a test for positive sign, what if t^* is negative and large (e.g., $t^* = -5.5$)? Intuitively, this casts doubt on the null hypothesis, but it surely does not lend support to the alternative. Formally, the conclusion is that we do not reject the null hypothesis that $\beta_j = 0$, and we might say that β_j is "not significantly positive." At this point it is tempting to redo the whole test—to make it a test for negative sign rather than for positive sign. The conclusion would be that the coefficient is significantly negative. Such a reformulation of the hypotheses after t^* has been observed is formally wrong, however, because it invalidates the level of significance and the probabilistic interpretation of the findings.

As is true of any significance test, a test for sign is going to lead us to make mistakes sometimes. Of course, we never know whether or not we have made a mistake; no flag is waved. However, we do know that the probability of our incorrectly rejecting the null hypothesis (i.e., of making a Type I error) is given by α, the level of significance. This is because a mistake of this type occurs if the outcome of t falls in the critical region when H_0 is in fact true. By (12.11) the probability of this occurring is α.

For example, we return to the earnings function (12.8). Suppose economic theory predicts that the impact of education on earnings is positive. A test of this theory can be made by testing the hypothesis that $\beta_1 > 0$ in our regression model. Our procedure forces us to set up the null hypothesis H_0: $\beta_1 = 0$ and use the theoretical prediction as the alternative. We are clearly interested in the test for positive sign. First, we state the hypotheses, as in (12.9). Second, we choose $\alpha = .05$, by convention. Third, given this α and noting that df $= 98$, we see from Table A.3 that $t^c = 1.660$ approximately. Thus, our decision rule is to reject H_0 if $t^* \geq 1.660$. Fourth, we compute the value of $t^* = \hat{\beta}_1^*/s^*(\hat{\beta}_1) = 0.797/0.128 = 6.23$. Fifth, seeing that t^* is in the critical region, we decide to reject the null hypothesis and we conclude that education has a significantly positive effect on earnings.

The two earnings function tests that we have used as examples provide a comparison of one- and two-tailed tests and raise the question of what hypotheses should be tested. A comparison can be made using the same level of significance in the two tests, for which the appropriate critical regions are shown in Figure 12.4. In our examples, $t_2^c = 1.984$ and $t_1^c = 1.660$ in the two-tailed and one-tailed tests, respectively. It turned out that $t^* = 6.23$, so the outcome is in the critical region for both cases. Suppose, instead, that t^* occurs in the range between t_1^c and t_2^c. In the basic significance test we would conclude that β_1 is not significant, whereas in the test for positive sign we would say that β_1 is significantly positive. Hence, our qualitative conclusion can be very sensitive to the initial specification of hypotheses. If t^* occurs in the unshaded range between $-t_2^c$ and t_1^c, we would not reject the null hypothesis in either case. Finally, if t^* occurs in the range below $-t_2^c$, the basic significance test would conclude that β_1 is significant whereas the test for sign would conclude that it is not significantly positive.

The case in which t^* falls between t_1^c and t_2^c can lead to an awkward situation. If a researcher begins an analysis with a theory that $\beta_1 > 0$, the test for positive sign would be appropriate and he would be satisfied with the conclusion that β_1

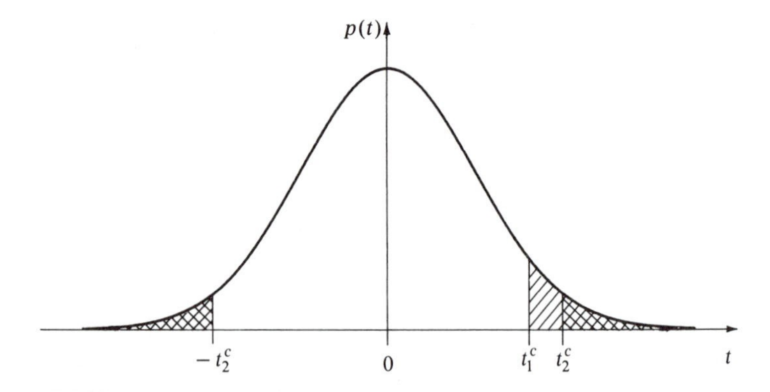

FIGURE 12.4 To compare the possible results of using a test for positive sign rather than the basic significance test, we use the same level of significance in both cases. The most important point of comparison to note is that if t^* happens to fall in the range between t_1^c and t_2^c, the tests give seemingly contradictory results. In the basic significance test we would conclude that the coefficient is not significant, whereas in the test for positive sign we would conclude that the coefficient is significantly positive. Since the conclusion of a test can be very sensitive to its technical specification, proper procedure calls for the hypotheses to be clearly stated before any results of the estimation are seen.

is significantly positive. On the other hand, if he begins with no particular theory and carries out a basic significance test, he would find that β_1 is not significant. But if he were alert at all, he would notice that a switch to a one-tailed test would lead him to significance. On formal grounds, however, this is wrong because it invalidates the meaning of the level of significance. But if the researcher were even more alert, he would notice that by doubling the level of significance the new t_2^c would equal the old t_1^c, so that the regression result would be significant in a two-tailed test. By extending this strategy, most any finding can be made to look significant!

For example, in Section 7.3 we reported a Phillips curve with a dummy variable allowing for a shift during the 1965–1970 period. The regression (7.36) is reported fully in Table 11.5. The t^* for the coefficient on *UINV*, the inverse of the unemployment rate, is $t^* = 1.886$. With df $= 12$, $t_1^c = 1.782$ and $t_2^c = 2.179$ at the 5 percent level of significance. The conclusion of the basic significance test is that the unemployment regressor is *not* significant, whereas the conclusion of the test for positive sign is that the regressor *is* significantly positive. In this case, economic theory and an understanding of the reciprocal specification naturally lead to the proposition that $\beta_1 > 0$, so the test for positive sign is appropriate. It should be noted that the unemployment regressor would be judged significant by both hypothesis tests if the level of significance were 10 percent.

The proper choice of hypotheses and levels of significance requires good judgment, both of economic and statistical matters. Although there are no hard and fast rules, prudence calls for (1) determining the hypotheses to be tested before seeing any results, (2) choosing a 5 percent level of significance, and (3) if appropriate, reporting additionally how changes in the hypotheses or the level of significance would alter the qualitative conclusion. *To specify the hypotheses after looking at the results of estimation, or to plan a sequence of tests that depend on the observed results, invalidates the formal validity of the hypothesis testing procedure.*

We have presented two different tests regarding the effect of education on earnings for the sake of providing two examples and to introduce the question of the choice of test. It is not a pattern to follow; usually only one test should be applied to any coefficient.

12.4 Tests for Specific Coefficient Values

Sometimes the interesting proposition that we wish to test involves some particular nonzero value, β_j^0. For example, if theory suggests that β_1 equals 1, we would test

$$\begin{aligned} H_0&: \beta_1 = 1 \\ H_1&: \beta_1 \neq 1 \end{aligned} \tag{12.14}$$

Similarly, if theory implies that β_1 is less than 1, we would test

$$H_0\colon \beta_1 = 1 \qquad H_1\colon \beta_1 < 1 \tag{12.15}$$

In both cases the null hypothesis is of the form $H_0\colon \beta_j = \beta_j^0$.

As with the other tests we have developed, the procedure is to examine the implications of the null hypothesis being true and to construct a decision rule that can lead to rejecting the hypothesis. To be general, if $\beta_j = \beta_j^0$, sampling theory tells us that $E[\hat{\beta}_j] = \beta_j^0$. Thus the implied estimation error is $\hat{\beta}_j - \beta_j^0$, and the statistic

$$t = \frac{\hat{\beta}_j - \beta_j^0}{s(\hat{\beta}_j)} \tag{12.16}$$

has a t distribution with $n - k - 1$ degrees of freedom ($n - 2$ for simple regression). On the basis of a particular regression, the outcome of t is

$$t^* = \frac{\hat{\beta}_j^* - \beta_j^0}{s^*(\hat{\beta}_j)} \tag{12.17}$$

Although t may take on any real value, it is unlikely that it would be very large or very small (negatively) because that would reflect an estimation error (i.e., a difference $\hat{\beta}_j - \beta_j^0$) that is very large relative to the standard error. Our logic is that if we have a finding that is unlikely to occur when the null hypothesis is true, the null hypothesis should be judged false. If the alternative hypothesis is that $\beta_j \neq \beta_j^0$, as in (12.14), we set up a two-tailed test in which finding t^* to be very large or very small (negatively) leads us to reject the null hypothesis. Similarly, if the alternative hypothesis is that $\beta_j < \beta_j^0$, as in (12.15), we set up a one-tailed test in which finding very small (negative) values of t^* leads us to reject the null hypothesis in favor of the alternative. The procedure of the tests is the same as those discussed previously, with identical forms for the decision rules.

For example, consider the aggregate consumption function reported in Chapter 6:

$$\begin{aligned} \widehat{CON}_i &= \underset{(6.973)}{0.568} + \underset{(0.010)}{0.907}DPI_i \\ R^2 &= .997 \qquad SER = 8.935 \end{aligned} \tag{12.18}$$

(the numbers in parentheses are the standard errors for the corresponding coefficients). As noted in Chapter 1, Keynes theorized that the marginal propensity to consume would be less than 1, which is a testable proposition about β_1. First, the proper hypotheses are given by (12.15), which calls for a one-tailed test.

TABLE 12.1 Summary of Regression *t* Tests

If the Alternative Hypothesis Is of the Form:	Then the Decision Rule Is of the Form:	Where t^c Is Determined by:
H_1: $\beta_j \neq \beta_j^0$	Reject H_0 if $\lvert t^* \rvert \geq t^c$	$\Pr(\lvert t \rvert \geq t^c) = \alpha$
H_1: $\beta_j > \beta_j^0$	Reject H_0 if $t^* \geq t^c$	$\Pr(t \geq t^c) = \alpha$
H_1: $\beta_j < \beta_j^0$	Reject H_0 if $t^* \leq -t^c$	$\Pr(t \geq t^c) = \alpha$

Notes:

1. In all cases, H_0: $\beta_j = \beta_j^0$.
2. The level of significance is α.
3. The degrees of freedom is df $= n - k - 1$, where k is the number of regressors.
4. In all cases, $t^* = (\hat{\beta}_j^* - \beta_j^0)/s^*(\hat{\beta}_j)$.

Second, we choose the conventional 5 percent level of significance. Third, noting that $\alpha = .05$ and df $= n - 2 = 23$, we see from Table A.3 that $t^c = 1.714$; thus, our decision rule is to reject H_0 if $t^* \leq -1.714$. Fourth, we compute the value of

$$t^* = \frac{\hat{\beta}_1^* - \beta_1^0}{s^*(\hat{\beta}_1)} = \frac{0.907 - 1.0}{0.010} = -9.3 \tag{12.19}$$

Fifth, seeing that t^* is in the critical region, we decide to reject the null hypothesis. We conclude that the marginal propensity to consume is significantly less than 1, supporting Keynes' theory.

If $\beta_j^0 = 0$ in the test for a specific coefficient value, we are back to the basic significance test or the test for sign. Thus all the tests presented in this chapter can be considered special cases of a general procedure. The most important distinction among the cases is not the value of β_j^0, but the specification of the alternative hypothesis. This distinction is made in Table 12.1, which summarizes the procedure for testing individual coefficients in regression models.

12.5 *P*-Values

The essence of hypothesis testing is a comparison of a test statistic calculated from a set of data with the probability distribution that would prevail for that statistic if the null hypothesis were true. In this comparison, a result that is very unlikely to have occurred casts doubt on the truth of the null hypothesis. In the formal test procedure described above, a level of significance is chosen and a clear decision rule for rejecting the null hypothesis is formulated. In Section 12.3 we saw that the conclusion can be very sensitive to the technical specification of the test—that is, to the choice of α and the alternative hypothesis. In a sense, the formality of the procedure interferes with our learning from the data.

Consider, for example, the test for positive sign, in which the hypotheses are

$$H_0: \beta_j = 0$$
$$H_1: \beta_j > 0 \tag{12.20}$$

and the decision rule is

$$\text{Reject } H_0 \text{ if } t^* \geq t^c \tag{12.21}$$

where t^c is determined from

$$\Pr(t \geq t^c) = \alpha \tag{12.22}$$

for a chosen level of α. If the null hypothesis is true, the probability distribution of

$$t = \frac{\hat{\beta}_j}{s(\hat{\beta}_j)} \tag{12.23}$$

is as illustrated in Figure 12.5a. In this figure, the critical regions corresponding to two levels of significance, .05 and .10, are shown by the shaded areas.

It is possible to reformulate the test procedure without changing its essence. Focusing on the t^* obtained in the data, we can compute a probability amount known as the ***P*-value.**

$$P\text{-value} = \Pr(t \geq t^*) \qquad \text{(one-tailed)} \tag{12.24}$$

under the supposition that the null hypothesis is true. Such a probability is given by the area under the distribution to the right of t^*, as shown in Figure 12.5b. The standard decision rule (12.21) can be reformulated as:

$$\text{Reject } H_0 \text{ if } P\text{-value} \leq \alpha \tag{12.25}$$

In the test for positive sign, whenever t^* is greater than t^c then the P-value is less than α; this can be verified by comparing (12.22) with (12.24).

The P-value, by itself, is the probability that the t statistic would be as large or larger than it actually is, when the null hypothesis is true. Thus the P-value is an inverse measure of the ''surprise'' we feel about the estimate of a coefficient. We are very surprised when the P-value is low, and this casts doubt on the null hypothesis.

This interpretation of P-values is consistent with the standard formulation of hypothesis tests. If the P-value is small, say $P = .035$, then the result is quite unlikely and leads to formally rejecting the null hypothesis at $\alpha = .05$. If the P-value is a bit larger, say $P = .075$, then the result is still fairly unlikely. It would not lead to formal rejection at $\alpha = .05$ but it would at $\alpha = .10$; this case is illustrated in Figure 12.5a. Finally, if $P = .125$, the result is moderately unlikely—there is only a 12.5 percent chance that t^* would be as large or larger

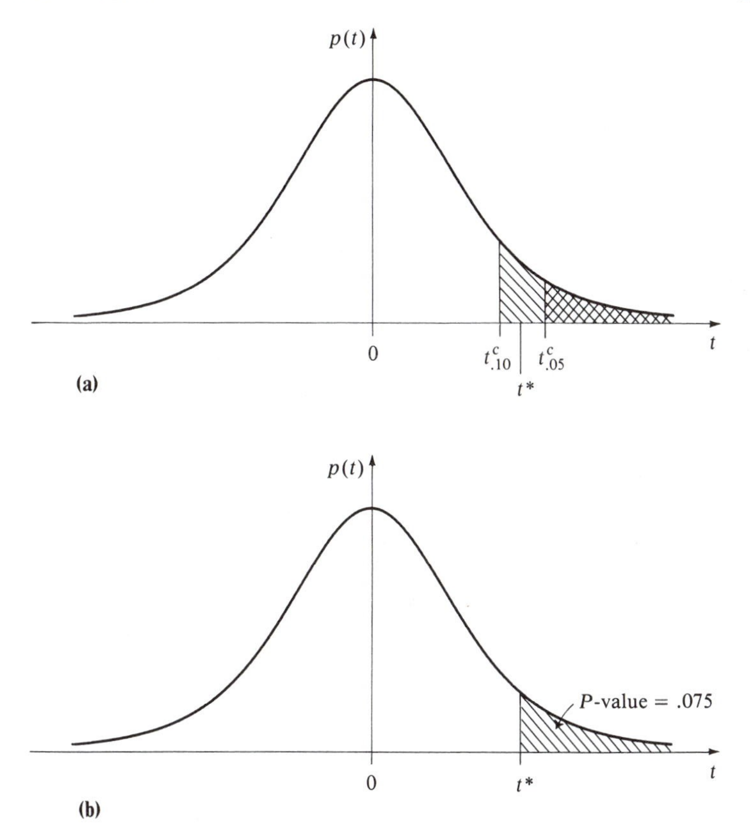

FIGURE 12.5 Part (a) shows the critical regions for the test of positive sign when $\alpha = .05$ and $\alpha = .10$. For the t^* indicated, the tests show significance at the 10 percent level but not at the 5 percent level. In (b), Pr $(t \geq t^*)$ is shown by the shaded area. This probability is known as the *P*-value of the test, and it embodies all the information needed for statistical inference without setting up the formal decision rule and choosing the level of significance. When the *P*-value equals .075, we realize that a one-tailed test would show significance at 10 percent, but not at the 5 percent level.

than it actually is, if H_0 were true—but it would not lead to significance at the conventional levels of significance.

Although rephrasing the test in terms of *P*-values permits us to carry out the formal procedure in just a slightly different language, it also permits us to proceed less formally and thereby to avoid the strictness involved with choosing the level of significance in the formal procedure. The difficulty with that strictness is that it invites inappropriate posturing: we have seen that a result can be made "significant" merely by changing the α, which serves as the formal criterion for significance. Also, if one has blind allegiance to the 5 percent level, a result for which $P = .06$ would be declared to be insignificant with no further analysis. Reporting the *P*-value of a test in addition to, or instead of, the formal decision is a way of concisely conveying some information about what is going on in the data.

For a two-tailed test, a two-tailed P-value can be calculated

$$P\text{-value} = \Pr(|t| \leq t^*) \qquad \text{(two-tailed)} \tag{12.26}$$

and compared with the formal level of significance or interpreted directly.

Although practice varies, some statisticians suggest that calculation and reporting of the P-value ought to be done without any reference to the alternative hypothesis—that is, that the P-value ought to be computed only for the single tail of the sampling distribution in which the t^* happens to fall. This practice breaks away from the formal procedure, but for a very practical reason: often the appropriate null hypothesis to consider is objectively clear to the researcher, but whether the alternative hypothesis should be one-tailed or two-tailed is unclear. Since it invalidates the formal procedure to choose H_1 on the basis of sample evidence for t^*, the researcher may be left in a quandary. The suggested P-value procedure is simply a probabilistic calculation involving the null hypothesis and the sample results. It is up to the researcher or the reader to provide the substantive interpretation, rather than to rely on the formal decision rule.

For example, in the Phillips curve discussed above, the t statistic for the unemployment variable is $t^* = 1.886$, having the anticipated positive sign. The one-tailed P-value is

$$P\text{-value} = \Pr(t \geq 1.886) \approx .04 \tag{12.27}$$

A computer is necessary to calculate this exactly, but the value here is interpolated from Table A.2 with df = 12. Thus, if the null hypothesis were true ($\beta_1 = 0$), there would be only a 4 percent chance of getting a t^* of 1.886 or larger. This casts doubt on the null hypothesis, and the coefficient might be judged fairly significant. We see immediately that on a formal one-tailed test it would be significant if $\alpha = .10$ or $\alpha = .05$, but not significant at $\alpha = .01$.

P-values are not commonly reported in econometric studies, perhaps because standard t tables do not contain sufficient detail. However, computers can calculate P-values with no trouble at all, and some regression programs routinely print them out. (Sometimes they are labeled "achieved level of significance.") Also, for very large samples the t distribution is practically the same as the standard normal, so a table like Table A.1 can be used.

A common practice among econometricians is to report t^* values and indicate if the coefficients are significant at any of the conventional levels of significance (e.g., 1, 5, and 10 percent). This conveys information similar to that of the P-value approach, but with less succinctness and less precision.

Finally, thinking in terms of P-values helps clarify the relation between the size of the t^* value in a formal hypothesis test and its meaning for the decision to reject the null hypothesis or not. Suppose that we consider a one-tailed test with 30 degrees of freedom. If $t^* = 1.0$, yielding $P = .163$, the null hypothesis would not be rejected at any of the commonly reported levels of significance

(.10, .05, and .01). If t^* moved up to 1.5 with $P = .072$, the result would be significant (i.e., H_0 would be rejected) at the .10 level but not the others. If t^* moved up to 2.0 with $P = .027$, the result would be significant at the .10 and .05 levels, but not at the .01 level. If t^* moved up to 2.5 with $P = .009$, the result would be significant at all these levels. Thus, relatively modest differences in t^* when it is near 2.0 in value can have important implications for the conclusions of our tests. By contrast, we see that there is no difference of practical consequence between getting a t^* value of 5 or a t^* of 50: in both cases it is nearly impossible that the sample value would have occurred if the null hypothesis were true, and therefore the coefficient is significant.

Problems

Section 12.1

★ **12.1** For each of the following, could it be the statement of a null hypothesis? Could it be an alternative hypothesis?
(a) $\beta_1 < 60$
(b) $\beta_1 \neq 60$
(c) $\beta_1 = 60$
(d) $\beta_1 > 60$

12.2 Set up the appropriate null and alternative hypotheses for testing the proposition that
(a) $\beta_1 = 0$
(b) $\beta_1 < 0$
(c) $\beta_1 \neq 100$
(d) $\beta_1 > 100$

Section 12.2

12.3 Carry out a complete basic significance test for the effect of *DPI* on *CON* in Equation (12.18).

★ **12.4** In the earnings function (7.33), reported fully in Table 11.1, does a person's race affect his earnings?

★ **12.5** In the earnings function (7.15), reported fully in Table 11.1, does experience affect earnings? Carry out the basic significance test at the 10 percent level and then do it again at the 5 percent level.

12.6 In the demand for money regression (7.51), reported fully in Table 11.6, does income have a significant effect?

12.7 Among the five earnings functions reported in Table 11.1, in which equations is the intercept significantly different from zero?

12.8 In the earnings function (7.41) reported fully in Table 11.1, assess the significance of each of the coefficients on the regional dummy variables. Discuss the implications of these three tests taken together.

12.9 Under what condition would the two-standard-error rule be precisely the same as a regular 5-percent significance test?

Section 12.3

12.10 In the consumption function (12.18), determine quickly if the intercept is significantly positive.

12.11 Is the rate of growth of *GNP* estimated in regression (6.45), and reported fully in Table 11.4, significantly positive?

★ **12.12** In the demand for money regression (7.51), reported fully in Table 11.6, is the interest elasticity significantly negative?

★ **12.13** For a 5 percent significance test with df = 100, determine the values that correspond to t_1^c and t_2^c in Figure 12.4.

12.14 In the earnings function (7.46) reported fully in Table 11.1, is the coefficient of *EXPSQ* significant at the 10 percent level? Based on general economic knowledge, what would be the appropriate test for sign in this case? Carry out this test at the 10 percent level and compare its conclusion with that of the basic significance test.

12.15 Suppose that H_0: $\beta_j = 0$ is true, and consider the following procedure. If the estimated coefficient $\hat{\beta}_j^*$ is positive, set up a 5 percent significance test for positive sign; if $\hat{\beta}_j^*$ is negative, set up a 5 percent significance test for negative sign. Now, viewing the situation before the estimate is calculated, what is the probability that this procedure will lead to a Type I error?

Section 12.4

12.16 In a study with a very large number of observations, the price elasticity of demand for gasoline was estimated to be −0.90 with a standard error of 0.06; is the estimated elasticity significantly different from −1.0?

12.17 In a study with a very large number of observations, the income elasticity of demand was estimated to be 0.70 with a standard error of 0.12. Is the elasticity significantly less than 1.0?

★ **12.18** In the semilog earnings function (7.52), reported fully in Table 11.2, is the elasticity of *EARNS* with respect to *MONTHS* worked significantly different from 1?

Section 12.5

12.19 Using Table A.2, determine (approximately) and interpret the one-tailed *P*-values for the tests in

(a) Problem 12.4
(b) Problem 12.5
(c) Problem 12.10
(d) Problem 12.11
(e) Problem 12.12

Estimation and Regression Problems

This chapter covers a variety of topics relating to the estimation of regression models. We first expand the concept of estimation by introducing confidence intervals, which are related to the hypothesis testing of Chapter 12. After a general consideration of the properties of estimators, we reconsider two regression problems discussed in Chapter 7: misspecification and multicollinearity.

13.1 Confidence Intervals for the Coefficients

Until now, our notion of the "estimation" of an unknown regression coefficient has been the calculation of a single value that serves as our best guess of the coefficient's true value. In statistical theory, this value is called a ***point estimate.*** Since sampling theory makes it clear that every estimate involves some estimation error, it is useful to provide information regarding the likely error when reporting the value of the point estimate. This can be accomplished to some extent by reporting the standard error of the coefficient.

A more structured approach is to specify a range of values that is likely to include the true value of the unknown coefficient. This range of values is an interval of the general form

$$\hat{\beta}_j^* \pm h \quad \text{or} \quad (\hat{\beta}_j^* - h) \text{ to } (\hat{\beta}_j^* + h) \tag{13.1}$$

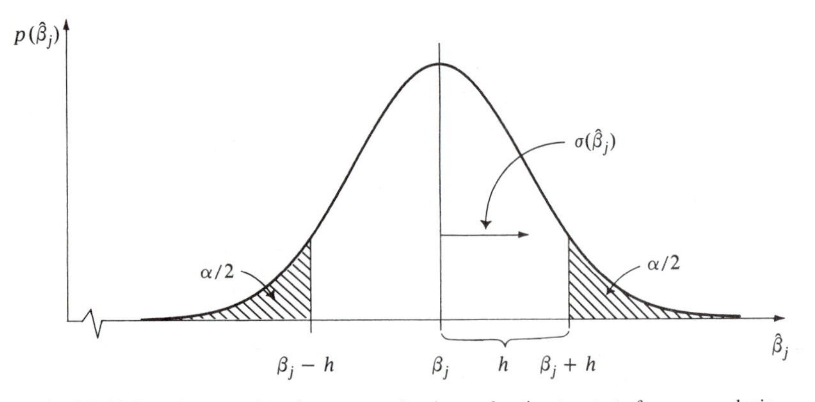

FIGURE 13.1 The procedure for constructing interval estimates starts from an analysis of the sampling distribution of $\hat{\beta}_j$, which arises because $\hat{\beta}_j$ depends on the disturbances. We consider here the case in which the standard error of $\hat{\beta}_j$ is known. The distance h, which runs from the mean β_j to the bounding point $\beta_j + h$, is determined by the probability amount $(\alpha/2)$ to be bounded in the tail. By construction, the probability of an outcome of $\hat{\beta}_j$ falling within h of the true β_j is $1 - \alpha$. Hence, the probability of $\hat{\beta}_j \pm h$ containing the true β_1 is $1 - \alpha$.

where h is the half-width of the interval. Sampling theory leads to a relatively simple method that allows us to determine the size of h in association with some probability amount so that we confidently can make statements like "the chance that the true β_1 is in the interval $\hat{\beta}_1^* \pm h$ is 95 percent."

To aid understanding, we start with the hypothetical case in which the true standard deviation of the disturbance, σ_u, is known. In this case, the sampling distribution of the estimator $\hat{\beta}_j$ is normal with an unknown mean β_j and a known standard deviation $\sigma(\hat{\beta}_j)$. Figure 13.1 illustrates this distribution. The value h is to be determined so that the probability that $\hat{\beta}_j$ will fall within h of β_j is $1 - \alpha$:

$$\Pr(\beta_j - h \leq \hat{\beta}_j \leq \beta_j + h) = 1 - \alpha \tag{13.2}$$

The value of h can be calculated directly from

$$\begin{aligned}\frac{\alpha}{2} &= \Pr(\hat{\beta}_j \geq \beta_j + h) \\ &= \Pr(Z \geq Z_k), \quad \text{where} \quad Z_k = \frac{(\beta_j + h) - \beta_j}{\sigma(\hat{\beta}_j)} = \frac{h}{\sigma(\hat{\beta}_j)} \qquad (13.3) \\ &= \Pr\left(Z \geq \frac{h}{\sigma(\hat{\beta}_j)}\right)\end{aligned}$$

In moving from the first to the second line we are starting with a probability statement involving a particular normal variable and setting up the equivalent statement involving a standard normal variable. Now

$$\frac{\alpha}{2} = \Pr(Z \geq Z^c) \tag{13.4}$$

defines the critical value Z^c that is equivalent to $h/\sigma(\hat{\beta}_j)$. Hence

$$h = Z^c\sigma(\hat{\beta}_j) \tag{13.5}$$

where Z^c depends on the given value of α, through (13.4). By construction [see (13.2)], the probability that $\hat{\beta}_j$ will fall within h of the true value β_j is $1 - \alpha$.

Suppose that we adopt the procedure of using the interval

$$\hat{\beta}_j^* \pm h, \qquad \text{where} \quad h = Z^c\sigma(\hat{\beta}_j) \tag{13.6}$$

to serve as an ***interval estimate*** of β_j. Referring to Figure 13.1, if the $\hat{\beta}_j^*$ value happens to fall in the range $\beta_j - h$ to $\beta_j + h$, β_j^* is clearly within h of the true β_j and β_j is within h of β_j^*—that is, β_j is in the interval $\hat{\beta}_j^* \pm h$. By (13.2), the probability of β_j^* being so situated is $1 - \alpha$. Continuing the reasoning, if $\hat{\beta}_j^*$ happens to fall outside the range $\beta_j - h$ to $\beta_j + h$, it is clear that $\hat{\beta}_j^*$ and β_j are farther than h away from each other—that is, β_j is not in the interval (13.6). The probability of this occurring is α.

Thus we can be confident that an interval constructed according to this method will contain the true β_j about $(1 - \alpha) \cdot 100$ percent of the times we make such an estimate. Because of this interpretation, the interval is known as a ***confidence interval,*** and the probability amount $1 - \alpha$ is known as the ***level of confidence.***

In the more realistic case when the standard deviation of the disturbance is unknown, we proceed along similar lines but use $s(\hat{\beta}_j)$ instead of $\sigma(\hat{\beta}_j)$. We can show that the interval

$$\hat{\beta}_j^* \pm h, \qquad \text{where} \quad h = t^c s^*(\hat{\beta}_j) \tag{13.7}$$

and where t^c is defined by

$$\frac{\alpha}{2} = \Pr(t \geq t^c) \tag{13.8}$$

can be used to construct an interval estimate of an unknown β_j at the $1 - \alpha$ level of confidence.

The validity of this procedure is more complex to establish than in the previous case, because both $s(\hat{\beta}_j)$ and $\hat{\beta}_j$ depend on the outcomes of the disturbances. We start with a probability statement that defines the critical t values, t^c and $-t^c$, for which

$$\Pr(-t^c \leq t \leq t^c) = 1 - \alpha \tag{13.9}$$

In regression, we know that the sampling statistic

$$t = \frac{\hat{\beta}_j - \beta_j}{s(\hat{\beta}_j)} \tag{13.10}$$

has a t distribution with $n - k - 1$ degrees of freedom. By substituting (13.10) into (13.9) we have a probability statement involving the random variables $\hat{\beta}_j$ and $s(\hat{\beta}_j)$, and the parameter β_j:

$$\Pr\left(-t^c \leq \frac{\hat{\beta}_j - \beta_j}{s(\hat{\beta}_j)} \leq t^c\right) = 1 - \alpha \tag{13.11}$$

This can be arranged to yield

$$\begin{aligned} 1 - \alpha &= \Pr\left(-t^c \leq \frac{\hat{\beta}_j - \beta_j}{s(\hat{\beta}_j)} \leq t^c\right) \\ &= \Pr\,(-t^c s(\hat{\beta}_j) \leq \hat{\beta}_j - \beta_j \leq t^c s(\hat{\beta}_j)) \qquad (13.12) \\ &= \Pr\,(\hat{\beta}_j - t^c s(\hat{\beta}_j) \leq \beta_j \leq \hat{\beta}_j + t^c s(\hat{\beta}_j)) \end{aligned}$$

In going from the second line to the third, we first subtract $\hat{\beta}_j$ from each of the three terms, then multiply through by -1 (which switches the direction of the inequalities), and then do a left-right swap to reswitch the direction of the inequalities. Now we focus on the end and beginning of (13.12):

$$\Pr\,(\hat{\beta}_j - t^c s(\hat{\beta}_j) \leq \beta_j \leq \hat{\beta}_j + t^c s(\hat{\beta}_j)) = 1 - \alpha \tag{13.13}$$

This is a probability statement about the ***random interval*** $\hat{\beta}_j \pm t^c s(\hat{\beta}_j)$, not about the parameter β_j. The interpretation of (13.13) is that the probability that an interval of the form $\hat{\beta}_j \pm t^c s(\hat{\beta}_j)$ will include the value β_j within it is $1 - \alpha$. In other words, we can be $(1 - \alpha) \cdot 100$ percent confident that an interval of the form (13.7) contains the true value β_j.

The procedure is simple to apply in practice. For example, consider again the aggregate consumption function

$$\begin{aligned} \widehat{CON}_i &= \underset{(6.973)}{0.568} + \underset{(0.010)}{0.907}DPI_i \\ R^2 &= .997 \qquad SER = 8.935 \end{aligned} \tag{13.14}$$

To construct a 95 percent confidence interval for estimating the marginal propensity to consume, we first determine the t^c such that

$$\Pr\,(t \geq t^c) = .025 \tag{13.15}$$

for a t distribution with 23 degrees of freedom. Table A.3 gives the value $t^c = 2.069$. Hence

$$h = t^c s^*(\hat{\beta}_1) = (2.069)(0.010) = 0.021 \tag{13.16}$$

and the interval is

$$\hat{\beta}_1^* \pm h = 0.907 \pm 0.021 \quad \text{or} \quad 0.886 \text{ to } 0.928 \tag{13.17}$$

That is, our best estimate of β_1 is that it is 0.907, and we are 95 percent confident that β_1 is between 0.886 and 0.928.

Sometimes confidence intervals are reported less formally, using a rule of thumb that sets $t^c = 2$. This produces a ***two-standard-error confidence interval,*** $\hat{\beta}_j^* \pm 2s^*(\hat{\beta}_j)$, which is approximately a 95 percent confidence interval except when the sample size is small. To see exactly how much approximation is involved, we note first that the exact level of confidence is given by

$$1 - \alpha = 1 - \Pr(|t| \geq 2) \tag{13.18}$$

From Table A.2 we see that when there are 10 degrees of freedom the exact level of confidence for a two-standard-error confidence interval is .926; when df = 25, $1 - \alpha = .944$; and when df = 100, $1 - \alpha = .952$.

Confidence Intervals and Hypothesis Testing

There is an important equivalence between creating a confidence interval and carrying out a two-tailed test of the hypotheses

$$\begin{aligned} H_0&: \beta_j = \beta_j^0 \\ H_1&: \beta_j \neq \beta_j^0 \end{aligned} \tag{13.19}$$

Suppose that the confidence interval and significance test are based on the same value of α. Whenever the confidence interval contains the value β_j^0, the null hypothesis is "accepted" (i.e., not rejected). And whenever the confidence interval does not contain β_j^0, the null hypothesis is rejected.

This equivalence can be understood informally along the following lines. If the hypothesis is rejected with a positive t^*, then

$$\frac{\hat{\beta}_j^* - \beta_j^0}{s^*(\hat{\beta}_j)} \geq t^c \tag{13.20}$$

Rearranging, we see that this implies that

$$\beta_j^0 \leq \hat{\beta}_j^* - t^c s^*(\hat{\beta}_j) \tag{13.21}$$

that is, β_j^0 lies to the left of (or on) the lower boundary of the confidence interval. Similarly, the conditions under which the null hypothesis is rejected with a negative t^* correspond to β_j^0 lying to the right of (or on) the upper boundary of the confidence interval.

For example, in Chapter 6 we estimated a Phillips curve (6.19) whose intercept can be interpreted as the limiting rate of inflation as the unemployment rate gets very high. The complete estimation, reported in Table 11.5, is

$$\begin{aligned} \widehat{RINF1}_i &= -\underset{(1.219)}{1.984} + \underset{(5.594)}{22.234}UINV_i \\ R^2 &= .549 \qquad SER = 0.956 \end{aligned} \tag{13.22}$$

Our best estimate of the limiting rate of inflation is -1.984 percent (i.e., a disinflation of roughly 2 percent). The 95 percent confidence interval for estimating this parameter is

$$\hat{\beta}_0^* \pm h = -1.984 \pm (2.160)(1.219) \quad \text{or} \quad (-4.62) \text{ to } (0.65) \qquad (13.23)$$

Since the confidence interval contains zero, a basic significance test with $\alpha =$ .05 leads to the conclusion that the intercept is not significantly different from zero. To verify this we see that $t^* = -1.628$ is not in the left-tail critical region that is bounded by $-t^c = -2.160$.

13.2 Confidence Intervals for Prediction

One use of regression models is to make predictions of the dependent variable for observations that are not included in the data used for estimation. In time-series work this usually means considering an observation that occurs after the time period covered by the data, and such a prediction is called a ***forecast.*** In cross-section work, this usually means considering an observation that is in the same population as the data but that was not selected in the data collection. In either case, we call it an ''out-of-sample'' observation.

We consider the case of simple regression and focus on an out-of-sample observation whose value for the explanatory variable is X_p. Presuming that the economic process that determines Y_p is the same as that which produced the data used for estimation (which is an assumption that we must make if we are to use the estimated regression), the actual value of Y_p will be

$$Y_p = \beta_0 + \beta_1 X_p + u_p \qquad (13.24)$$

In the prediction context, Y_p is unknown (and perhaps not yet determined even) but if we know the value of X_p we predict Y_p to be

$$\hat{Y}_p = \hat{\beta}_0 + \hat{\beta}_1 X_p \qquad (13.25)$$

Associated with any prediction is a ***prediction error,***

$$e_p = Y_p - \hat{Y}_p = (\beta_0 - \hat{\beta}_0) + (\beta_1 - \hat{\beta}_1)X_p + u_p \qquad (13.26)$$

We see that there are two sources of error in the prediction we are making. First, the actual Y_p depends on the outcomes of the disturbance, but this is neglected in making the prediction. Second, there will be some estimation errors involved with the coefficients.

It turns out that the expected prediction error is equal to zero:

$$E[e_p] = 0 \qquad (13.27)$$

so we say that the predictions from (13.25) are unbiased.

In any particular situation the prediction error will not be zero except by coincidence, and an estimate of the typical error is given by

$$s_p = SER\sqrt{1 + \frac{1}{n} + \frac{(X_p - \overline{X})^2}{\sum (X_i - \overline{X})^2}} \qquad (13.28)$$

which is the ***standard error of prediction.*** Regarding (13.28), we see that the terms under the square-root sign sum to greater than 1, so the typical error of an out-of-sample prediction (s_p) is always greater than the typical error of fit (SER). Also, the typical error of prediction depends on $X_p - \overline{X}$, which is the difference between the value of X for the observation being predicted and the mean of X for the observations in the data set. If the new observation is like the typical observation in the data (i.e., if X_p is close to $\overline{X}$), the typical prediction error is relatively small. However, if $|X_p - \overline{X}|$ is large, the typical prediction error is relatively large. This is unfortunate for time-series applications, because $X_p - \overline{X}$ is likely to be relatively large for the observation being predicted.

Confidence intervals for the prediction are of the form

$$\hat{Y}_p \pm h, \qquad \text{where} \quad h = t^c s_p \qquad (13.29)$$

where t^c is determined from

$$\frac{\alpha}{2} = \Pr\,(t \geq t^c) \qquad (13.30)$$

For example, DPI in 1981 was 1040.2 billion dollars. Based on the consumption function (13.14) estimated for 1956–1980,

$$\hat{Y}_p = 0.568 + (0.907)(1040.2) = 944.03 \qquad (13.31)$$

and

$$s_p = 8.935\sqrt{1 + \frac{1}{25} + \frac{(1040.2 - 705.7)^2}{875{,}270}} = 9.656 \qquad (13.32)$$

We predict with 95 percent confidence that Y_p is in the interval 944.03 ± (2.069)(9.656) = 944.03 ± 19.98, or between 924.05 and 964.01. (The actual value was $Y_p = 959.1$, which lies within the interval.)

Since the width of the confidence interval depends on s_p, then for any given level of confidence the confidence interval is wider the greater is the difference $|X_p - \overline{X}|$. This is illustrated in Figure 13.2, which shows schematically the 95 percent confidence bands for making predictions in a simple regression model. For an observation with X_p, the prediction of $\hat{Y}_p$ is read off the fitted regression line and the confidence interval extends vertically from the lower confidence band to the upper one.

The prediction methods discussed here take the X_p value as given, or known with certainty. This may be appropriate in cross-section applications, but for time-series forecasting it often is not. When X_p also must be predicted, the actual errors involved in predicting Y_p are more complex. However, if the regressor in a time-series model is lagged variable, X_p need not be predicted and we are in a position to make predictions of Y_p with (13.25).

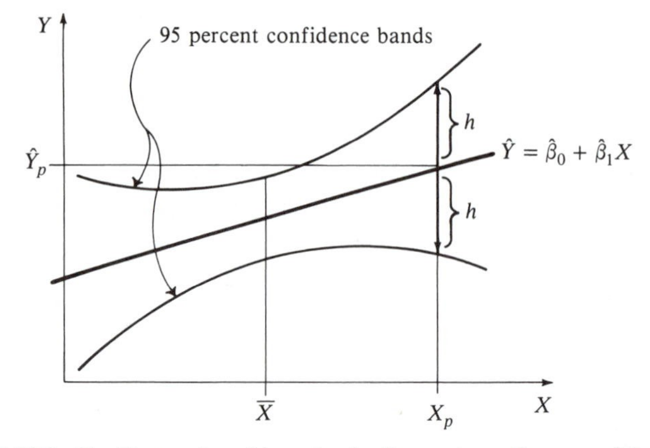

FIGURE 13.2 The illustrated confidence bands allow us to see the range of the 95 percent confidence interval for predictions made with the estimated regression model. For example, when $X = X_p$, the prediction interval runs from $\hat{Y}_p - h$ to $\hat{Y}_p + h$. As illustrated, the width of the confidence interval increases with the distance of X_p from $\bar{X}$.

13.3 Comparing Alternative Estimators

In all our work with regression so far, the method used to estimate the coefficients has been ordinary least squares (OLS) and our sampling theory has presumed that the estimators are applied to a correctly specified model. We should realize, however, that there can be alternative estimators for the coefficients and thus that the same data can yield more than one estimated value for a given parameter.

Alternative estimators arise from two sources. First and foremost, a method other than ordinary least squares may be adopted as the criterion for deriving the estimator of a coefficient. The method will lead to a rule or formula for determining an estimate of the coefficient from observed data; this rule or formula is the estimator. For example, in Chapter 5 we noted that one approach to determining the best-fitting line is to measure the error of fit from a plotted point to the fitted line by determining the perpendicular distance. (By contrast, in OLS we determine the vertical distance, $Y_i - \hat{Y}_i$.) For any given set of data, the "perpendicular estimates" of the coefficients will be different from the OLS estimates, and the corresponding sampling distributions of the estimators will be different also.

A second source of alternative estimators is the use of some estimator in a model specified differently from the one for which it was derived. For example, if we were to use the regular simple regression OLS estimator for β_1 to estimate the coefficient of the first explanatory variable in a multiple regression, we would obtain an estimate for the impact of that variable. Whether this should be considered to be a misapplication of OLS or the application of an "alternative estimator" is perhaps semantic, and we will return to this in Section 13.4.

For the purpose of discussion here, suppose that we are interested in estimating a particular coefficient, β_j, in a properly specified normal regression model. One way to make the estimate is to apply the appropriate OLS estimator, which we denote as usual by $\hat{\beta}_j$. A second way is to apply an alternative estimator, which we denote by $\tilde{\beta}_j$. What difference might it make whether we use $\hat{\beta}_j$ or $\tilde{\beta}_j$ to estimate β_j? To answer this question, we need to compare the sampling distributions of the estimators.

In Chapter 11 we noted the features of $p(\hat{\beta}_j)$, the sampling distribution of $\hat{\beta}_j$: it is normal, its expected value is equal to β_j, and its standard deviation is a known function of σ_u and the values of the explanatory variable(s). The sampling distribution $p(\tilde{\beta}_j)$ of the alternative estimator $\tilde{\beta}_j$ may be quite different: its form may be not-normal, its mean may be greater or less than β_j, and its standard deviation may be greater or less than $\sigma(\hat{\beta}_j)$. Figure 13.3 illustrates one possible situation. As compared with $p(\hat{\beta}_j)$, the sampling distribution $p(\tilde{\beta}_j)$ of the alternative estimator has a smaller mean and a smaller standard deviation.

In the general analysis of an estimator, statistical theory usually focuses on certain characteristics of the sampling distribution in relation to β_j; these are known as ***properties*** of the estimator. The statistical properties of estimators fall into two classes: those that hold true regardless of the sample size, and those that hold true only as the sample size approaches infinity. The latter are called ***large-sample*** properties, and by contrast the former are called ***small-sample*** properties even though they hold true in large samples also. We are concerned with small-sample properties relating to the mean and to the variance of the sampling distribution and with one large-sample property that combines the two.

If the expected value of an estimator is equal to the unknown parameter whose value is being estimated, the estimator is said to be ***unbiased;*** otherwise, it is said to be ***biased.*** For some estimator $\tilde{\beta}_j$ of β_j,

$$\text{bias of } \tilde{\beta}_j = E[\tilde{\beta}_j - \beta_j] = E[\tilde{\beta}_j] - \beta_j \tag{13.33}$$

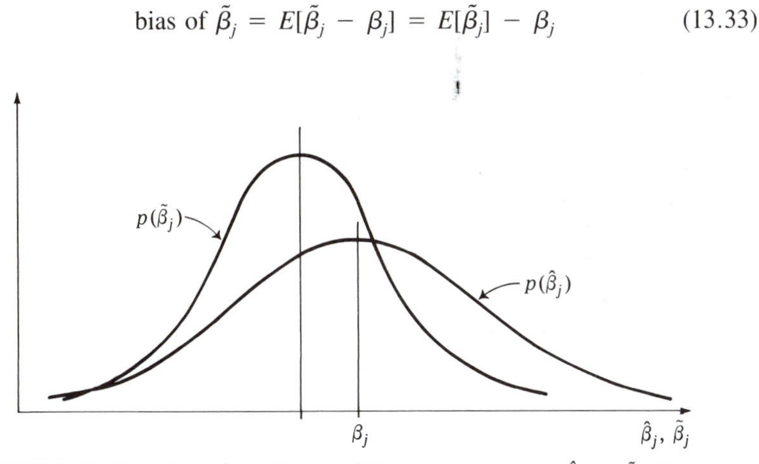

FIGURE 13.3 Two alternative estimators of β_j are denoted here by $\hat{\beta}_j$ and $\tilde{\beta}_j$. The estimator $\tilde{\beta}_j$ is negatively biased, since $E[\tilde{\beta}_j] < \beta_j$. However, $\tilde{\beta}_j$ has a smaller variance (i.e., smaller standard error) than $\hat{\beta}_j$, which is unbiased. When choosing between these two estimators, we are forced to balance off the desirability of unbiasedness against that of minimum variance.

In Figure 13.3 we see that $\tilde{\beta}_j$ has a negative bias, because $E[\tilde{\beta}_j] - \beta_j < 0$. By contrast, for the OLS estimator,

$$\text{bias of } \hat{\beta}_j = E[\hat{\beta}_j - \beta_j] = E[\hat{\beta}_j] - \beta_j = 0 \qquad (13.34)$$

and we see that it is unbiased. The essence of unbiasedness is that on average (i.e., in expectation) the point estimate is right on target; a biased estimator yields point estimates that tend to be too large or too small. Unbiasedness is a desirable property for an estimator to have, but it is not essential.

It is desirable also that the sampling distribution of an estimator have a small standard deviation or variance, so that large estimation errors are not likely to occur. In Figure 13.3 it seems clear that $\tilde{\beta}_j$ has a smaller variance than $\hat{\beta}_j$. Comparing all possible estimators of β_j, the one with the smallest variance is called the ***minimum variance*** estimator. When we compare all unbiased estimators, the one with the smallest variance is said to be ***efficient.***

If we had to choose between $\hat{\beta}_j$ and $\tilde{\beta}_j$, we would have to balance off our desire for unbiasedness against that for minimum variance. One criterion for doing this is to choose the estimator with the smaller ***mean squared error*** (*MSE*). It turns out that the *MSE* is simply the sampling variance of an estimator plus the square of its bias. For any estimator, say $\tilde{\beta}_j$,

$$MSE = E[(\tilde{\beta}_j - \beta_j)^2] = \sigma^2(\tilde{\beta}_j) + (E[\tilde{\beta}_j - \beta_j])^2 \qquad (13.35)$$

The square root of the *MSE* is a measure of the typical estimation error that would result from using $\tilde{\beta}_j$ as an estimator of β_j. The *MSE* criterion amounts to saying that the best estimator is the one that yields the smallest typical error. In Figure 13.3, it is not obvious whether $\hat{\beta}_j$ or $\tilde{\beta}_j$ has the smaller *MSE*.

In statistical theory, as the sample size increases toward infinity an estimator is said to be ***consistent*** if its sampling distribution becomes totally concentrated at the value of the unknown parameter it seeks to estimate. In other words, if $\tilde{\beta}_j$ is a consistent estimator of β_j, both $E[\tilde{\beta}_j - \beta_j]$ and $\sigma^2(\tilde{\beta}_j)$ head toward zero as the sample size increases. Consistency is a desirable property for an estimator to have. However, in small or finite-size samples, a consistent estimator may be biased or it may be inefficient. Conceivably, in small samples a consistent estimator may be less preferred than one that is inconsistent. A reason for being interested in large-sample properties is that for some techniques it is difficult to determine the small-sample properties. In these cases, the large-sample properties provide some basis for choosing among alternative estimators.

Although it is desirable for an estimator to have good properties, they need not be ideal in order for the estimator to be useful. For example, the standard error of regression (*SER*) can be used to estimate the standard deviation of the disturbances (σ_u), but it is a biased estimator and its sampling distribution is not symmetric. The square of *SER* is an unbiased estimator of the variance of the disturbance, but its sampling distribution is not symmetric either (it has a chi-square distribution with $n - k - 1$ degrees of freedom).

Under the assumptions of the normal regression model, the ordinary least squares coefficient estimators are unbiased and consistent. Also, among possible estimators that are unbiased and that are defined as linear combinations of the observed Y values, the OLS estimators have the smallest variance. Technically, the OLS estimators are the "best linear unbiased estimators" for the regression coefficients. This result is known as the ***Gauss–Markov Theorem,*** and it contributes to the high regard in which OLS is held.

13.4 Multicollinearity

As discussed in Section 7.6, the term ***multicollinearity*** names the situation that arises when there is a substantial degree of linear dependence among the regressors in the data we are using. This may result from there being high correlations among some of the regressors. Such a situation frequently occurs in macroeconomic time-series models, in which many of the variables tend to rise together over time. It also may occur in cross-section models—either naturally or as a result of specification, such as when we include a variable and its square as regressors.

The consequence of multicollinearity, especially when it is severe, is that the sampling distributions of the coefficient estimators have relatively large standard errors. As we know, the larger is the standard error, the greater is the probability that the estimated coefficient will be "far" from its expected value. In other words, multicollinearity makes the estimates imprecise.

To gain some insight into this problem, we consider an exploratory situation in which we can take more than one sample (set of data) from the same economic process. For example, suppose that among some population of workers the true process determining earnings is given by

$$EARNS_i = \beta_0 + \beta_1 ED_i + \beta_2 EXP_i + u_i \qquad (13.36)$$

We choose the samples so that they are characterized by different degrees of multicollinearity and then compare the properties of the estimators in the samples. We focus on a cross-section model in this exploration, because planned sample selection is more practical with it than with a time-series model.

The first sample is a selection among workers who are about 30 years old. In this sample there will be a strong negative correlation between ED and EXP, because experience is measured approximately by the number of years since leaving school. Those workers with more-than-average years of education will tend to have fewer-than-average years of experience. The second sample is one in which a wide variety of ages are represented. This will result in a smaller correlation between ED and EXP. (The two sets of data should be drawn with certain technical similarities so that differences are due essentially to the different correlations between the explanatory variables.)

It should be clear that both sets of data are appropriate for estimating the coefficients of the true process (13.36). The same OLS estimators are used in both cases, but the corresponding sampling distributions will differ because their standard errors depend on the values of the regressors in the different data sets.

Since the estimators in the two sets of data are appropriate OLS estimators, we know that each one is unbiased and has a normal sampling distribution. It turns out, however, that the standard errors of the sampling distributions are larger in the first sample—the one with a high correlation between the explanatory variables. Figure 13.4 illustrates this for any coefficient in the model. The key point is that a greater correlation between the regressors leads to larger standard errors for the estimated coefficients.

One consequence of larger standard errors is that the confidence intervals we make are wider; this is another way of saying that the estimators are less precise.

The consequence for hypothesis testing is more subtle. Consider the basic significance test with hypotheses

$$\begin{aligned} H_0&: \beta_j = 0 \\ H_1&: \beta_j \neq 0 \end{aligned} \tag{13.37}$$

Whether the standard error is large or small does not affect the validity of the procedure or the decision rule that is set up in terms of critical t values. Also, the probability of making a Type I error is unaffected, because it is equal to the chosen level of significance.

However, a larger standard error leads to a greater probability of making a Type II error: that is, there will be a greater probability of not rejecting the null hypothesis when in fact it is false. Thus, considering a regressor that does have some effect on the dependent variable, a severe situation of multicollinearity can make it unlikely that we will correctly conclude that the associated coefficient is significant (i.e., significantly different from zero). In this sense, multicollinearity makes it hard to find significance.

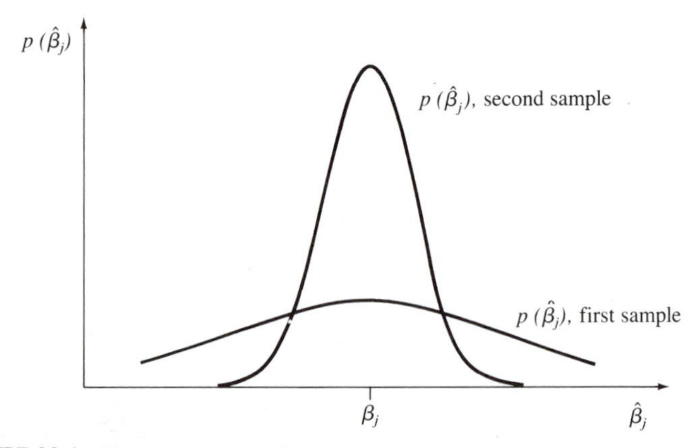

FIGURE 13.4 The consequences of multicollinearity for estimating any coefficient in the regression model (13.36) are shown by comparing the sampling distributions in two samples of data: the first has a high correlation between the regressors, and the second has a low correlation. A greater correlation between the regressors leads to a larger standard error for the estimator.

Since there usually is some degree of linear dependence among the regressors, one might say that multicollinearity is usually present in regression models. It is standard, however, to reserve the term for situations that are especially severe and in which the conclusions of statistical inference are somewhat disappointing.

What should be done if multicollinearity is present? Unfortunately, there is not much one can do about it easily. It is sometimes suggested that we should delete one of the correlated variables in order to reduce the multicollinearity, but this generally does not make sense because it leads to a specification error that biases the estimates of the remaining coefficients (see Section 13.5).

Since the degree of multicollinearity is a sample phenomenon, we might seek to reduce it by sample selection. It is helpful to make the sample as large as possible. However, in time-series work—where the problem is most acute—there are usually no more data to be had unless one is willing to wait for more history to happen. In cross-section work, there are occasions when control can be exercised in sample selection and the potential for severe multicollinearity can be reduced.

A difficulty of living in a multicollinear world is that some regression results are disappointing because the parameters of interest cannot be estimated with much precision. We must interpret our results with appropriate caution.

13.5 Misspecification

In Section 7.6 we considered the consequences of misspecifying the list of variables to be included in a linear multiple regression model. The previous discussion is extended here by incorporating the ideas of sampling theory. Two situations are distinguished: one is when a variable that belongs in the model is omitted from the specification of the estimated regression, and the other is when a variable that does not belong in the model is included. We present two related models for comparison, choosing one and then the other as the correct specification.

The first model supposes that a certain economic process is correctly described by

$$Y_i = \beta_0 + \beta_1 X_{1i} + \beta_2 X_{2i} + u_i \tag{13.38}$$

with u_i satisfying the assumptions of the normal regression model. The Gauss–Markov Theorem asserts that the OLS estimators (7.8)–(7.10) are the best linear unbiased estimators of the β_j. In this model the true impact of X_1 on Y is given by the coefficient β_1. Its proper OLS estimator is

$$\hat{\beta}_1 = \frac{\left(\sum x_1 y\right)\left(\sum x_2^2\right) - \left(\sum x_2 y\right)\left(\sum x_1 x_2\right)}{\left(\sum x_1^2\right)\left(\sum x_2^2\right) - \left(\sum x_1 x_2\right)^2} \tag{13.39}$$

using the deviation-from-mean notation of (7.7) for simplicity.

The second model supposes that the process is correctly specified by

$$Y_i = \gamma_0 + \gamma_1 X_{1i} + v_i \tag{13.40}$$

with the disturbance v_i satisfying the regular assumptions. In this model γ_1 gives the true impact of X_1 on Y, and its proper OLS estimator is

$$\hat{\gamma}_1 = \frac{\sum x_1 y}{\sum x_1^2} \tag{13.41}$$

which is equivalent to (5.53) rewritten in deviation form. Applying the Gauss–Markov Theorem again, this is the best linear unbiased estimator of γ_1.

The first type of misspecification that we consider occurs when a relevant variable is omitted from the estimated model. For this analysis we regard (13.38) as the correct specification of the process determining Y. Our attention is focused on the task of estimating the true impact of X_1 on Y, which is given by β_1. The OLS estimator of β_1 in the correctly specified model is (13.39).

Suppose, however, that we make the mistake of ignoring X_2 and carry out an estimation of a simple regression of Y on X_1, as though the correct model were (13.40). The OLS estimator for the slope coefficient on X_1 is given by $\hat{\gamma}_1$ in (13.41), but since we are interested in estimating the true (*ceteris paribus*) impact of X_1 on Y we can also denote this estimator by $\tilde{\beta}_1$. In other words, the OLS estimator (13.41) for the slope coefficient in a simple regression can be interpreted as an alternative estimator for the coefficient on X_1 in (13.38). Accordingly, $\tilde{\beta}_1$ is sometimes called an "omitted-variable estimator."

How do $\hat{\beta}_1$ and $\tilde{\beta}_1$ compare as estimators of β_1, the true impact of X_1 on Y? To answer this question we must examine and compare the two sampling distributions. With regard to their means, we know from sampling theory that

$$E[\hat{\beta}_1] = \beta_1 \tag{13.42}$$

so that $\hat{\beta}_1$ is unbiased. It turns out that

$$E[\tilde{\beta}_1] = \beta_1 + \beta_2 \frac{S(X_1, X_2)}{S^2(X_1)} \tag{13.43}$$

Thus $\tilde{\beta}_1$ is biased, unless the second term in (13.43) happens to be zero. Whether the bias, which is equal to the second term, is positive or negative depends on the sign of the true parameter β_2 and the sign of the covariance $S(X_1, X_2)$ in the data we have. Because of this bias, the procedures of hypothesis testing and interval estimation are no longer valid.

For example, in Section 7.6 we examined earnings functions involving *ED* and *EXP* as possible explanatory variables. Recasting these in terms of (13.38) and (13.40), X_1 plays the role of *ED* and X_2 plays the role of *EXP*. In the previous discussion, β_2 was presumed to be positive and the covariance was discovered to be negative in the data. Thus if the multiple regression specification were correct, the estimated impact of education on earnings in the simple regression would tend to be an underestimate of its true impact; that is, $\tilde{\beta}_1$ would be negatively biased.

Further understanding of the bias comes from noting that the ratio $S(X_1, X_2)/S^2(X_1)$ is equal to the slope of an OLS regression of X_2 on X_1. Thus (13.43) shows that the expected value of $\tilde{\beta}_1$ is equal to the true impact of X_1 on Y plus a combination of the true impact of X_2 on Y and the relation between X_2 and X_1 in the particular data we use.

A hypothetical comparison of the sampling distributions of the two estimators is given in Figure 13.5a for a case that corresponds to the earnings function example. In general, the shape of $p(\tilde{\beta}_1)$ is not normal, and its variance may be greater or less than that of the sampling distribution of $\hat{\beta}_1$. That the variance of $\tilde{\beta}_1$ might be less than that of $\hat{\beta}_1$ opens the door to the possibility that the biased omitted-variable estimator may have a smaller mean square error than the proper OLS estimator. Whether this possibility represents a practical opportunity in any particular case is not easy to determine, and the sure bias of the omitted-variable estimator leads econometricians to avoid it in practice.

These results generalize easily. When a relevant variable is excluded from the specification of some economic process, OLS applied to the misspecified model yields biased estimators for all the coefficients. This invalidates the procedures of statistical inference in the estimated model.

The second type of misspecification that we consider occurs when an irrelevant variable is included in the estimated model. Turning around the comparisons just made, we regard the simple regression (13.40) as the correct model, but we suppose that (13.38) is chosen as the regression specification to be estimated. For example, it might be that the true process determining earnings depends only on education, but that experience is added to the regression specification by mistake or to test whether it has an effect.

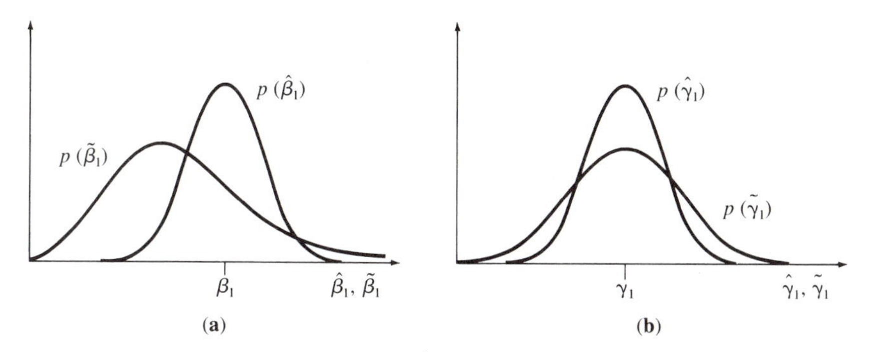

FIGURE 13.5 In Part (a), the estimator $\hat{\beta}_1$ is the OLS estimator of β_1 in a correctly specified model, while $\tilde{\beta}_1$ is an omitted-variable estimator of the same coefficient. In the case illustrated, $\tilde{\beta}_1$ is negatively biased and has a larger variance than $\hat{\beta}_1$. In part (b), the estimator $\hat{\gamma}_1$ is the OLS estimator of γ_1 in a correctly specified model, while $\tilde{\gamma}_1$ is an irrelevant-variable estimator of the same coefficient. Both estimators are unbiased, but the irrelevant-variable estimator has a larger standard error.

The true impact of X_1 on Y is given by γ_1 in the correct simple regression, and its proper OLS estimator is given by (13.41). If instead the multiple regression specification (13.38) is estimated by OLS, the estimator of the coefficient on X_1 is given by the right-hand side of (13.39). This can be considered an alternative estimator of γ_1 and denoted by $\tilde{\gamma}_1$. Like the proper OLS estimator $\hat{\gamma}_1$, this "irrelevant-variable estimator" $\tilde{\gamma}_1$ has a sampling distribution that is normal and it is unbiased. However, the standard error for $\tilde{\gamma}_1$ is greater than that for $\hat{\gamma}_1$, and thus the irrelevant-variable estimator is inefficient. A hypothetical comparison of the two sampling distributions is given in Figure 13.5b. It should be noted that the coefficient estimates calculated from the same data are different.

In this situation, when (13.40) is the correct model, (13.38) could be considered "correct" if the restriction that $\beta_2 = 0$ is added and somehow taken into account during estimation. The regular OLS estimator $\hat{\beta}_2$, which is given by (7.8) but does not take the restriction into account, has a normal sampling distribution with an expected value of zero. Of course, the actual estimate $\hat{\beta}_2^*$ will not be equal to zero, except by coincidence. In this misspecified model (13.38), a basic significance test for β_2 will usually conclude appropriately that the coefficient is not significant. However, a Type I error can occur (with probability α, equal to the level of significance), in which case we would mistakenly conclude that the irrelevant variable is significant.

The results of this special case generalize easily. When irrelevant variables are added to the correct regression specification of some economic process, OLS applied to the misspecified model yields estimators that are unbiased and have normal sampling distributions. These are inefficient estimators of the true coefficients in the correctly specified model, but the methods of hypothesis testing and interval estimation can be applied without alteration.

In comparing the two types of misspecification, we note first that including an irrelevant variable has relatively mild statistical effects. The standard errors of the coefficients are made greater than they should be, thereby decreasing the precision of our estimates. By contrast, excluding a relevant variable seems more serious; it leads to biased estimators and thereby undermines the validity of our statistical inference procedures. It should be remembered, also, that both types of misspecification can lead to substantial misinterpretation of reality.

Specification Strategy

Much research is guided by an explanatory strategy, in which various models are estimated and their results are tested for significance. Eventually the models must be compared, and perhaps one is chosen as the "best." In following this strategy, which is a reasonable one overall, the results of the hypothesis tests must be handled with special care. If a model is misspecified—and if several models are estimated most of them must be misspecified—the hypothesis testing procedure can yield misleading results.

In comparing various models or in proceeding along a search strategy, there is a tendency to act as though a variable that is found to be insignificant in a regression truly does not belong there. Often the next step taken is estimating a new version of the regression with the offending variable deleted. This is not always a wise procedure, however. Even in a correctly specified model, pure sampling variability can lead a relevant variable to be judged insignificant. This is a Type II error in our testing procedure. One factor determining the chance of this is the level of significance: the lower is α, the greater is the chance of judging a relevant variable to be insignificant. A second, more general, factor is that the data may be such that the coefficients cannot be estimated with very much precision (i.e., the standard errors may be appropriately large). This leads to small t^* values, and insignificance. The problem might be that there are too few observations, in which case a larger sample from the same economic process might show significance for the coefficient in question. Alternatively, it might be that the regression suffers from a severe case of multicollinearity, whose technical consequence is that the standard errors tend to be large.

In assessing a regression, if theory suggests that a variable belongs in the model and if the insignificance of the variable reasonably can be ascribed to a problem with the data or with the strictness of conventional levels of significance, good judgment calls for leaving it in. The ill consequences of mistaken exclusion are often worse than mistaken inclusion. Of course, if a variable does not belong in the specification of the model, there is no reason for having it there. The presence of an irrelevant variable imposes a statistical cost, and it may mislead the readers of the research report. With no strict rules to work by, the researcher needs to exercise care and judgment. This is the art of econometrics.

Problems

Section 13.1

* **13.1** Construct a 95 percent confidence interval for estimating the impact of education on earnings, based on the regression (12.8).

13.2 Construct a 90 percent confidence interval for estimating the marginal propensity to consume, based on the regression (13.14). Now construct a 90 percent confidence interval for the same parameter, and compare the two interval estimates.

* **13.3** It is important for the U.S. Treasury to have a good estimate of the marginal propensity to consume, in order to recommend tax changes. Would a 100 percent confidence interval be ideal?

13.4 Use a confidence interval to test the hypothesis that the income elasticity of the demand for money, estimated in regression (7.51) and reported in Table 11.6, is equal to 1.0.

13.5 Use a confidence interval to assess whether experience affects earnings, based on regression (7.15) as reported in Table 11.1.

⋆ **13.6** Suppose that we perform a statistical experiment, taking many samples of the same size from a given economic process. In each sample we construct a 95 percent confidence interval for estimating the coefficient β_j.

(a) In roughly what proportion of the samples does the confidence interval contain the value β_j?

(b) Assume that β_j is unknown. If we look just at one sample, can we tell from the confidence interval whether in fact it contains the value β_j?

13.7 In a simple regression, what happens to the interval estimate for the slope if the data are selected with a lot of variation in the X values rather than with a smaller amount?

Section 13.2

⋆ **13.8** Based on the information in Section 13.2, construct a 95 percent confidence interval for predicting the value of aggregate consumption when disposable income reaches 2 trillion dollars.

13.9 Based on the earnings function reported as regression (12.8), construct a 90 percent confidence interval for predicting the earnings of men with exactly 16 years of education. (Note that in the data sample the mean ED is 11.58 and its standard deviation is 3.44; also, note that the predictions are for earnings measured in 1963 dollars.)

⋆ **13.10** Suppose that we are omniscient and know the true values of β_0, β_1, and σ_u in an economic process described by a simple regression model. How can we construct a 90 percent confidence interval for predicting the Y_p associated with a given X_p?

Section 13.3

13.11 Suppose that we are seeking to estimate a coefficient β_j with two alternative estimators, $\tilde{\beta}'_j$ and $\tilde{\beta}''_j$. In addition, suppose that it is very important to us that our estimate fall within the range $\beta_j \pm \delta$, where δ is some positive constant. Draw a figure showing the sampling distributions of the two estimates consistent with this: $\tilde{\beta}'_j$ is biased while $\tilde{\beta}''_j$ is unbiased, but $\tilde{\beta}'_j$ is definitely preferred over $\tilde{\beta}''_j$. Explain.

Section 13.4

13.12 Examine the estimated regression (7.29) reported in Table 11.3 for the possible effects of multicollinearity.

Section 13.5

13.13 Suppose that aggregate income and wealth both increase over time, as does aggregate consumption. Simple Keynesian theory assumes that consumption depends on income alone, whereas life cycle theory assumes that it depends positively on both income and wealth. What are the statistical implications of these theories for estimation of aggregate consumption functions?

13.14 Suppose that we have control over the selection of our sample, so that we can have either a high correlation or no correlation between X_1 and X_2 in Equation (13.38). What impact would this choice have on the consequences of excluding X_2 and estimating a regression of Y on just X_1?

13.15 Compare the two Phillips curves reported in Table 11.5 with regard to what might be expected on the basis of possible misspecification.

Topics in Econometrics

Autocorrelation and Heteroscedasticity

In Chapter 11 we saw that it is necessary to make certain assumptions about the disturbances in a regression model in order to determine the sampling distributions of the coefficient estimators. These sampling distributions, in turn, provide the basis for making statistical inferences and for assessing the properties of the ordinary least squares technique. In particular, it is assumed that for a given observation, the disturbance is a random variable with a probability distribution that is normal with a mean of zero and a certain standard deviation, σ_u. Among the observations, it is assumed both that the disturbances are independent and that all the standard deviations are equal: $\sigma(u_i) = \sigma_u$, for all i.

In this chapter we examine alternatives to the last two assumptions. If the disturbances are related in a certain way, rather than independent of each other, they are said to exhibit autocorrelation. If the disturbances do not have equal standard deviations, they are said to exhibit heteroscedasticity. These alternative assumptions are more plausible than the regular ones in certain cases, and we investigate their consequences for regression estimation. Both alternative assumptions point the way toward modifying the basic OLS estimation technique in order to get better estimates.

14.1 Autocorrelation

Consider the simple regression specification

$$Y_i = \beta_0 + \beta_1 X_i + u_i \tag{14.1}$$

In time-series applications the observations are usually successive periods, and the index i indicates the time period. (Often t is used as the index, instead of i, to highlight this.) Under the assumptions of the normal regression model, successive periods' disturbances are independent, and a typical time plot of the outcomes of the disturbances is given in Figure 14.1a. Given a set of X values, a typical set of data generated for Y through (14.1) looks like Figure 14.1b. Essentially, the observed values for Y are randomly scattered around the true regression line.

Autocorrelation (or, ***serial correlation***) is the situation in which successive disturbances are related to each other rather than independent of each other. The circumstances surrounding time-series regression models make this a plausible occurrence in many cases. Recall that one of the factors contributing to the disturbance term in any regression model is measurement error for the dependent variable. It seems possible that measurement errors may be serially correlated because of repetition in data-gathering techniques. A second factor usually con-

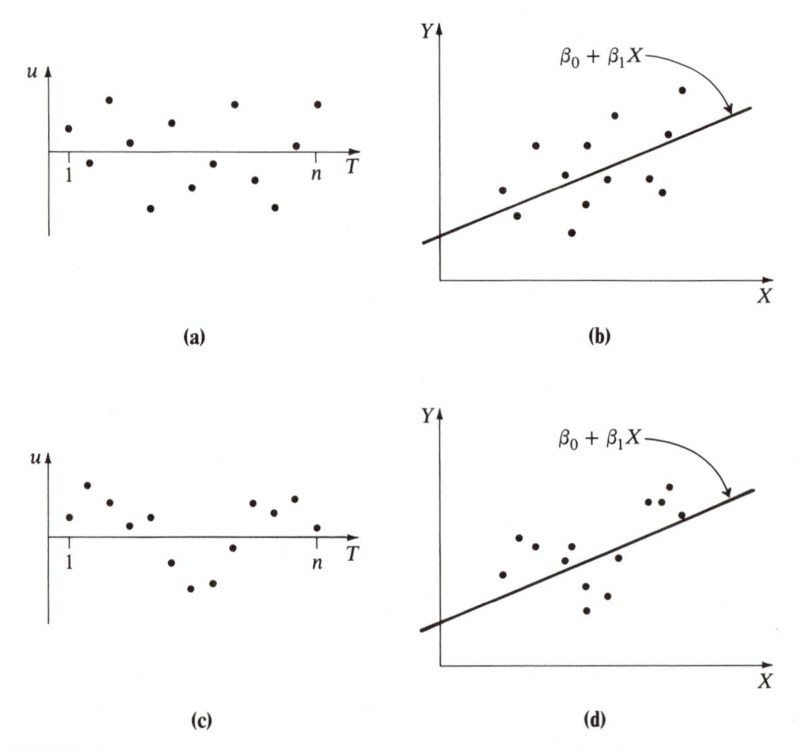

FIGURE 14.1 In the normal regression model the disturbances for different observations are independent. In (a), a typical time plot of disturbances shows a random pattern. In (b), these disturbances also show a random pattern around the true regression line. With positive first-order autocorrelation, successive disturbances are correlated, and a typical time plot of disturbances is given in (c). If the values of X increase over time, these disturbances show a snakelike pattern around the true regression line, as in (d).

tributing to the disturbance term is the exclusion of some relatively unimportant explanatory variables. Each of these is likely to vary systematically with time, and the combination of the unimportant variables may be serially correlated. In both of these cases, the autocorrelation model presented below is a straightforward way of taking these factors into account.

When autocorrelation occurs, the time plot of the actual disturbances may resemble Figure 14.1c. In the figure, successive disturbances appear to be small deviations from the previous value, and the overall pattern is snakelike. If the values of X increase over time, which is common in time series, the Y values generated through the simple regression model (14.1) will snake around the true regression line, as illustrated in Figure 14.1d. In order to understand the consequences of this phenomenon and to take corrective measures, we need to develop a formal model of autocorrelation. There are a variety of realistic forms in which the disturbances may be related, and each requires separate analysis. We examine only the most common, which is known as first-order autocorrelation.

The Model

The simplest model of ***first-order autocorrelation*** starts with the regression specification (14.1). In contrast to the regular model, however, the disturbance term u_i is assumed to be related to the previous period's disturbance, u_{i-1}, according to

$$u_i = \rho u_{i-1} + \epsilon_i, \qquad 0 < \rho < 1 \tag{14.2}$$

(ρ is "rho" and ϵ is "epsilon"). This specifies the case of *positive* autocorrelation, which is its most common form. When (14.2) holds with $-1 < \rho < 0$, the case is that of *negative* autocorrelation. We focus on the positive case and leave some aspects of the negative case as problems. Much of the analytics is the same in both cases.

In this autocorrelation model, the outcome of each observation's disturbance is a random deviation from a portion of the previous disturbance. The ϵ_i are assumed to be random variables having the same set of assumptions that were previously made for the u_i: the ϵ_i are independent and have identical normal distributions with $E[\epsilon_i] = 0$ and $\sigma(\epsilon_i) = \sigma_\epsilon$. Given these assumptions regarding ϵ_i, the properties of the disturbances u_i can be determined. It turns out that all the u_i have unconditional normal distributions with $E[u_i] = 0$ and

$$\sigma(u_i) = \sigma_u = \frac{\sigma_\epsilon}{\sqrt{1 - \rho^2}} \tag{14.3}$$

the same for all i. For any particular observation, the disturbance u_i has a greater standard deviation than the random term ϵ_i; if ρ is large (close to 1 in absolute value), σ_u is much larger than σ_ϵ.

Notice that when we make the regular "disturbance" assumptions about ϵ in (14.2), all of them carry over to u_i in (14.1) except independence: the u_i are normal, with means of zero and identical standard deviations. But, it can be shown that the u_i have a joint probability distribution in which the correlation between adjacent observations is ρ in (14.2). In time-series work, adjacent observations are sequential, so that ρ is the correlation between one period's disturbance and the next, or previous, one.

The impact of positive autocorrelation on the determination of Y is illustrated in Figure 14.2. Suppose that the actual outcome of the disturbance term for the first observation is u_1^*. For the second observation, substituting (14.2) into (14.1), we see that Y_2 is determined from

$$Y_2 = \beta_0 + \beta_1 X_2 + \rho u_1^* + \epsilon_2 \tag{14.4}$$

The random term ϵ_2 causes there to be a (conditional) probability distribution for Y_2 that has a mean of $\beta_0 + \beta_1 X_2 + \rho u_1^*$ and a standard deviation of σ_ϵ. One outcome from this distribution occurs, and the difference between this outcome Y_2 and the unconditional mean $(\beta_0 + \beta_1 X_2)$ is the disturbance u_2^* (which is not shown). Similarly, for the third period (not shown) there is a conditional probability distribution for Y_3 that has a mean of $\beta_0 + \beta_1 X_3$

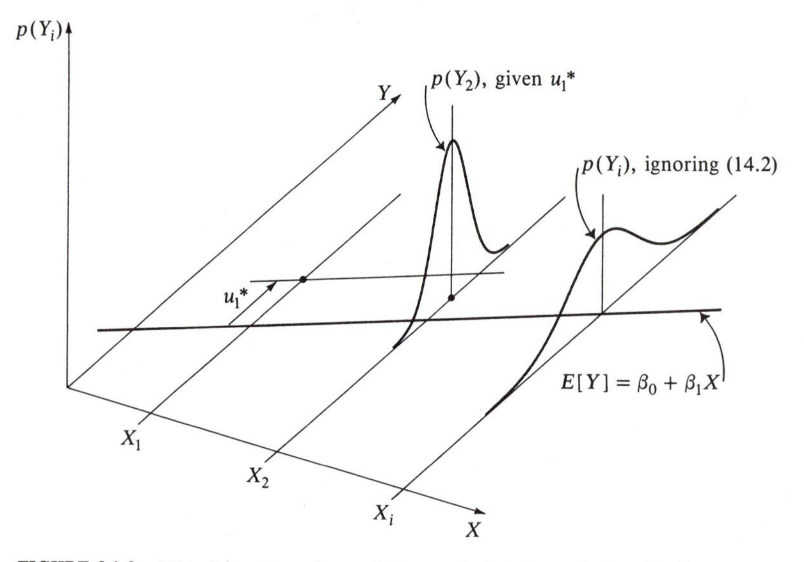

FIGURE 14.2 With first-order autocorrelation, each disturbance is thought of as a random variable whose mean depends on the actual disturbance of the previous observation. Thus, for the second observation, Y_2 is thought of as an outcome from "$p(Y_2)$, given u_1^*." If we try to apply OLS directly, this is tantamount to assuming that each Y_i is determined from a probability distribution like "$p(Y_i)$, ignoring (14.2)," which has a larger standard deviation than the true distribution of Y_i. Thus OLS is no longer efficient. The determination of Y in the case of autocorrelation should be compared with that in the normal regression model, illustrated in Figure 11.2.

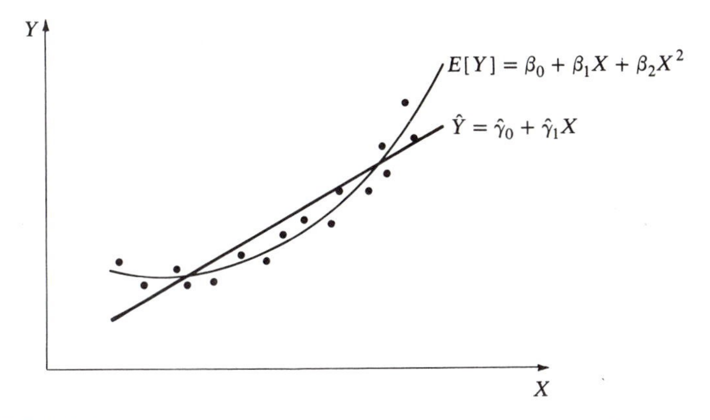

FIGURE 14.3 If an estimated model is misspecified, the pattern of residuals around the fitted regression line may resemble the pattern of autocorrelation. Here, the true regression model is quadratic, but a simple linear model is fit to the data. The proper procedure is to correct the misspecification, not to treat the situation as one of autocorrelation. Of course, it is not easy to know when the estimated model is misspecified, so some exploration may be appropriate.

$+ \rho u_2^*$ and a standard deviation of σ_ϵ; its outcome includes a disturbance u_3^*, which helps determine the mean of the conditional distribution for Y_4. The process continues in this way, and for $0 < \rho < 1$ it leads to a set of Y values that may snake around the true regression line, as in Figure 14.1d.

The process of true autocorrelation should be distinguished from one in which a misspecification of the model makes the disturbances appear to be serially correlated, when in fact they are not. For example, suppose that Y is a second-degree polynomial function of X,

$$Y_i = \beta_0 + \beta_1 X_i + \beta_2 X_i^2 + u_i \tag{14.5}$$

with regular disturbances. Figure 14.3 shows how the data generated from this process might look. If we mistakenly estimate a simple regression of Y on X,

$$\hat{Y}_i = \hat{\gamma}_0 + \hat{\gamma}_1 X_i \tag{14.6}$$

the residuals will seem to snake around the fitted line. We might think that autocorrelation is present when in fact the real problem is the exclusion of an important explanatory variable. The proper course of action when misspecification is present is to correct the specification—it is not appropriate to assign the problem to the disturbance and treat the situation as one of autocorrelation.

Consequences of Autocorrelation for OLS

To assess the consequences of first-order autocorrelation for the problem of estimation, we reconsider the model determining the values of Y, as illustrated in Figure 14.2. As explained previously, the observed values of Y are generated from a series of probability distributions like that of "$p(Y_2)$, given u_1^*," which has a standard deviation σ_ϵ. The means of these distributions sequentially shift according to the outcome of the previous disturbance. However, if we ignore the structure of the disturbances (14.2) and simply apply OLS to (14.1), this is tantamount to assuming that the values of Y are generated from a series of probability distributions like that of "$p(Y_i)$, ignoring (14.2)" in Figure 14.2. The standard deviation of this distribution is given by (14.3), and it is larger than the standard deviation of "$p(Y_2)$, given u_1^*."

The consequence of this is that OLS applied directly to (14.1) is no longer efficient. Intuitively, a technique that does not take the disturbance structure (14.2) into account will yield larger sampling errors for its estimates than a technique that does. Further, and very important, when $\rho > 0$ and the values of X are increasing over time, the direct application of OLS to (14.1) tends to underestimate the appropriate standard errors, meaning that the t ratios based on these calculations are erroneously large. This tends to cause our hypothesis tests to reject a correct null hypothesis with a probability much greater than the chosen level of significance. In other words, our testing procedure is now invalid, and it errs in the direction of finding too much significance.

The presence of autocorrelation does not affect the unbiasedness and consistency of OLS under the other regular assumptions. However, if one of the regressors is a lagged value of the dependent variable, the OLS estimates are biased and inconsistent.

When true first-order autocorrelation seems to be present, there are several methods of estimation that are preferable to OLS. If these methods cannot be used, OLS estimates are still worth examining. However, reports of the significance tests must be discounted.

Testing for Autocorrelation

In practice, one way to determine whether or not autocorrelation is present is to examine the time plot of the residuals. If they seem to cycle or snake around the time axis, positive autocorrelation should be suspected; if they are fairly random, the problem might be considered absent.

A rigorous approach to doing this is provided by the ***Durbin–Watson test.*** The null hypothesis is that there is no autocorrelation, specified as H_0: $\rho = 0$ in the basic autocorrelation model. The alternative hypothesis can be specified in various ways, depending on the situation. Most commonly, the alternative hypothesis is H_1: $0 < \rho < 1$, which is the case of positive autocorrelation.

The sampling statistic for the test is

$$d = \frac{\sum_{i=2}^{n} (e_i - e_{i-1})^2}{\sum_{i=1}^{n} e_i^2} \tag{14.7}$$

The statistic d ranges from 0 to 4. If the null hypothesis is true, then the sampling distribution of this statistic has a mean of about 2. If positive autocorrelation is present, differences like $e_i - e_{i-1}$ in the numerator of (14.7) tend to be smaller than the typical value e_i (neglecting sign) in the denominator, and d tends to be small. Hence, small values of d (close to 0) are evidence against the null hypothesis.

The decision rule for the test should be of the form "reject H_0 if $d^* \leq d^c$," where d^c is the critical value of d that bounds exactly α probability in the left-hand tail of the sampling distribution of d when $\rho = 0$. Unfortunately, the critical value d^c cannot be determined exactly because the sampling distribution depends on the values of the regression variables. However, Durbin and Watson were able to derive upper and lower bounds for d^c, so that $d_l \leq d^c \leq d_u$ as illustrated in Figure 14.4. These upper and lower bounds depend on the number of regressors (k), the number of observations (n), and the level of significance for the test (α).

In practice, the decision rule for testing against the alternative of positive autocorrelation is

$$\text{If:} \quad \begin{array}{ll} 0 < d^* < d_l, & \text{reject } H_0 \\ d_l \leq d^* \leq d_u, & \text{the result is indeterminate} \\ d_u < d^*, & \text{do not reject } H_0 \end{array} \tag{14.8}$$

For this one-tailed test, Table A.7 gives the d_l and d_u bounds at the .05 level of significance. Most econometric computer programs routinely calculate the actual value of the test statistic, d^*.

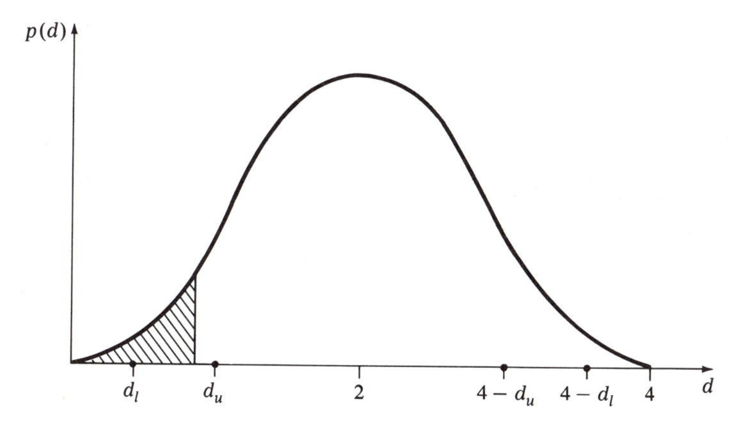

FIGURE 14.4 The exact sampling distribution of the Durbin-Watson statistic (d) depends on the values of the regression variables. Hence, the critical value d^c cannot be generally determined. However, lower and upper bounds have been derived, so we know that $d_l \leq d^c \leq d_u$. If the actual value of the test statistic, d^*, falls between these bounds, the conclusion of the hypothesis test is indeterminate.

When we suspect negative autocorrelation, the alternative hypothesis is H_1: $-1 < \rho < 0$. When negative autocorrelation is present, d tends to be large, and large values of d (close to 4) are evidence against the null hypothesis. The decision rule is

$$\text{If:}\quad \begin{array}{rl} 4 - d_l < d^* < 4, & \text{reject } H_0 \\ 4 - d_u \le d^* \le 4 - d_l, & \text{the result is indeterminate} \\ d^* \le 4 - d_u, & \text{do not reject } H_0 \end{array} \tag{14.9}$$

When there is no particular belief regarding the sign of ρ, a two-tailed test is appropriate. In this case, the critical and indeterminate regions are combinations of the corresponding regions for the one-tailed tests. Table A.8 presents the d_l and d_u bounds for two-tailed tests at the .05 level of significance.

For example, we consider again the simple aggregate consumption function (6.10) reported in Table 11.3

$$\begin{array}{rl} \widehat{CON}_i = & \underset{(6.973)}{0.568} + \underset{(0.010)}{0.907}DPI_i \\ R^2 = .977 & SER = 8.935 \end{array} \tag{14.10}$$

for which $d^* = 0.66$. With $n = 25$ and $k = 1$, we see from Table A.7 that $d_l = 1.29$ and $d_u = 1.45$. Since $d^* < d_l$, we reject the null hypothesis of no autocorrelation and suspect that positive autocorrelation is present.

It should be noted that calculation of the test statistic d^* demands that the observations be ordered from 1 to n corresponding to the passing of time. If the observations are shuffled, or ordered some other way, then the calculated d^* will be a different value. Shuffling the observations causes the d^* value to be close to 2, leading to a conclusion of no autocorrelation. It should be clear, however, that shuffling the data does not remove autocorrelation: it merely invalidates the test procedure.

Finally, the validity of the Durbin–Watson test depends on the regressors being truly fixed. If a lagged value of the dependent variable is specified as a regressor, this assumption is no longer tenable. In this case d is biased toward 2. We note without much elaboration that in this situation a useful test for autocorrelation is based on the ***Durbin h statistic,*** which can be calculated as:

$$h = \left(1 - \frac{d}{2}\right)\sqrt{\frac{n}{1 - n \cdot V}} \tag{14.11}$$

where d is the Durbin–Watson statistic and V is the square of the standard error of the coefficient on the lagged dependent variable. The test can be used only if $n \cdot V < 1$. In large samples, h has approximately a standard normal distribution if the null hypothesis of no autocorrelation is true, and Table A.1 can be used to determine the critical value of the test statistic.

Estimation Procedures

Suppose that the Durbin–Watson statistic indicates that autocorrelation is present. What should we do?

The first-order autocorrelation model for simple regression, given earlier as (14.1) and (14.2) is

$$Y_i = \beta_0 + \beta_1 X_i + u_i \tag{14.12}$$

$$u_i = \rho u_{i-1} + \epsilon_i \tag{14.13}$$

in which ϵ_i satisfies the regular disturbance assumptions. In order to estimate (14.12) efficiently we begin with a hypothetical question: if we know the true value of ρ, how can we estimate β_0 and β_1? A technique known as ***generalized differencing*** provides an avenue for us to apply OLS to a modification of the original model and data.

The basic model (14.12) applies to every observation, including that for period $i - 1$, and the equation statement for that period can be multiplied through by ρ to yield

$$\rho Y_{i-1} = \rho\beta_0 + \rho\beta_1 X_{i-1} + \rho u_{i-1} \tag{14.14}$$

Subtracting (14.14) from (14.12) yields

$$[Y_i - \rho Y_{i-1}] = \beta_0(1 - \rho) + \beta_1[X_i - \rho X_{i-1}] + \epsilon_i \tag{14.15}$$

which holds validly for $i = 2, 3, \ldots, n$. Note that $u_i - \rho u_{i-1} = \epsilon_i$ by (14.13). If we consider the terms in square brackets to be a regressand and regressor, respectively, then (14.15) specifies a simple regression in which the intercept is $\beta_0(1 - \rho)$ and the slope is β_1.

Equation (14.15) provides the key to estimation. The fruit of its derivation is that we end up with a regression specification that has a regular disturbance, because ϵ_i satisfies all the regular assumptions. Now, since ρ is assumed known, we can construct the indicated regressand and regressor and apply OLS to this equation to get nearly efficient estimates of the parameters. From a practical viewpoint, the difficulty with this technique is that ρ is truly unknown. Several popular procedures that let the data help determine the value of ρ are in common use, and we outline two of them.

The ***Hildreth–Lu procedure*** closely mimics the generalized differencing technique. Since ρ is not known, a set of trial values that span the relevant range for ρ are used, one at a time, to construct the regressand and regressor and estimate (14.15). Then the estimated equation having the smallest sum of squared residuals is designated the final one. For example if positive autocorrelation is indicated, we could let $\rho^* = 0, .1, .2, \ldots, .9$, and 1.0 in successive trial estimations of (14.15). If more precision for the value of ρ is desired, more values of ρ^* can be used. Effectively, the procedure scans a grid of possible values of ρ and chooses the most promising candidate.

The ***Cochrane–Orcutt procedure*** begins by estimating (14.12) with OLS using the original data. The residuals from this estimation, e_i for $i = 1, 2, 3, \ldots, n$, are a set of data that can be used to estimate

$$e_i = r \cdot e_{i-1} + \epsilon_i' \tag{14.16}$$

by OLS using observations 2 through n (the first is lost in creating the lagged variable). This equation, which has no constant term, is a model of the first-order autocorrelation process specified by (14.13), except that we use residuals rather than actual disturbance values as variables. The estimate of r is taken as a proxy for ρ, and we denote this value by ρ^*.

The next step of the Cochrane–Orcutt procedure uses ρ^* as though it were the true ρ, and transforms the data on Y and X to get their generalized differences $[Y_i - \rho^* Y_{i-1}]$ and $[X_i - \rho^* X_{i-1}]$. Then the OLS estimates $\hat{\beta}_0$ and $\hat{\beta}_1$ from (14.15) are computed. This could be our end result, or we could continue the procedure by repeating all the steps: based on the parameter estimates $\hat{\beta}_0$ and $\hat{\beta}_1$, we calculate the residuals $e_i = Y_i - \hat{\beta}_0 - \hat{\beta}_1 X_i$; then we estimate a regression of the form (14.16); then using the estimate of r as ρ^*, we construct the new generalized difference regressand and regressor; finally, we use OLS on (14.15) to compute new $\hat{\beta}_0$ and $\hat{\beta}_1$. The procedure can be repeated until successive values of ρ^* converge.

Both the Hildreth–Lu and Cochrane–Orcutt procedures lead to consistent estimates of the regression parameters, as does OLS. In large samples these new procedures are more efficient than OLS. In small samples the new procedures are biased, while OLS is not, and they may have a larger mean squared error than OLS when $|\rho|$ is small. The new procedures provide a better basis for carrying out hypothesis tests, although the validity of the tests strictly holds only for large samples.

These procedures are illustrated by the reestimation of the simple consumption function (14.10), for which the Durbin–Watson statistic suggests positive autocorrelation. A series of Cochrane–Orcutt estimations yields $\rho^* = .6791, .6795, .6795, \ldots$, with no further change in ρ^* beyond the third iteration. The final estimated equation is

$$\begin{aligned}[Y_i - \rho^* Y_{i-1}] = \underset{(17.89)}{-2.21}(1 - \rho^*) + \underset{(0.023)}{0.912}[X_i - \rho^* X_{i-1}]\end{aligned} \tag{14.17}$$

The results of a coarse Hildreth–Lu scan are shown in Table 14.1. For this set of trial ρ values, the lowest *SSR* occurs with $\rho^* = .7$, and the corresponding equation is

$$\begin{aligned}[Y_i - \rho^* Y_{i-1}] = \underset{(19.19)}{-2.25}(1 - \rho^*) + \underset{(0.024)}{0.912}[X_i - \rho^* X_{i-1}]\end{aligned} \tag{14.18}$$

A second grid search for $\rho^* = .60, .61, \ldots, .70$ yields the smallest *SSR* for $\rho^* = .68$. In this case both procedures lead to essentially the same result; however, this is not always the case in practice.

TABLE 14.1
Hildreth–Lu Scan

ρ^*	SSR
.0	1836.3
.1	1619.9
.2	1437.6
.3	1289.4
.4	1175.4
.5	1095.7
.6	1050.5
.7	1040.1
.8	1065.0
.9	1122.7
1.0	1158.1

Since the coefficient estimates in the generalized difference equations are estimates of the underlying behavioral model (14.12), it is common and appropriate to report (14.17) as

$$\begin{aligned} \widehat{CON}_i &= \underset{(17.89)}{-2.21} + \underset{(0.023)}{0.912 DPI_i} \\ R^2 &= .997 \qquad SER = 9.010 \qquad \rho^* = .6795 \end{aligned} \tag{14.19}$$

The R^2 and SER here are based on the fit of (14.19) to the original 25 observations; these are different from the goodness-of-fit measures for (14.17), which has a different regressand. It should be noted that the estimated coefficients in (14.19) differ from those in (14.10), which ignored the autocorrelation problem. Also, we note that the standard errors in (14.19) are larger than those in (14.10). Presuming that the true model is the simple regression with first-order autocorrelation, we interpret the low standard errors in (14.10) to be invalid as a consequence of ignoring the autocorrelation.

These procedures extend directly to multiple regression. The standard model

$$Y_i = \beta_0 + \beta_1 X_{1i} + \beta_2 X_{2i} + \cdots + \beta_k X_{ki} + u_i \tag{14.20}$$

is adopted in conjunction with the first-order autocorrelation specification (14.13). The generalized differencing technique leads to the creation of the regressand $[Y_i - \rho Y_{i-1}]$ and the regressors $[X_{ji} - \rho X_{j(i-1)}]$, for $j = 1$ to k. The estimates of β_1 through β_k come directly from the OLS regression of the regressand on the regressors, and β_0 is recovered from the constant $\beta_0(1 - \rho)$. The Hildreth–Lu and Cochrane–Orcutt procedures can be used in practice.

In the two procedures outlined above, ρ is effectively estimated by scanning or iterating. A simpler two-step procedure suggested by Durbin involves: (1) estimating the original model by OLS, and taking $\hat{\rho} = 1 - d/2$ as an estimate of the correlation ρ, and (2) estimating a generalized difference model using this $\hat{\rho}$ to create the regressand and the regressors.

14.2 Heteroscedasticity

Heteroscedasticity is the situation occurring when the standard deviations of the disturbance terms are not identical for all observations. This most often arises in the analysis of cross-section data, although it may be present in time-series work also. Heteroscedasticity is of equal stature with autocorrelation as a theoretical and empirical problem, but in practice it is given considerably less attention. In large part this may be due to the wide variety of forms that it may take on and the resulting lack of standard procedures to deal with it.

The Model and Its Consequences

Consider the simple regression model

$$Y_i = \beta_0 + \beta_1 X_i + u_i \tag{14.21}$$

in which the disturbance u_i satisfies all but one of the regular assumptions of the normal regression model. We assume now that the standard deviation of the disturbance may take on a different value for each observation

$$\sigma(u_i) = \sigma_i \tag{14.22}$$

It could be that the value of σ_i is purely idiosyncratic, but only if there is a systematic relation between σ_i and some characteristics of the observation will it be possible to make econometric headway. For our discussion, it may be reasonable to suppose that σ_i is related to X_i, the value of the explanatory variable.

For example, consider a cross-section family consumption function, in which Y is consumption expenditures and X is family income. In order to clarify the presentation, temporarily suppose that only four different values of X are included in the sample. The theoretical determination of Y is illustrated in Figure 14.5a. As X increases, the standard deviation (σ_i) of the relevant $p(Y)$ distribution increases also. This makes sense for the example, because families have much greater latitude in making expenditure decisions when incomes are high than when they are low. Now, among families with income level X_1, the observed values of Y are (theoretically) a set of outcomes generated from "$p(Y)$ for X_1," like those illustrated by the points having X_1 in Figure 14.5a. For greater values of X, the observations on Y will be less concentrated around their mean, and the process will generate a set of data resembling that in Figure 14.5b.

If we apply OLS directly to the estimation of (14.21) while ignoring the assumption (14.22), the consequences are similar to those of autocorrelation. The OLS estimators are still unbiased and consistent. However, the estimators are no longer efficient, and the regular OLS calculations of the now-appropriate standard errors are wrong. This causes the hypothesis tests to be invalid. In

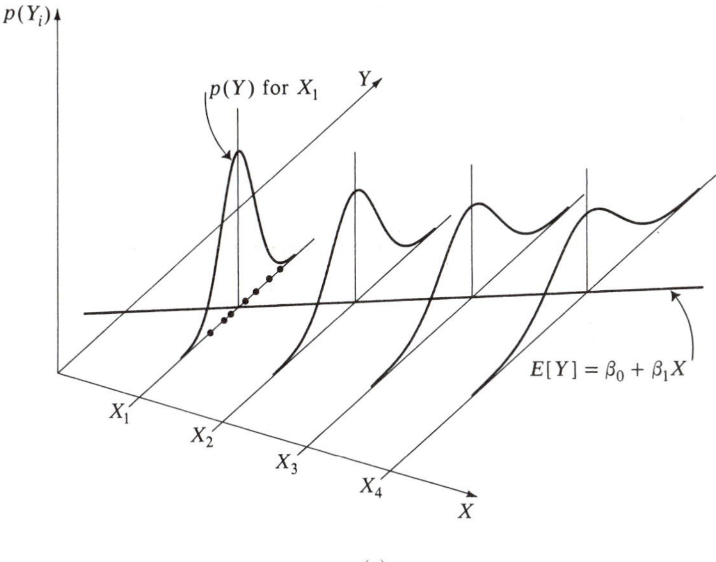

(a)

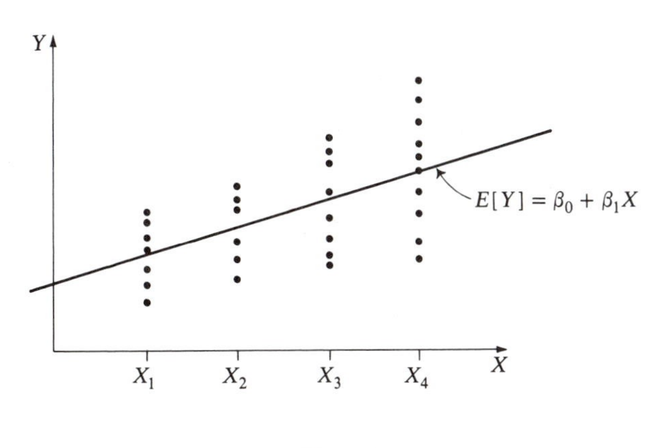

(b)

FIGURE 14.5 If the standard deviation of the disturbance increases with the value of X, we have a case of heteroscedasticity. The determination of Y in this case is illustrated in (a), which should be compared with Figure 11.2. The observed values of Y for several observations having X_1 in common are thought of as a set of outcomes from "$p(Y)$ for X_1." When only four distinguishable values of X are observed, this process leads to a set of data illustrated in (b).

practice, OLS can be applied and its results interpreted cautiously, but better procedures can be followed in many cases.

Detection

The essence of detecting heteroscedasticity involves estimating the basic model (14.21) with OLS and then examining the residuals for evidence regarding the relation between $\sigma(u_i)$ and the characteristics of the observations. It need not be true that the relevant characteristics are variables that appear in the model, but this is usually the case. In any event, it may be difficult to ascertain what the relation is.

Sometimes it is possible to proceed with an informal, graphical analysis of the residuals. For example, returning to the cross-section consumption function, suppose that (14.21) is estimated and then the residuals are plotted against the values of X, as in Figure 14.6. (We now suppose that X takes on many values, not just four.) It appears that the typical absolute value of e, $|e|$, is roughly proportional to X. This suggests that the standard deviation of the residuals associated with a given value of X is also roughly proportional to X. With this finding, we might be willing to say that (14.22) should be specified as

$$\sigma(u_i) = \sigma X_i \tag{14.23}$$

where σ now stands simply as a constant of proportionality.

This graphical method can be formalized by using regression methods to examine the relation between the residuals obtained from estimating (14.21) and

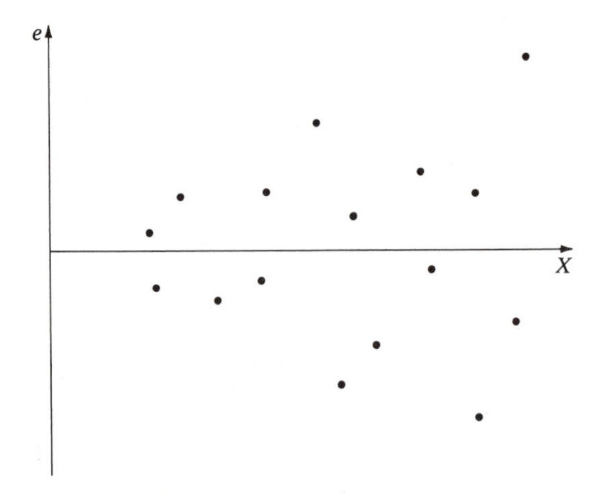

FIGURE 14.6 A heteroscedastic process in which $\sigma(u_i)$ increases with X leads to an estimated set of residuals like this. It appears that the typical absolute value of e increases with X.

the explanatory variable X. One suggestion is to explore a variety of alternative functional forms, such as

$$|e_i| = \beta_1 X_i + v_i \tag{14.24}$$

and

$$|e_i| = \beta_1 \sqrt{X_i} + v_i \tag{14.25}$$

In these second-stage regressions, the regressand is a data variable consisting of the absolute values of the residuals from the original regression. The regressor is simply the explanatory variable X or some transformation of it. The specification (14.24) or (14.25) would be appropriate if the plot of $|e|$ against X looks like Figure 14.7a or b, respectively; of course, not both of them could be correct in the same situation. If β_1 is statistically significant in these or other second-stage regressions, then we conclude that σ_i is not independent of X_i (i.e., that the disturbances are heteroscedastic). This procedure, known as the ***Glejser test,*** has the additional benefit of giving an indication of just how σ_i is related to X_i.

A similar formal method is known as the ***Park test*** for heteroscedasticity. To examine the relation between the residuals from the original regression and the explanatory variable X, Park suggests the functional form

$$\ln e_i^2 = \beta_0 + \beta_1 \ln X_i + v_i \tag{14.26}$$

As discussed in Chapter 6, the log-linear specification permits a wide variety of relations between e^2 and X to be discovered. If β_1 is statistically significant, we conclude that the disturbances are heteroscedastic.

The procedures suggested by Glejser and Park provide a formal structure of what is essentially an exploratory analysis. They serve as useful guidelines in uncharted territory, and the required second-stage regressions can be set up and estimated with only a moderate amount of effort in many econometric computer programs. From a statistical viewpoint, however, the functional forms are likely

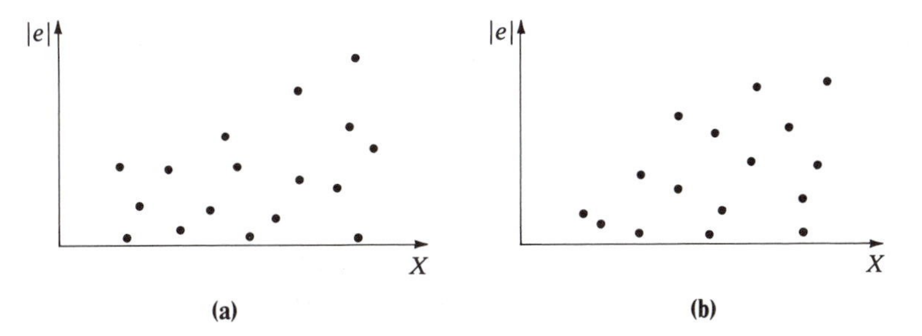

FIGURE 14.7 Various cases of heteroscedasticity show different patterns when the absolute value of the residual is graphed against X. In (a), $|e|$ tends to increase with X and the pattern may be specified by Equation (14.24). In (b), $|e|$ tends to increase with the square root of X, and the pattern may be specified by Equation (14.25).

to be misspecified and the disturbances v_i in (14.24)–(14.26) are themselves likely to be heteroscedastic. Hence we should think of these tests as giving indications rather than formal conclusions.

A more general procedure for detecting heteroscedasticity is the ***Breusch–Pagan test.*** This allows for more than one variable to be considered as the source of heteroscedasticity, but it does not attempt to estimate a specific form for the relation. As in the other tests, the first stage is to estimate the basic model (14.21) with OLS and then to calculate each of the residuals. The second stage is to estimate a linear regression in which the regressand is $e_i^2/(\Sigma\, e_i^2/n)$ and the k regressors are whatever variables are thought to possibly affect σ_i. If the null hypothesis of no heteroscedasticity is true, a test statistic equal to one-half of the explained variation of Y has a chi-squared distribution with k degrees of freedom in large samples. Values of the test statistic greater than a critical value indicate heteroscedasticity. [The "explained variation of Y," which was defined in connection with R^2 in (5.24), is equal to $\Sigma\,(\hat{Y}_i - \overline{Y})^2$.]

Estimation

Suppose we have concluded that the disturbances in the regression model (14.21) are heteroscedastic as specified by (14.23). In this particular case the technique for estimating, or reestimating, the coefficients in (14.21) is straightforward to derive and apply. The object is to respecify the original model in such a way that the resulting disturbances are ***homoscedastic*** (i.e., free from heteroscedasticity).

The first step is to divide through the original model (14.21) by X_i, the measure to which $\sigma(u_i)$ is proportional in the specification (14.23). Letting $\epsilon_i = u_i/X_i$, this yields

$$\left[\frac{Y_i}{X_i}\right] = \beta_0 \left[\frac{1}{X_i}\right] + \beta_1 + \epsilon_i \qquad (14.27)$$

which is a regression specification with regressand and regressor given in brackets. Note, however, that the original intercept β_0 in (14.21) is the slope coefficient in (14.27), and the original slope now appears as the intercept. It can be shown that

$$\sigma(\epsilon_i) = \sigma \qquad (14.28)$$

where σ is the constant proportionality factor from (14.23). Thus (14.27) specifies a regression model in which the disturbances satisfy all the regular assumptions, and OLS can be used to make estimates of β_0 and β_1 that are unbiased, efficient, and consistent. The regressand and regressor in (14.27) are easy to construct, and therefore the technique is easy to apply.

Other cases result when the particular specification of heteroscedasticity is different from (14.23), and the resulting final models differ from (14.27). The general procedure is to divide through the original model by the measure to which $\sigma(u_i)$ is proportional. This is known as ***weighted least squares,*** because the transformation procedure is similar to attaching a separate weight to each observation. The purpose of the transformation is to respecify the original model to yield a homoscedastic regression specification, because this permits the proper application of OLS to the task of estimating the coefficients of the original model.

Sometimes the best way to deal with heteroscedasticity is to think about it while formulating the original model. For example, if we are interested in family consumption behavior we might focus on the consumption-income ratio, rather than the consumption level, as the variable of interest. We would expect that the disturbances affecting this ratio would be relatively free of heteroscedasticity, especially in comparison with the disturbances affecting the consumption level. Note that (14.27) is a model explaining the consumption–income ratio; its reciprocal specification is consistent with the consumption level being linearly related to income. Sometimes it is suggested that a relation be specified in log-linear rather than linear form, in order to reduce the apparent heteroscedasticity. However, the choice of functional form should be made predominantly on the basis of other considerations.

Finally, it is useful to contrast our treatments of heteroscedasticity and autocorrelation. In dealing with autocorrelation we worked only with the first-order case, and alternative estimation procedures focused on estimating the parameter ρ. With heteroscedasticity, the difficult part is determining the case: in our example we let $\sigma(u_i) = \sigma X_i$, but $\sigma(u_i) = \sigma\sqrt{X_i}$ might be equally reasonable. The estimation procedures for these cases are similar, and there is no need to estimate σ beforehand. In other words, with heteroscedasticity the basic specification is always at issue and there is not much of an estimation problem, whereas with autocorrelation most researchers adopt the same specification and its estimation is the issue.

Problems

Section 14.1

14.1 Suppose that $\rho = .9$ in the first-order autocorrelation specification (14.2). If $u_1^* = 20$, what is the expected value of u_2 (i.e., $E[u_2]$)? Assuming that $\sigma(\epsilon_i) = 2$, name two values, u_2^* and u_3^*, that serve as good examples of results from this process.

⋆ **14.2** Repeat Problem 14.1 for $\rho = -.9$, a case of negative autocorrelation.

14.3 Draw a diagram in the spirit of Figure 14.1d that shows hypothetical data generated from a model with negative autocorrelation.

⋆ **14.4** Looking only at Table A.7, describe what happens to the sampling distribution of the Durbin–Watson statistic as the number of observations increases.

14.5 For $d^* = 1.50$, is positive autocorrelation indicated on a one-tailed test at the .05 level of significance if:
(a) $k = 1, n = 20$?
(b) $k = 4, n = 20$?
(c) $k = 1, n = 60$?
(d) $k = 4, n = 60$?

⋆ **14.6** Suppose that a time-series regression was estimated and it found $d^* = 1.60$; then a Hildreth–Lu scan determined that $\rho^* = .20$. Only after all this computing was it realized that the original d^* did not indicate autocorrelation. What results should be reported from this research?

Section 14.2

⋆ **14.7** Suppose that measurement error in economic data has decreased over time. How might this affect the task of estimating a simple time-series regression model?

14.8 In the derivation leading to Equation (14.27), show that the disturbance ϵ is homoscedastic.

14.9 Draw a figure illustrating the following heteroscedastic specifications:
(a) $|e_i| = \beta_1 X_i^2 + v_i$, for $\beta_1 > 0$
(b) $|e_i| = \beta_1(1/X_i) + v_i$, for $\beta_1 > 0$
(c) $|e_i| = \beta_0 + \beta_1 X_i + v_i$, for $\beta_1 > 0, \beta_0 > 0$
(d) $|e_i| = \beta_0 + \beta_1 X_i + v_i$, for $\beta_1 < 0, \beta_0 > 0$

⋆ **14.10** In the functional forms suggested by Glejser (and those in Problem 14.9), is it correct to assume that the disturbance v_i has a normal distribution?

⋆ **14.11** Suppose that $\sigma(u_i) = \sigma\sqrt{X_i}$. Starting from Equation (14.21), derive a final estimating equation with homoscedastic disturbances that is suitable for estimation by ordinary least squares.

14.12 Draw figures showing the relation between e_i^2 and X_i indicated by the following results of the Park test:
(a) $\hat{\beta}_1 = -0.5$
(b) $\hat{\beta}_1 = 0.5$
(c) $\hat{\beta}_1 = 2.0$

14.13 Suppose that we are formulating a model that explains the profitability of different firms. Is the specification more likely to be heteroscedastic if the dependent variable is the level of profits or the percentage rate of return on equity?

More Testing and Specification

This chapter extends our basic knowledge of regression in two separate directions. In the first section we explain a procedure for testing hypotheses about more than one coefficient. In the second section we analyze models that have a dummy variable representing the outcome of the economic process.

15.1 *F* Tests in Multiple Regression

Standard t tests regarding single β_j coefficients are by far the most commonly examined hypothesis tests in regression analysis. Sometimes, however, we are concerned with more than one coefficient, and the t tests are inadequate for our needs.

For example, Table 15.1 reports the results of five variations of our earnings function with the dependent variable in its logarithmic form, *LNEARNS*. In this table the results of each regression are shown in separate columns and the presence or absence of values in particular rows indicates whether or not a particular regressor is included in the equation. Regression (1), which replicates (6.40), includes only *ED* as an explanatory variable; regression (2) adds *EXP* in order to examine the linear effect of experience, and regression (3) lets experience enter quadratically; regression (4) adds a dummy variable for race; and regression (5) adds three dummy variables for regional location.

TABLE 15.1 Earnings Functions
(Dependent Variable: *LNEARNS*, $n = 100$)

	(1)	(2)	(3)	(4)	(5)
Constant	0.673	0.418	−0.0781	0.0746	0.119
	(0.158)	(0.289)	(0.390)	(0.3891)	(0.408)
ED	0.107	0.116	0.118	0.110	0.111
	(0.013)	(0.016)	(0.016)	(0.016)	(0.016)
EXP		0.00651	0.0542	0.0522	0.0500
		(0.00620)	(0.0263)	(0.0258)	(0.0260)
EXPSQ			−0.00103	−0.00101	−0.000942
			(0.00055)	(0.00054)	(0.000546)
DRACE				−0.381	−0.306
				(0.175)	(0.179)
DNCENT					0.0550
					(0.116)
DSOUTH					−0.135
					(0.119)
DWEST					−0.134
					(0.133)
R^2	.405	.412	.433	.460	.479
SER	0.446	0.446	0.440	0.432	0.431
SSR	19.5033	19.2844	18.6085	17.7241	17.0904

The significance of *ED* can be tested in each regression using a regular t test, as can the effect of *EXP* in regression (2). However, the effect of experience in regressions (3), (4), and (5) enters through two regressors: *EXP* and *EXPSQ*. The null hypothesis that experience has no effect on (the logarithm of) earnings, together with its contradictory alternative, is of the form

$$\begin{aligned} &H_0\colon \beta_2 = 0 \text{ and } \beta_3 = 0 \\ &H_1\colon \beta_2 \neq 0 \text{ and/or } \beta_3 \neq 0 \end{aligned} \tag{15.1}$$

which we have not seen before.

To preview the appropriate procedure for testing this null hypothesis, suppose that regression (3) in Table 15.1 is the basic specification that we believe characterizes the earnings function. The underlying regression model is

$$LNEARNS_i = \beta_0 + \beta_1 ED_i + \beta_2 EXP_i + \beta_3 EXPSQ_i + u_i \tag{15.2}$$

If the null hypothesis that $\beta_2 = \beta_3 = 0$ is true, the underlying regression model can be restated validly as

$$LNEARNS_i = \beta_0 + \beta_1 ED_i + u_i \tag{15.3}$$

In other words, the null hypothesis serves to restrict some of the coefficients in the underlying regression model. Equation (15.3) is referred to as the ***restricted*** form of the model, and (15.2) is the ***unrestricted*** form.

The essence of the test of the null hypothesis involves comparing the estimates of these two forms. In each regression, the sum of squared residuals (SSR) measures the overall error of fit. If the restricted form has a much poorer fit than the unrestricted form, this suggests that the restrictions (i.e., the implications of the null hypothesis) are quite different from the reality that generated the data. By contrast, if the error of fit of the restricted form is not much different from that of the unrestricted form, this suggests that the restrictions are fairly consistent with reality.

Theory

We start formally by considering a multiple regression model with regular assumptions:

$$Y_i = \beta_0 + \beta_1 X_{1i} + \beta_2 X_{2i} + \cdots + \beta_k X_{ki} + u_i \qquad (15.4)$$

Consider a null hypothesis stating that several of the coefficients take on specific values or that there exists a certain relation among the coefficients (to be elaborated by example below). The alternative hypothesis is the denial of any one or more of the elements of the null.

Now we consider estimating two regressions and comparing the sums of squared residuals

$$SSR = \sum_{i=1}^{n} e_i^2 \qquad (15.5)$$

from them. The first regression is of the full model (15.4) under consideration and the resulting sum of squared residuals is denoted by SSR_U, indicating that the estimation is unrestricted. The second regression results from somehow imposing a set of restrictions on the estimation in such a way that the null hypothesis is necessarily fulfilled. For example, in the earnings function (15.2) the null hypothesis that $\beta_2 = 0$ and $\beta_3 = 0$ is necessarily fulfilled by estimating (15.3). In general, other types of null hypotheses lead to other types of restriction. The restricted regression is estimated, yielding a certain sum of squared residuals denoted by SSR_R.

It turns out that SSR_R is always greater than SSR_U, unless they are equal by coincidence. Loosely speaking, this is because the unrestricted form allows the OLS technique to assign values freely to the $\hat{\beta}_j^*$ to make SSR as small as possible. By contrast the restricted form gives the technique less freedom to minimize the SSR, and thereby leads to a higher value for it. If the null hypothesis is true, SSR_R is likely to be only slightly greater than SSR_U because the estimated coefficients in the unrestricted form will differ from the values (perhaps implicit) in the restricted form only because of sampling variation. On the other hand, if the null hypothesis is false, SSR_R is likely to be much greater than SSR_U because the estimation of the restricted form must include (perhaps implicitly) some coefficients that are incorrect, leading to greater errors of fit.

Let n be the total number of observations, let k be the number of regressors in the unrestricted form, and let r be the number of restrictions. Generally, the number of restrictions, r, is equal to the number of fewer coefficients that are estimated in the restricted form as compared with the unrestricted form. We accept without derivation or proof the fact that if the null hypothesis is true, then the sampling statistic

$$F = \frac{(SSR_R - SSR_U)/r}{SSR_U/(n - k - 1)} \tag{15.6}$$

has an F distribution with r degrees of freedom in the numerator and $n - k - 1$ in the denominator (see the appendix to Chapter 10).

The value of F reflects the increase in SSR resulting from estimating the restricted form of the model rather than the unrestricted form, as can be seen from (15.6). Following standard testing procedures, we choose a critical region consisting of values of F that seem unlikely to occur if H_0 is true but that would occur because of sampling variability with probability α, which is the level of significance. If the value of the statistic F^* is in the critical region, the null hypothesis is rejected. The logic is that if the restrictions inherent in the null hypothesis are wrong, F^* will tend to have a high value. Hence, large values of F^* cast doubt on the null hypothesis, and the critical region consists of values of F greater than a critical value, F^c. This is illustrated in Figure 15.1. We

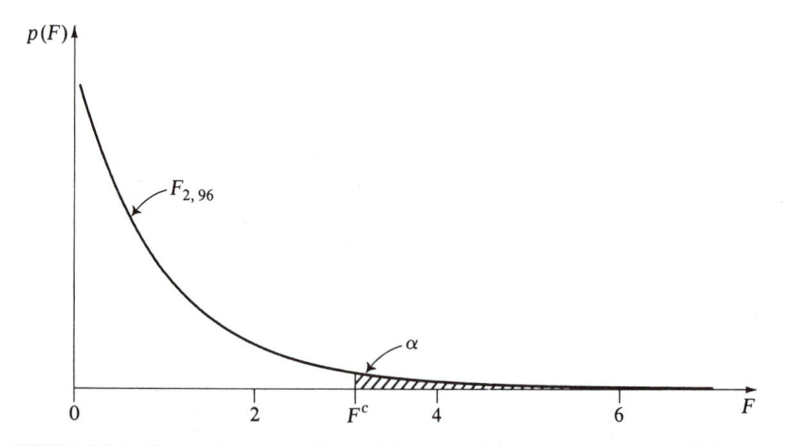

FIGURE 15.1 In carrying out an F test of linear restrictions on regression coefficients, the decision rule is to reject the null hypothesis if $F^* \geq F^c$. The critical value, F^c, is determined from Tables A.5 and A.6, for $\alpha = .05$ and $\alpha = .01$, respectively. The value of the test statistic is calculated according to Equation (15.9).

recognize that there is a chance of mistakenly deciding to reject the null hypothesis when in fact it is true: this chance is equal to α. Appropriately enough, this whole procedure is known as an ***F* test.**

In brief, the decision rule is of the form

$$\text{Reject } H_0 \text{ if } F^* \geq F^c \tag{15.7}$$

where the critical value F^c is determined from

$$\Pr(F \geq F^c) = \alpha \tag{15.8}$$

and where the test statistic is

$$F^* = \frac{(SSR_R - SSR_U)/r}{SSR_U/(n - k - 1)} \tag{15.9}$$

The specification of the hypotheses and formulation of the restricted and unrestricted regressions needed to calculate F^* vary from application to application. As in the case of t tests, it is a good idea to carry out the following five steps in conducting the test:

1. State the hypotheses clearly.
2. Choose the level of significance, α.
3. Construct the decision rule.
4. Determine the value of the test statistic, F^*.
5. State the conclusion of the test.

Most of the work comes in step 4.

Applications

The most common application of the F test involves the null hypothesis that in the regression model under consideration *all* the coefficients except the intercept are equal to zero:

$$\begin{aligned} &H_0\text{: } \beta_1 = \beta_2 = \cdots = \beta_k = 0 \\ &H_1\text{: } \beta_j \neq 0 \qquad \text{for at least one } j,\ j = 1, \ldots, k \end{aligned} \tag{15.10}$$

Essentially, H_0 says that all the regressors taken together have no ability to explain the dependent variable. This test is something of a straw man, in that we expect the null to be rejected in all but the most worthless of regression models.

The unrestricted form is the basic regression model under consideration and its sum of squared residuals is denoted by SSR_U. The restricted form, incorporating k restrictions, is simply

$$Y_i = \beta_0 + u_i \tag{15.11}$$

This regression need not be specially estimated because some algebraic manipulation shows that

$$F = \frac{(SSR_R - SSR_U/k}{SSR_U/(n - k - 1)} = \frac{R^2/k}{(1 - R^2)/(n - k - 1)} \quad (15.12)$$

where R^2 is from the unrestricted form. Accordingly, this common F test is sometimes said to be a test of the significance of R^2. It should be noted that it is possible for a regression to include at least one variable that is significant by its own t test and still not lead to rejection of the null hypothesis that the coefficients of all the regressors are equal to zero.

A second type of application involves a null hypothesis that *some* of the coefficients are equal to zero. For example, we return to the example of testing experience in the polynomial form of the earnings function. The relevant hypotheses are

$$\begin{aligned} H_0&: \beta_2 = \beta_3 = 0 \\ H_1&: \beta_2 \neq 0 \quad \text{and/or} \quad \beta_3 \neq 0 \end{aligned} \quad (15.13)$$

Note that H_0 is specific and that it formalizes the statement that experience has no effect on earnings in this model. Also, H_1 is vague and it is the simplest denial of H_0. We choose the .05 level of significance. The unrestricted form of the model is (15.2) and the restricted form is (15.3). There are $r = 2$ restrictions imposed in the restricted form, there are $k = 3$ regressors in the unrestricted form, and there are the same $n = 100$ observations in both forms. The decision rule is to reject H_0 if $F^* \geq 3.1$ approximately, as determined from Table A.5. In Table 15.1 the restricted form is reported as regression (1) and the unrestricted form is regression (3). Thus the value of the test statistic is

$$F^* = \frac{(19.5033 - 18.6085)/2}{18.6085/96} = 2.308 \quad (15.14)$$

Since F^* is not in the critical region we do not reject the null hypothesis, and we conclude that experience has an insigificant effect on the logarithm of earnings.

However, the P-value is

$$P\text{-value} = \Pr\,(F_{2,96} \geq 2.308) = .105 \quad (15.15)$$

(a computer is necessary to calculate this) which indicates that experience just misses being significant at the .10 level. This suggests that we should not simply discard experience as a relevant variable.

The results of an F test may seem to contradict those for t tests on the individual coefficients. Despite the fact that in regression (3) *EXP* is significant at the .05 level and *EXPSQ* is significant at the .10 level, the F test has shown that experience is not significant at the .10 (or .05) level. The F test is different from a simple combination of two t tests, and the results can be different. Note, however, that all the test statistics here are fairly close to the critical values and the differences between the statistical findings of the F and t tests are not really very great.

A similar test procedure can be used to test whether there are differences among the different categories of a single categorical variable. For example, we can test whether the region of residence is a significant determinant of earnings. The hypotheses are formulated as

$$\begin{aligned} &H_0\colon \beta_5 = \beta_6 = \beta_7 = 0 \\ &H_1\colon \beta_5 \neq 0 \quad \text{and/or} \quad \beta_6 \neq 0 \quad \text{and/or} \quad \beta_7 \neq 0 \end{aligned} \tag{15.16}$$

The unrestricted form of the model must include the three regional dummy variables; the restricted form is the same except that the three dummy variables must be excluded. In Table 15.1, if regression (5) is considered the relevant model, then (4) is its restricted form. Tests of the significance of dummy variables are related to the analysis of variance, which is discussed in Chapter 19.

Also, it should be noted that the simple set of hypotheses

$$\begin{aligned} &H_0\colon \beta_j = 0 \\ &H_1\colon \beta_j \neq 0 \end{aligned} \tag{15.17}$$

can also be handled in the F test framework. Of course, we also recognize this as the basic significance test involving a t test. It turns out, nicely, that in this case F values correspond to the square of t values, and the conclusions of the two tests are identical.

A third type of application involves testing a hypothesis that specifies a relation among coefficients but not particular values for them. For example, suppose that we believe in a linear aggregate consumption function but want to examine whether or not the marginal propensity to consume labor income is the same as that for property income. We might have in mind a regression model of the form

$$CON_i = \beta_0 + \beta_1 LABINC_i + \beta_2 PROPINC_i + u_i \tag{15.18}$$

The null hypothesis and its alternative are formulated as

$$\begin{aligned} &H_0\colon \beta_1 = \beta_2 \\ &H_1\colon \beta_1 \neq \beta_2 \end{aligned} \tag{15.19}$$

If the null hypothesis is true, then $\beta_1 = \beta_2$, and we can let β_1 stand for their common value. Imposing this restriction on the basic model, we see that

$$\begin{aligned} CON_i &= \beta_0 + \beta_1 LABINC_i + \beta_1 PROPINC_i + u_i \\ &= \beta_0 + \beta_1 [LABINC_i + PROPINC_i] + u_i \end{aligned} \tag{15.20}$$

or

$$CON_i = \beta_0 + \beta_1 TOTINC_i + u_i \tag{15.21}$$

The restricted form of the model is simply the regression of *CON* on a newly constructed regressor equal to *LABINC* plus *PROPINC* for every observation. The number of restrictions, r, is equal to 1 because there is one fewer coefficient to estimate in (15.21) than in (15.18). We proceed with the F test as before, comparing SSR_R from (15.21) with SSR_U from (15.18).

All the F tests described in this section come under the general heading of ***tests of linear restrictions*** on the coefficients, and still other types of application are possible. For example, one might be interested in testing the proposition that $\beta_1 + \beta_2 = 1$ in some regression model. Given our current knowledge, all we need to do is formulate a restricted version of the model that incorporates this hypothesis.

The Chow Test

A fourth type of application is known as the ***test for equality of coefficients*** or the ***Chow test*** (after its originator, Gregory Chow). When the observations under study naturally fall into two or more groups, the question arises as to whether or not all the groups are subject to the same economic process. This, of course, is the essential criterion governing the selection of observations to be used in the estimation of a model. For example, in time-series work it might be believed that the whole consumption function permanently changed in a certain year; this establishes two groups, the ''before'' and the ''after'' observations. In cross-section work it might be believed that the earnings function is substantially different for blacks and whites; this establishes two groups based on race.

We outline the test when the number of groups is two, but it may be extended to any larger number. If we estimate a multiple regression model (15.4) for a complete set of observations, we implicitly restrict the intercept and the effect of each of the regressors to be the same for all observations. But if we estimate the same model specification separately for each group, we allow the intercept and the effect of each of the regressors to differ between the groups. The test for equality of coefficients considers the situation of possibly unequal coefficients for the groups to be the general, unrestricted model. The null hypothesis that every β_j has the same value for both groups is imposed on the general model by estimating the model just once, with the complete set of observations; this is the restricted form of the model.

In carrying out the test, SSR_U is obtained by adding the SSRs from the two separate regressions, and the corresponding number of degrees of freedom is equal to the number of observations minus the number of parameters estimated (counting both equations). The SSR_R is obtained directly from the combined regression, and the number of restrictions is equal to the number of fewer parameters in the restricted regression as compared with the total number of parameters in the unrestricted regressions. An underlying assumption of the test is that the disturbances are identical for the groups, even if the β_j coefficients are not.

For example, we may accept the general validity of the earnings function (15.2) but wonder whether the coefficients are the same for blacks and whites. The unrestricted form, allowing for differences, consists of two estimated regressions: for blacks,

$$\begin{aligned}\widehat{LNEARNS_i} = & -\underset{(2.282)}{1.408} + \underset{(0.083)}{0.129}ED_i + \underset{(0.184)}{0.123}EXP_i \\ & - \underset{(0.00377)}{0.00220}EXPSQ_i \end{aligned} \tag{15.22}$$

$$R^2 = .562 \qquad SER = 0.473 \qquad SSR = 0.6713$$

and for whites,

$$\begin{aligned}\widehat{LNEARNS_i} = & \underset{(0.401)}{0.104} + \underset{(0.016)}{0.110}ED_i + \underset{(0.0264)}{0.0508}EXP_i \\ & - \underset{(0.000559)}{0.000995}EXPSQ_i \end{aligned} \tag{15.23}$$

$$R^2 = .391 \qquad SER = 0.437 \qquad SSR = 16.9793$$

Thus, $SSR_U = 0.6713 + 16.9793 = 17.6506$. The restricted form is regression (3) in Table 15.1. The coefficients seem to differ substantially, but this may be due just to sampling variability. The test statistic is

$$F^* = \frac{(18.6085 - 17.6506)/4}{17.6506/92} = 1.25 \tag{15.24}$$

The critical value at the .05 level of significance is $F^c \approx 2.47$, so the null hypothesis is not rejected. We judge that the earnings functions are not significantly different for the two races. It should be noted, however, that regression (4) in Table 15.1 shows a significant effect for the dummy variable *DRACE*.

In time-series work, the Chow test serves as a ***test of structural stability.*** The notion of stability here corresponds to our usual presumption that the true coefficients in the structural relation maintain the same values over all the time spanned by the data set. Such stability is of fundamental importance for our regression methods, for only if the relation is stable could all the data come from the same process. Of course, it is reasonable that economic processes might change over time (i.e., that they might be unstable), and whether a given set of data satisfy the essential criterion is always open to question and testing.

The most common form of the test divides the sample into two subperiods. The null hypothesis is that the coefficients of the basic structural relation are the same in both subperiods, and this leads to a restricted regression. The alternative hypothesis is that the corresponding coefficients are different in the two subperiods, and this leads to the unrestricted form. (Note that this is a very special characterization of the lack of stability: we are not considering a drift or a randomness for the coefficients.)

For example, consider the proposition that the coefficients of the simple logarithmic demand for money function

$$LNM_i = \beta_0 + \beta_1 LNGNP_i + u_i \tag{15.25}$$

were different in the 1971–1980 period from what they were in the 1956–1970 period. The null hypothesis is that there was structural stability:

$$\begin{aligned} H_0&: \beta_0^{1956-1970} = \beta_0^{1971-1980} \text{ and } \beta_1^{1956-1970} = \beta_1^{1971-1980} \\ H_1&: \beta_0^{1956-1970} \neq \beta_0^{1971-1980} \text{ and/or } \beta_1^{1956-1970} \neq \beta_1^{1971-1980} \end{aligned} \tag{15.26}$$

The restricted form of the model, which embodies the null hypothesis, was first reported as (6.33) and is given again as regression (1) in Table 15.2. The unrestricted form of the model consists of two separate regressions, one for the 1956–1970 period and the other for 1971–1980. These are reported as regressions (2) and (3) in Table 15.2, and SSR_U is the sum of their *SSR*s.

In this test the number of restrictions is $r = 2$ and there are a total of $25 - 4 = 21$ degrees of freedom in the unrestricted form. Choosing $\alpha = .05$ as the level of significance, the critical value for $F_{2,21}$ is 3.47. From Table 15.2 we see that $SSR_R = 0.02140$ while $SSR_U = 0.004727 + 0.006198 = 0.01092$. Hence the test statistic is

$$F^* = \frac{(SSR_R - SSR_U)/r}{SSR_U/(n - k - 1)} = \frac{(0.02140 - 0.01092)/2}{0.01092/21} = 10.08 \tag{15.27}$$

TABLE 15.2 Demand for Money—Chow Test (Dependent Variable: *LNM*)

	(1) $i = 1$ to 25	(2) $i = 1$ to 15	(3) $i = 16$ to 25	(4) $i = 1$ to 25
Constant	3.948	3.426	6.138	3.426
	(0.165)	(0.193)	(0.695)	(0.230)
LNGNP	0.215	0.292	−0.0910	0.292
	(0.024)	(0.029)	(0.0968)	(0.034)
D				2.712
				(0.614)
$D \cdot LNGNP$				−0.383
				(0.086)
R^2	.780	.890	.100	.888
SER	0.0305	0.0191	0.0278	0.0228
SSR	0.02140	0.004727	0.006198	0.01092
n	25	15	10	25

Since $F^* > F^c$, the null hypothesis is rejected and we conclude that the relation was not stable over the whole period.

In assessing this result, we note that it could arise because of instability of the coefficients in (15.25), or it could arise because that equation does not give the true relation. Whatever the cause, the results of the Chow test indicate that the estimated regression (6.33) should be viewed with caution.

Although in this example and the previous one the unrestricted form of the model was estimated with two equations, it is possible to estimate this form in a single equation with the help of dummy variables. In the demand for money example, let $D_i = 1$ for the latter period and $D_i = 0$ for the earlier period. Consider the regression specification

$$LNM_i = \beta_0 + \beta_1 LNGNP_i + \beta_2 D_i + \beta_3 [D_i \cdot LNGNP_i] + u_i \quad (15.28)$$

where $[D_i \cdot LNGNP_i]$ is a single regressor that is constructed as the product of two other regression variables. When this specification is estimated for the whole sample period ($n = 25$), the fit is precisely the same as that of the two separate regressions of the unrestricted form. The results are reported as regression (4) in Table 15.2. We see that $\hat{\beta}_0^*$ and $\hat{\beta}_1^*$ in regression (4) are precisely the same as the estimated intercept and slope in the separate regression (2) for 1956–1970. Also, $(\hat{\beta}_0^* + \hat{\beta}_2^*)$ from regression (4) is the same as the estimated intercept in regression (3) for 1971–1980, and $(\hat{\beta}_1^* + \hat{\beta}_3^*)$ from regression (4) is the same as the corresponding estimated slope in regression (3). Further, SSR in regression (4) is equal to the sum of the SSRs from regressions (2) and (3), so it is precisely the SSR_U that is required for the Chow test.

This dummy variable estimation procedure can be used for any Chow test. In the demand for money example, the restricted form (15.25) is obtained from the unrestricted form (15.28) by imposing the null hypothesis H_0: $\beta_2 = \beta_3 = 0$. Thus, the Chow test is another variant of the tests for linear restrictions. Note that the proper counting of the degrees of freedom for the F statistic should conform to the dummy variable procedure whether or not it is used in the Chow test; the use of n and k may be confusing when the unrestricted form is estimated as two separate regressions.

15.2 Dummy Dependent Variables

It sometimes is the case that the economic process under study leads to outcomes that are categorical rather than continuous in nature. For example, we might be interested in the process determining whether a person takes private or public transportation to work, whether a firm goes bankrupt or not, or whether a high school senior goes to college or not. In each of these cases the outcome of the process can be coded as a dummy variable, with one outcome arbitrarily assigned the value 0 and the other assigned the value 1.

It is natural to try to represent this behavioral process in a regression model.

As a simple example, we consider a model that explains the labor force participation decision of individual married women on the basis of the wage available to them. Suppose that we have sufficient data to create two variables as follows. Let $Y = 1$ if the woman is in the labor force, 0 if she is not. Let X be the wage she does or could command in the market. The model

$$Y_i = \beta_0 + \beta_1 X_i + u_i \tag{15.29}$$

can be estimated by ordinary least squares.

If we look at a hypothetical scatter plot of the data and a graph of the fitted regression line, given in Figure 15.2, we see immediately that something special is at work. First, while the observed values of Y are all either 0 or 1, the fitted values take on a continuum of values from less than 0 to more than 1. Clearly, we need a special framework to explain the model and interpret the results. Second, it is apparent that the disturbance terms in the regression are of a very special form; certainly they are not normally distributed. Hence the properties of OLS and the possibilities for statistical inference must be reassessed.

Recall from Section 11.1 that the normal regression model is built up from an analysis of the determination of Y_i for an individual. As before, we temporarily let Y_i denote both the name of the random variable and its outcome in data. In the normal regression model, Y_i is assumed to have a normal distribution with a mean of $\beta_0 + \beta_1 X_i$ and a standard deviation σ_u. Clearly, the assumption of normality is no longer tenable when the dependent variable takes on only the values 0 and 1.

The usual interpretation given to (15.29) in the present context is that it is a ***linear probability model.*** Since Y_i can take on only the values 0 and 1, it has a ***Bernouilli distribution***—a binomial distribution with $n = 1$ (see Section 9.3).

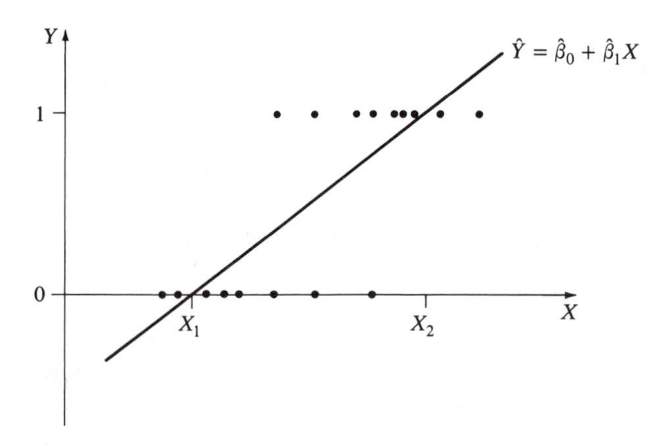

FIGURE 15.2 When the dependent variable is a dummy variable, which takes on values of only 0 or 1, the scatterplot of the data takes on this special form. The regression is interpreted as a linear probability model, and $\hat{Y}$ values for X between X_1 and X_2 are thought of as predicted probabilities. This interpretation breaks down, however, when $X < X_1$ or $X > X_2$.

Let P_i denote Pr $(Y_i = 1)$. In the linear probability model interpretation of (15.29), P_i is given by the systematic part of the behavioral process

$$P_i = E[Y_i] = \beta_0 + \beta_1 X_i \tag{15.30}$$

The slope coefficient β_1 shows the impact of a unit change in the wage on the probability of labor force participation. If β_1 is positive, for example, then the higher the wage the more likely it is that a woman will seek to work. A difficulty with this interpretation can be seen in Figure 15.2. For values of X below X_1 or above X_2, the predicted probabilities are outside the usual 0 to 1 range.

Returning to thinking of Y_i as having a Bernouilli distribution, we see that since $E[Y_i] = P_i$, then by (9.16) $\sigma(Y_i) = \sqrt{P_i(1 - P_i)}$. Now, the disturbance in (15.29) is simply $u_i = Y_i - E[Y_i]$, so the standard deviation of u_i is the same as that of Y_i. Since $\sigma(Y_i)$ depends on X_i through P_i, so does $\sigma(u_i)$. Hence the disturbances have different standard deviations: they are heteroscedastic.

Also, since the disturbances clearly are not normal, we cannot apply our usual methods of statistical inference in small samples. Despite these difficulties arising from the nature of the disturbance, it is fairly common to estimate linear probability models with ordinary least squares.

Another approach to estimating a model of a categorical economic process can be introduced by grouping the original data on X in a manner similar to that in Chapter 4. For each group, the values of X are now represented by the class mark, X_k. Also for each group, we calculate the proportion of its observations for which $Y_i = 1$, and denote this proportion by Y_k. In our example, Y_k is the labor force participation rate in each wage-defined group of women. For each group, the proportion Y_k can be thought of as an estimate of the probability P_i for each of its members.

As a result of this grouping we have m observations (m is the number of groups), each of the form (X_k, Y_k), and these data are plotted in Figure 15.3. Our task is to determine the structural relation between Y and X. One possibility is the simple regression model

$$Y_k = \beta_0 + \beta_1 X_k + u_k \tag{15.31}$$

This has many of the same pitfalls as the linear probability model (15.29). Also, in Figure 15.3a we see that the linearity of the systematic part of (15.31) does not very well describe the snakelike pattern of the data. Clearly, some nonlinear functional form would be preferable, but none that we have studied so far takes on this shape.

Logit Models

A functional form that is well suited to this problem is the ***logistic function,*** which is given by

$$Y = \frac{1}{1 + \exp\left[-(\beta_0 + \beta_1 X)\right]} \tag{15.32}$$

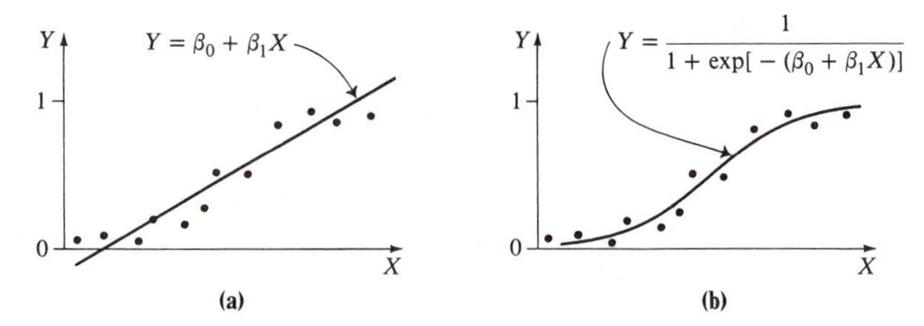

FIGURE 15.3 When observations with dummy dependent variables are grouped, Y takes on values between 0 and 1. We consider a process in which the systematic relation between Y_k and X_k takes on a snakelike pattern. In (a), a simple linear function is drawn through the points, but it does not describe the pattern very well. In (b), a better fit results from a logistic specification of the relation between Y_k and X_k.

This function has a minimum value of 0 and a maximum of 1, and it can take on the shape underlying our hypothetical data, as in Figure 15.3b. Equation (15.32) can be rearranged to yield

$$\ln\left(\frac{Y}{1 - Y}\right) = \beta_0 + \beta_1 X \tag{15.33}$$

The term on the left-hand side of (15.33) is known as the ***logit*** of Y. Since we have data on Y for each group, we can construct the values of the regressand

$$LOGITY_k = \ln\left(\frac{Y_k}{1 - Y_k}\right) \tag{15.34}$$

and formulate a regression model

$$LOGITY_k = \beta_0 + \beta_1 X_k + u_k \tag{15.35}$$

which includes a disturbance term for econometric reality. As in earlier chapters, we explicitly denote the regressand in (15.35) by a symbolic name in order to stress that it is a constructed variable, but the expression on the right-hand side of (15.34) is used commonly instead. This regression model can be estimated with ordinary least squares. Note that if $Y_k = 1$ or $Y_k = 0$, the logit is not defined; one possibility is to regroup the data into larger groups to avoid this.

For example, suppose that X is the hourly wage, measured in dollars. A hypothetical estimation of the logistic labor force participation rate model (15.35) is

$$\widehat{LOGITY_k} = -4.5 + 0.75X_k \tag{15.36}$$

To interpret the meaning of this regression it is necessary to examine the predictions that it yields.

The predicted values of $LOGITY_k$ and the corresponding values of the predicted labor force participation rates are given in Table 15.3 for wages in the range from 2 to 10 dollars. The predicted values of $LOGITY_k$ are derived from

TABLE 15.3 Labor Force Participation Rate—Predictions from the Logic Model

X_k	$LOGITY_k$	$\hat{Y}_k$
2.00	−3.00	0.047
3.00	−2.25	0.095
4.00	−1.50	0.182
5.00	−0.75	0.321
6.00	0.00	0.500
7.00	0.75	0.679
8.00	1.50	0.818
9.00	2.25	0.905
10.00	3.00	0.953

(15.36) and the values of Y_k from (15.34) or (15.32). Looking at these predictions, we see that the impact on Y resulting from equal-sized increments in X is small when X is small, large when X is middle-sized, and small again when X is large. A one-dollar increase in the wage leads to a 0.048 increase in $\hat{Y}_k$ when $X_k = 2$, a 0.179 increase in $\hat{Y}_k$ when $X_k = 6$, and a 0.087 increase in $\hat{Y}_k$ when $X_k = 8$. The numerical findings in this hypothetical example are not fully plausible, in part because labor force participation depends on more than just the wage rate. However, it serves to illustrate that the technical features of the logistic functional form make it appropriate for many economic processes whose outcomes are categorical.

We can think now of trying to use this logit model with individual observations rather than grouped data. In other words, we can formulate a logistic probability model

$$\ln\left(\frac{Y_i}{1 - Y_i}\right) = \beta_0 + \beta_1 X_i + u_i \tag{15.37}$$

Figure 15.4 shows what the individual data are like and what the fit would be. Unfortunately, ordinary least squares cannot be used to estimate the parameters

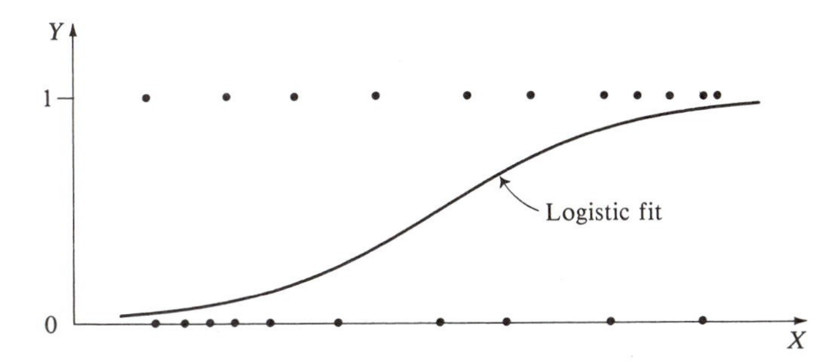

FIGURE 15.4 When observations with dummy dependent variables are treated individually, we might believe that the systematic part of the relation between Y and X is a logistic function. We cannot use OLS to estimate the transformed linear logit model, however, because the logit is not defined when $Y_i = 0$ or 1. More advanced estimation methods are available for this purpose.

β_0 and β_1 in the model, because the regressand cannot be constructed. Since each $Y_i = 0$ or 1, calculating the logit involves taking the logarithm of zero or infinity. Estimation methods that skirt this difficulty are available, but they are beyond the scope of this book. The results of these more advanced methods are often quite similar to those obtained by application of OLS to the linear probability model, and this allows a fair amount of confidence to be placed in its use.

Problems

Section 15.1

15.1 Use an F test to judge whether the explanatory variables in regression (2) of Table 15.1 help determine (the logarithm of) earnings.

⋆ **15.2** Based on regressions (5) and (4) in Table 15.1, test whether the region of residence significantly affects (the logarithm of) earnings.

15.3 Use an F test to judge whether *EXP* has a significant effect on (the logarithm of) earnings in regression (2) of Table 15.1.

15.4 Use a basic t test (two-tailed) to judge whether *EXP* has a significant effect on (the logarithm of) earnings in regression (2) of Table 15.1. Compare the squares of t^* and t^c with the corresponding F values in Problem 15.3.

⋆ **15.5** Suppose that we wish to test the hypothesis that the marginal propensity to consume out of property income is equal to 90 percent of the marginal propensity to consume out of labor income. Formulate the hypotheses and set up the procedure for carrying out this test.

15.6 Suppose that a dummy variable technique is used to combine the estimates of Equations (15.22) and (15.23). Explain how this regression is set up and determine each of its estimated coefficients.

⋆ **15.7** Explain intuitively why the standard errors in a dummy variable regression used for estimating the unrestricted form in a Chow test might be different from the corresponding standard errors in separate regressions. (Notice that they *are* in Table 15.2.)

Section 15.2

15.8 Consider the disturbance u_i in a linear probability model. Graph its probability distribution. Using the basic methods for discrete probability distributions, show that if $P_i = \beta_0 + \beta_1 X_i$, then $E[u_i] = 0$.

⋆ **15.9** On one set of axes, graph the probability distributions of the disturbances corresponding to two different values of X_i in the linear probability model.

★ **15.10** In the grouped data problem analyzed with a linear model in Figure 15.3a, do the disturbances appear to be continuous or discrete? Might they be normal?

★ **15.11** Consider the linear estimation illustrated in Figure 15.3a. Is this a case of autocorrelation?

15.12 Derive the logit formulation in Equation (15.33) from the logistic function in Equation (15.32).

15.13 Based on the estimated logit model in Equation (15.36), what is the predicted labor force participation rate of women whose wage would be $5.50 per hour?

Regression and Time Series

This chapter examines two topics that apply to regression models of time-series processes. The first section, on distributed lags, deals with the specification of regression models in which the effects of the explanatory variable are spread over more than one period. The second section covers the use of regression for estimating trend and seasonality patterns in time series.

16.1 Distributed Lags

An important element in the specification and estimation of regression models is that the value of the dependent variable is determined by the values of the regressors and the disturbance for that observation. For time-series work, in which observations are usually sequential time periods, this means that the dependent variable is determined by contemporaneous values of the regressors. This would seem to rule out intertemporal patterns of causation, but in this section we see that this is not the case.

The General Distributed Lag

In economic theory it makes sense that a given time-series variable should depend on previously occurring as well as contemporaneous values of its determinants. A process of decision making might well take into account the whole history of the determinant variables. For example, if persons are deciding how much to spend on consumption goods, they might take into account previous periods' incomes as well as the current period's amount. Sometimes economic processes have a gestation period that is longer than the time interval used to create the observations. For example, the production and sale of certain machine tools and other capital goods takes a long time; often the producing firms have backlogs of orders. Thus the firm's decision to purchase such equipment may occur a year or more before the actual purchase and delivery. This lag between decision and its measured fulfillment must be taken into account in modeling investment behavior.

Formally, we let Y be the dependent variable and X be its determinant. Suppose that the effects of X on Y do not all occur immediately, but that the total effect is distributed over several time periods. One way to specify the determination of Y is

$$Y_i = \beta_0 + \beta_1 X_i + \beta_2 X_{i-1} + \beta_3 X_{i-2} + \cdots + \beta_k X_{i-(k-1)} + u_i \quad (16.1)$$

which is known as a general ***distributed lag*** model. In this notation the current time period is i, the previous period is $i - 1$, and so on. Thus, (16.1) specifies that in period i, Y is determined by the contemporaneous value of X as well as the $k - 1$ previous values of X. In other words, the effect of X on Y is distributed over k periods.

Let us pursue a simple case, in which only two lagged values occur:

$$Y_i = \beta_0 + \beta_1 X_i + \beta_2 X_{i-1} + \beta_3 X_{i-2} + u_i \quad (16.2)$$

The effect of X on Y occurs partly with a lag, and it is distributed over three periods. For example, the idea that consumers' expenditures depend on their permanent incomes is sometimes formulated by hypothesizing that consumers have a three-period decision-making horizon: they view their "permanent incomes" as being a weighted average of three sequential years' incomes. In this example, Y is expenditures and X is income.

Some hypothetical data on Y and X are given in Table 16.1. As written, the model specified as (16.2) seems to involve just two variables, Y and X, but with multiple values of X. For example, the value $Y_3 = 415$ is determined by the contemporaneous value $X_3 = 475$, and by the previous values $X_2 = 425$ and $X_1 = 390$. This is not in regular regression form, which specifies a contemporaneous relation between a dependent variable and a set of regressors. The leap between the two specifications is made by constructing two new regressors, *XLAG1* and *XLAG2*, that contain the information of what the value of X was one and two periods previously, respectively. In Table 16.1 the values of *XLAG1* are constructed simply as the previous period's X value, and those of *XLAG2* are the values from two periods before. In (16.2), X_{i-1} corresponds to the ith value of *XLAG1* defined in Table 16.1, and X_{i-2} corresponds to $XLAG2_i$. Thus, for practical purposes we can think of rewriting (16.2) as

$$Y_i = \beta_0 + \beta_1 X_i + \beta_2 XLAG1_i + \beta_3 XLAG2_i + u_i \quad (16.3)$$

TABLE 16.1 Hypothetical Data on Y and X

i	Y	X	$XLAG1$	$XLAG2$
1	350	390	—	—
2	380	425	390	—
3	415	475	425	390
4	445	505	475	425
.				
.				
$i-2$	Y_{i-2}	X_{i-2}	$XLAG1_{i-2}$	$XLAG2_{i-2}$
$i-1$	Y_{i-1}	X_{i-1}	$XLAG1_{i-1}$	$XLAG2_{i-1}$
i	Y_i	X_i	$XLAG1_i$	$XLAG2_i$
.				
.				
n	Y_n	X_n	$XLAG1_n$	$XLAG2_n$

Notice that we start with n observations on Y and X, but that we have only $n - 2$ observations with which to estimate (16.3): the first two observations do not have values for $XLAG2$, and hence they cannot be used in the estimation. In this sense, constructing lagged variables leads to a loss of observations.

The actual reformulation of a distributed lag specification as (16.3) usually is understood implicitly in writing a model with notation like (16.2). With this understanding, equations like (16.2) should be viewed as properly specified regression models.

The relative importance of the current and lagged values of X in the determination of the current value of Y is given by the relative magnitudes of their regression coefficients. For example, consider a general distributed lag model like (16.1) with $k = 5$. A convenient way to compare the coefficients is in a graph like Figure 16.1. Here the values of the coefficients, β_j, are plotted

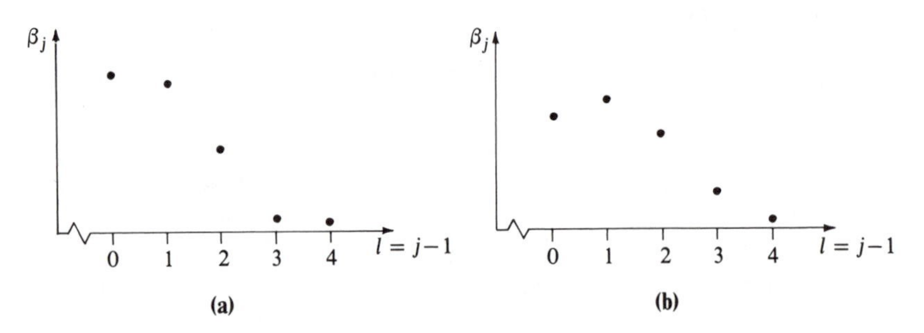

FIGURE 16.1 A graphical display of the values of the coefficients in a distributed lag model facilitates interpretation of the lag structure. The horizontal axis identifies the length of the lag, l. In (a), the one-period lagged value of X has almost as much impact on Y as the current-period X does, the second-period lag has half as much, and the third and fourth lags have much less. In (b), the one-period lag effect is greater than the current period's, but the importance of the effect smoothly decreases beyond there.

vertically. They are spaced and identified along the horizontal axis by the length of the lag, l, which is given simply by $l = j - 1$. This is not a plot of data, but a convenient graphical depiction of the values of the coefficients. In Figure 16.1a, we have a case in which the current ($l = 0$) and one-period lag ($l = 1$) values of X have nearly equal impacts on Y, and the impact of the second-period lag is only about half as much. The third and fourth lags have almost no effect compared with the current value of X. In Figure 16.1b, we have a different situation: the current effect is smaller than the first lag effect, but after that the importance of the effect smoothly decreases with the length of the lag.

A distributed lag model like (16.1) or (16.2) can be estimated with the technique of ordinary least squares, yielding estimates of the coefficients that have the usual properties. Two difficulties arise, however. First, the regressors are likely to be highly correlated, since it is usually the case that X_i and X_{i-1} (for example) both increase over time. This results in a situation of multicollinearity that can be extremely severe. Second, if we specify that the effect of X on Y is distributed over a large number of periods, the consequent loss of usable observations compounds the difficulty of making precise estimates.

For example, adopting the permanent income theory's idea that current consumption depends on a three-year span of incomes, we estimate

$$\begin{aligned}\widehat{CON}_i = \ & \underset{(0.084)}{0.735} + \underset{(0.159)}{0.886 DPI_i} + \underset{(0.2390)}{0.0116 DPILAG1_i} \\ & + \underset{(0.1762)}{0.0100 DPILAG2_i} \qquad (16.4)\end{aligned}$$

$$R^2 = .997 \qquad SER = 9.825$$

for the 23 observations 1958–1980. We see that the lagged impacts of disposable income are of negligible importance in comparison with the current impact, casting some doubt on the permanent income theory. Further, the current effect is clearly significant, whereas the lagged effects are clearly insignificant. However, we note that the presence of multicollinearity makes it difficult to precisely estimate the coefficients.

These difficulties of estimation are ameliorated to some extent if further specification for the distributed lag model is adopted, as discussed in the following sections. However, these specifications lead to other estimation difficulties. Also, if the adopted specification is not a reasonable characterization of the economic process, it will have created a misspecification problem.

The Koyck Model

The most common enhancement of the distributed lag model involves specifying that the coefficients decline geometrically in value as the length of the lag increases. Consistent with (16.1), the specification

$$\beta_j = \beta_1 \lambda^{(j-1)} \qquad 0 < \lambda < 1, \quad j = 1, \ldots, k \qquad (16.5)$$

puts this formally, so that $\beta_2 = \beta_1\lambda^1$, $\beta_3 = \beta_1\lambda^2$, $\beta_4 = \beta_1\lambda^3$, and so on. The parameter λ is a constant that reflects the relative importance of lag effects one period apart. To make things simple, we assume that k is infinite. Although this hardly seems reasonable or simple, it actually is: the factor λ^{j-1} gets very small as we push farther back in time (e.g., $0.5^{10} \approx 0.001$), so we are not claiming anything of substance regarding the distant past.

Now, letting the length of the lag be infinite, (16.1) can be written using (16.5) as

$$Y_i = \beta_0 + \beta_1(X_i + \lambda X_{i-1} + \lambda^2 X_{i-2} + \cdots) + u_i \tag{16.6}$$

This relation holds true every period, and we can write it down for period $i - 1$:

$$Y_{i-1} = \beta_0 + \beta_1(X_{i-1} + \lambda X_{i-2} + \lambda^2 X_{i-3} + \cdots) + u_i \tag{16.7}$$

Next, we multiply (16.7) through by λ to get a still-valid equation

$$\lambda Y_{i-1} = \lambda\beta_0 + \beta_1(\lambda X_{i-1} + \lambda^2 X_{i-2} + \lambda^3 X_{i-3} + \cdots) + \lambda u_{i-1} \tag{16.8}$$

and subtract (16.8) from (16.6) to get

$$Y_i - \lambda Y_{i-1} = \beta_0 - \lambda\beta_0 + \beta_1 X_i + u_i - \lambda u_{i-1} \tag{16.9}$$

Finally, moving λY_{i-1} to the right-hand side and letting $v_i = u_i - \lambda u_{i-1}$, we arrive at

$$Y_i = (1 - \lambda)\beta_0 + \beta_1 X_i + \lambda Y_{i-1} + v_i \tag{16.10}$$

This equation stands as the combination of the geometric lag assumption (16.5) with the infinite version of the general distributed lag model (16.1).

The regression model (16.10) can be estimated directly by OLS, with some difficulties to be discussed later. The specification of the geometric lag pattern enables us to replace an infinite stream of variables with just two regressors: the current value of X and the one-period lag of Y, the dependent variable. Only one observation is lost in the construction. This respecification has the beneficial effect of reducing the potential for severe multicollinearity that exists in the general distributed lag model. Notice that the coefficient on the lag of Y is λ, the geometric weight, which is to be estimated. The coefficient on the current value of X is β_1. Given (16.5), all the coefficients in the infinite distributed lag can be determined from these two values. This formulation is known as the ***Koyck model,*** after its originator; it is also known generically as the ***geometric lag model.***

For example, consider the assumption that the importance of previous years' incomes in the determination of current consumption expenditures declines geometrically with the length of the lag. The Koyck model specification of the aggregate consumption function is

$$\widehat{CON}_i = \underset{(0.349)}{2.611} + \underset{(0.121)}{0.623}DPI_i + \underset{(0.136)}{0.320}CONLAG1_i \tag{16.11}$$
$$R^2 = .998 \qquad SER = 8.483$$

estimated for the 23 observations, 1958–1980. It should be noted that 24 observations are available here, but we have kept the same sample as (16.4) for comparison. The estimated coefficient on current income is $\hat{\beta}_1^* = 0.623$. The rate of geometric decline is estimated to be $\hat{\lambda} = 0.320$, implying an estimated impact of $(0.623)(0.320) = 0.199$ for the one-period lag of income, $(0.623)(0.320)^2 = 0.064$ for the two-period lag, and so on. The estimated structure of the distributed lag coefficients is illustrated in Figure 16.2.

The Koyck lag model (16.11) provids a dynamic specification of the process by which income affects consumption. In a given year, a 1 billion dollar increase in *DPI* leads to a 0.623 billion dollar increase in predicted *CON*; this is the *short-run* marginal propensity to consume. If the increase in *DPI* is maintained for a second year, predicted *CON* in that period will be higher than it otherwise would have been for two reasons: because *DPI* is higher and because CON_{i-1} is higher. From the viewpoint of the second year, CON_{i-1} is 0.623 billion dollars higher, leading CON_i to be $(0.320)(0.623) = 0.199$ higher through the lag effect. This, coupled with the current-period impact of the higher *DPI*, leads to a total *two-year* marginal propensity to consume of 0.822. The equilibrium, or long run, impact of a maintained increase in *DPI* can be determined from

$$\Delta\widehat{CON}_i = 0.623\Delta DPI_i + 0.320\Delta CONLAG1_i \tag{16.12}$$

which is derived from (16.11). In equilibrium, the change in consumption above what it otherwise would have been is the same in each period: the effect has

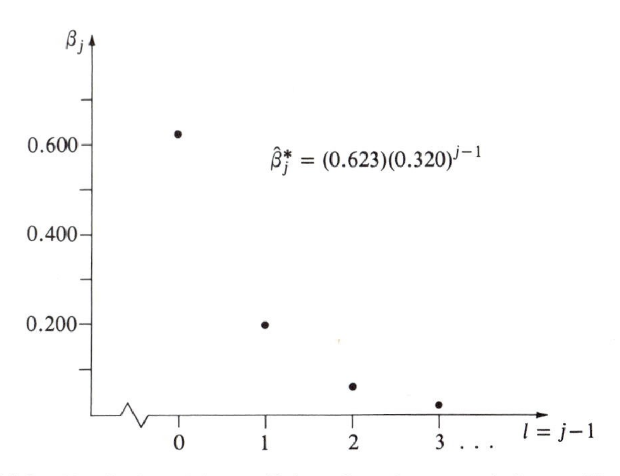

FIGURE 16.2 The distributed lag coefficients from the geometric lag specification of the consumption function, Equation (16.11), decline sharply with the length of the lag. The parameters β_1 and λ are estimated to be 0.623 and 0.320, respectively, and the implied value of each lag coefficient is determined from Equation (16.5).

reached a plateau. Thus, $\Delta CON = \Delta CONLAG1$, and (16.12) can be solved to yield $\Delta CON = 0.916\Delta DPI$. In this context, the *long-run* marginal propensity to consume is estimated as 0.916. In the general specification (16.10) of the Koyck model, the short-run impact of X on Y is given by β_1 and the long-run impact is given by $\beta_1/(1 - \lambda)$.

Partial Adjustment and Adaptive Expectations

We now briefly consider two other formulations that lead to estimating equations similar to (16.10). The first is the ***partial adjustment*** or ***stock adjustment*** model. The theory underlying this presumes that the behaviorally desired level of Y in period i is an unobservable variable Y^* and that this desired level is determined as

$$Y_i^* = \beta_0 + \beta_1 X_i + u_i \qquad (16.13)$$

Further, it is assumed that because of inertia, habit, costs, or other constraints, actual behavior each period is to close only part of the gap between last period's actual Y_{i-1} and this period's desired Y_i^*:

$$Y_i - Y_{i-1} = \gamma(Y_i^* - Y_{i-1}), \quad 0 < \gamma < 1 \qquad (16.14)$$

A little substitution and algebra combines (16.13) and (16.14) to yield

$$Y_i = \gamma\beta_0 + \gamma\beta_1 X_i + (1 - \gamma)Y_{i-1} + v_i \qquad (16.15)$$

where $v_i = \gamma u_i$. We recognize that the coefficient on Y_{i-1} identifies the adjustment factor, γ. Given this, the combined coefficient on X_i yields β_1, the impact of X on Y^*. The similarity of (16.15) and (16.10) should be noted.

The second formulation is the ***adaptive expectations*** model, which characterizes a certain behavior by saying that the actual value of Y is determined by the expected (or, desired) level of X:

$$Y_i = \beta_0 + \beta_1 X_i^* + u_i \qquad (16.16)$$

The expected level of X is the unobservable variable X^* that is theorized to be a weighted average of last period's X^* and this period's actual X:

$$X_i^* = \delta X_i + (1 - \delta)X_{i-1}^*, \quad 0 < \delta < 1 \qquad (16.17)$$

Substituting (16.17) into (16.16), we get

$$Y_i = \beta_0 + \beta_1(\delta X_i + (1 - \delta)X_{i-1}^*) + u_i \qquad (16.18)$$

Now, X_{i-1}^* depends on X_{i-1} and X_{i-2}^* in the same fashion as (16.17), and further substitutions lead us to an infinite lag specification of Y_i on X. We end up with the specification

$$Y_i = \delta\beta_0 + \delta\beta_1 X_i + (1 - \delta)Y_{i-1} + v_i \qquad (16.19)$$

where $v_i = u_i - (1 - \delta)u_{i-1}$. Notice that δ plays the same role as γ in the partial adjustment model, and each of these corresponds to $(1 - \lambda)$ in the Koyck lag model.

Since the partial adjustment and adaptive expectations models both lead to the same specification of regressand and regressors as the Koyck model, these models provide alternative frameworks for interpreting a regression with a lagged dependent variable. For example, consider the consumption function (16.11). Viewed through the partial adjustment model, the adjustment factor is estimated to be 0.680: this portion of the gap between desired consumption in a given year and actual consumption in the previous year is closed. We conclude that in the short run consumers adjust their actual consumption expenditures fairly completely to changes in desired expenditures. Also, using (16.15) we determine the marginal propensity to desire consumption [i.e., β_1 in (16.13)] to be $0.623/0.680 = 0.916$. It is no coincidence that this is the same as the long-run marginal propensity to consume [i.e., $\beta_1/(1 - \lambda)$ from (16.10)] in the Koyck interpretation. The formal correspondence between the estimating equations (16.10) and (16.15) leads to the correspondence between interpretations.

The results of (16.11) can be analyzed also through the framework of the adaptive expectations model. Expected income is fairly dependent on current actual income ($\delta = 0.680$), which is to say that through (16.17) expectations adjust rapidly to current reality. The marginal propensity to consume out of expected income is 0.916, although the mpc out of actual income is less, 0.623.

Estimation Problems

When the Koyck, partial adjustment, and adaptive expectations models are applied to the same economic process, they all lead to regression models having the same dependent variable and the same set of regressors. However, the error structures are not identical. To start with, we assume that the disturbance u_i in all the models satisfies the regular set of assumptions.

The situation is simplest for the partial adjustment model. The disturbance in the estimating regression (16.15) is $v_i = \gamma u_i$. Since v_i is proportional to u_i, it has the same properties. It can be shown that the OLS technique leads to consistent estimates of the parameters, although in small samples they will be biased because of the presence of the lagged dependent variable.

The Koyck model leads to a disturbance in the estimating regression (16.10) of the form $v_i = u_i - \lambda u_{i-1}$. Similarly, in the adaptive expectations model, the final (16.19) disturbance is $v_i = u_i - (1 - \delta)u_{i-1}$. This creates two difficulties. First, v_i will be autocorrelated even if u_i is not. Unfortunately, the Durbin-Watson statistic cannot be used to test whether v_i is autocorrelated, because the presence of the lagged dependent variable biases d toward 2. Instead, we use Durbin's h statistic, defined by (14.11).

The second and more troublesome difficulty is that the inclusion of the lagged dependent variable Y_{i-1} together with the autocorrelated disturbance all but guarantees that these two terms will be correlated. This condition invalidates one of the assumptions used to show that OLS leads to unbiased estimates of the unknown coefficients, leaving us with the conclusion that OLS will be biased. Further, this bias will not disappear even in very large samples, so the estimates will be inconsistent. Some advanced estimation methods may be helpful here, but we do not cover them.

Other Distributed Lag Specifications

The geometric lag specification (16.5) is fairly flexible because various values of λ are consistent with various degrees of importance for the lagged variables. However, the specification restricts the values of β_j to decline geometrically with j, as in Figure 16.2. Although this pattern of decline is appropriate for many economic processes, it is not universally so.

A variety of forms for specifying the structure of the distributed lag have been developed, and each is appropriate in certain circumstances. Some of these structures are illustrated in Figure 16.3. In Figure 16.3a the relative importance of the coefficient decreases linearly with the length of the lag; in Figure 16.3b the importance first increases and then decreases.

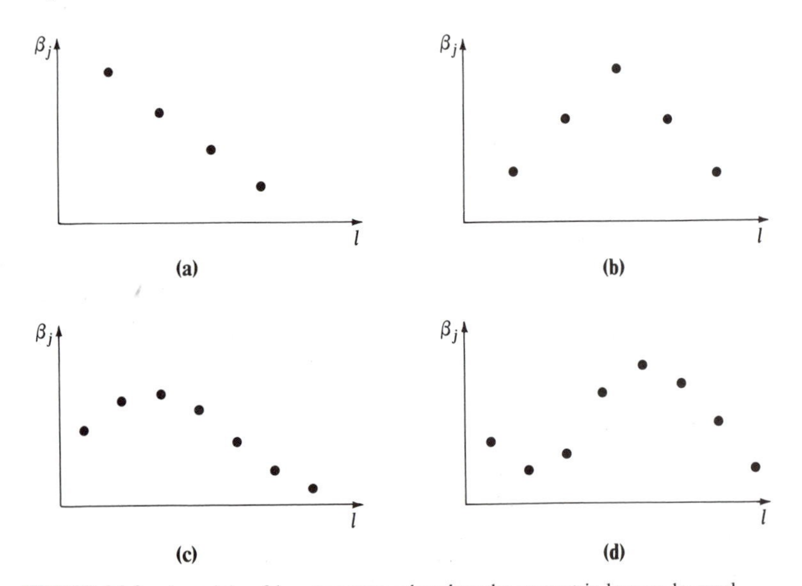

FIGURE 16.3 A variety of lag structures other than the geometric lag can be used. In (a), which is sometimes called the triangular lag, the relative importance of the coefficients decreases linearly with the length of the lag. In (b), the inverted-V lag specifies a special symmetric pattern for the coefficients. Parts (c) and (d) show two cases of the Almon lag structure, in which the lag coefficient values are specified to lie along a polynomial function of l, the length of the lag. These structures impose constraints on the relative values of regression coefficients and lead to the construction of a set of regressors that permit estimation by OLS.

Figure 16.3c and d show cases of the ***polynomial distributed lag,*** which is usually called the ***Almon lag*** (after its originator, Shirley Almon). In the Almon lag scheme, the β_j values are specified to lie along a polynomial function of the length of the lag, l. A particular Almon lag structure is parameterized by the maximum length of the lag and the degree of the polynomial (quadratic, cubic, etc.). Since there is often no firm basis for choosing the right specification a priori, the possibility of misspecification is always present.

The choice of a particular lag structure effectively imposes a set of restrictions on the relative impacts of the lagged variables. These lead to the construction of a set of regressors, which permit ordinary least squares to estimate all the necessary parameters. The procedures are not conceptually difficult, but they are sufficiently detailed to lie beyond our scope.

16.2 Regression Analysis of a Time Series

If we simply look at the plot of a time-series variable it is often possible to see a regularity that seems to permit forecasting into the near future. For example, Figure 16.4 plots the values of *GNP* from the time-series data set in Chapter 2 for the years 1956–1980. Intuition suggests that some method of extrapolation will be able to produce forecasts of *GNP* with relatively small error. If we are unable to specify and estimate a true structural model that determines *GNP* with predictive accuracy, then concentration on only the single time series may be the best thing to do. In this section we explore how regression techniques can be used for this purpose. This discussion only scratches the surface of a large and growing set of methods for the time-series analysis.

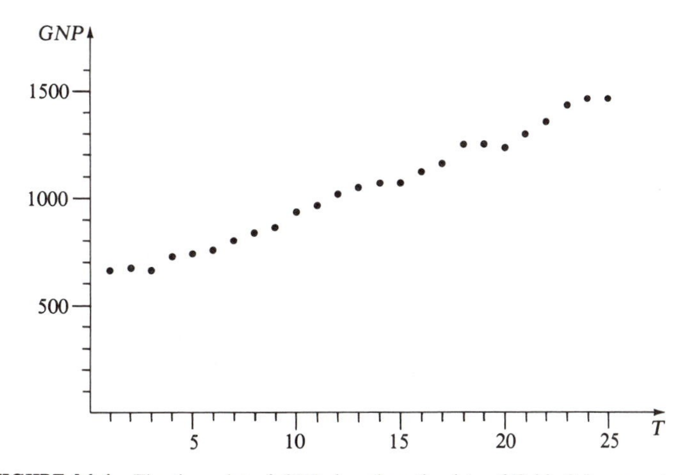

FIGURE 16.4 The time plot of *GNP*, based on the data of Table 2.3, suggests that extrapolation can be used to produce trend forecasts of *GNP*. A variety of regression specifications can be used to estimate this trend.

Trends

For the example of *GNP* reported in annual data, the simplest procedure is to estimate the regression model

$$GNP_i = \beta_0 + \beta_1 T_i + u_i \tag{16.20}$$

where T is the time period number. The fitted regression line

$$\widehat{GNP}_i = \hat{\beta}_0 + \hat{\beta}_1 T_i \tag{16.21}$$

is interpreted as the ***trend line*** around which actual observations occur. Sometimes a nonlinear trend line is assumed, and a regression of *GNP* on T and T^2 is estimated. In either case forecasts of *GNP* are made simply by determining the value of predicted *GNP* that corresponds to the specified time period along the trend line. The choice of the proper functional form depends on theoretical considerations and on an analysis of the residuals that result from the regression.

Another type of trend model is based on the semilog specification. In Chapter 6 we estimated

$$\begin{aligned} \widehat{LNGNP}_i &= \underset{(0.012)}{6.456} + \underset{(0.00082)}{0.0354}T_i \\ R^2 &= .988 \qquad SER = 0.0297 \end{aligned} \tag{16.22}$$

which is based on the theoretical premise that *GNP* tends to grow at a constant proportional rate. This regression can be used to make predictions of *LNGNP* directly. However, the best prediction of *GNP* is not simply the antilog of predicted *LNGNP*, but rather it is that amount multiplied by the antilog of $SER^2/2$.

To see why any adjustment is needed, we modify (6.42) to restate the model underlying the estimation of (16.22),

$$GNP_i = GNP_0(1 + r)^{T_i} e^{u_i} \tag{16.23}$$

where e is the base of natural logarithms. For a specific T_i,

$$E[GNP_i] = GNP_0(1 + r)^{T_i} E[e^{u_i}] \tag{16.24}$$

Now, if u_i has a normal distribution, it turns out that e^{u_i} has a ***lognormal distribution*** with

$$E[e^{u_i}] = \exp(\sigma_u^2/2) \tag{16.25}$$

which is greater than 1. Thus, $E[GNP_i]$ is systematically greater than $GNP_0(1 + r)^{T_i}$. In order to make our prediction a better guess of $E[GNP_i]$, we make the adjustment suggested above using *SER* as an estimate of σ_u.

For example, from (16.22) the prediction of *LNGNP* for 1981 ($T = 26$) is 7.376, yielding an antilog of 1597.2 billion dollars. This is then multiplied by the antilog of $(0.0297)^2/2$, which is about 1.0004, yielding a final *GNP* prediction of 1597.8 billion dollars. In cases when *SER* is relatively large, the adjustment is more dramatic. (This adjustment procedure can be applied to making predictions from any regression model with a logarithmic dependent variable.)

Seasonality

All the time-series variables that we have dealt with so far are measured on an annual basis. Shorter measurement periods are used for many actual time series. For example, the National Income Accounts data are available on a quarterly basis (January to March is the first quarter, April to June is the second, etc.), national unemployment rates are determined on a monthly basis, basic money supply data are determined on a weekly basis, and so on. Economists work with these data when short-period predictions are needed or when the dynamic behavior being studied displays itself naturally in these short periods.

Experience has shown that many variables are subject to regular patterns of variation during the course of a year, and this pattern is known as ***seasonality.*** For example, retail sales regularly go up in December because of holiday shopping, and the unemployment rate goes up in June because students leave school and enter the labor force searching for a job. Each time-series variable has its own seasonality pattern. Economic statisticians, who help prepare and analyze data, deal with seasonality in two years. First, they may use a variety of seasonal adjustment procedures to adjust the data before any analysis: such data are said to be ''seasonally adjusted.'' These procedures are complex and interesting, but we do not deal with them. Second, they may leave the data in their original form and take care of the seasonality as part of the analysis.

Table 16.2 displays quarterly data on expenditures for new plant and equipment, *NPE*, aggregated for all U.S. industries, 1966–1968. This short period

TABLE 16.2 *NPE* and Seasonal Dummies

Obs.	*T*	*NPE*	*Q1*	*Q2*	*Q3*	*Q4*
1966-1	1	12.77	1	0	0	0
-2	2	15.29	0	1	0	0
-3	3	15.57	0	0	1	0
-4	4	17.00	0	0	0	1
1967-1	5	13.59	1	0	0	0
-2	6	15.61	0	1	0	0
-3	7	15.40	0	0	1	0
-4	8	17.05	0	0	0	1
1968-1	9	14.25	1	0	0	0
-2	10	15.86	0	1	0	0
-3	11	16.02	0	0	1	0
-4	12	17.95	0	0	0	1

Source: Business Statistics (a supplement to the *Survey of Current Business*), 17th edition (1969), p. 9.

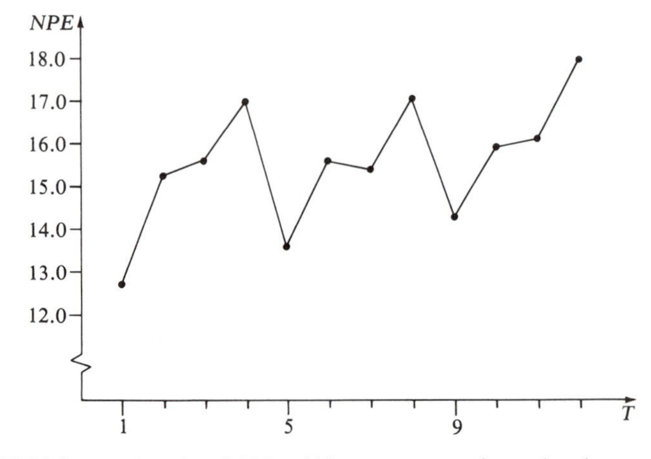

FIGURE 16.5 The time plot of *NPE*, which measures new plant and equipment expenditures on a quarterly basis, is based on the data in Table 16.2. The seasonal pattern around a positive trend stands out clearly. A regression model that includes a time trend and three seasonal dummy variables can be used to estimate the time pattern of *NPE*.

exhibited relatively steady growth with a low rate of unemployment, and therefore a simple trend model may be appropriate. However, as evident in the data and as shown in Figure 16.5, there is a strong seasonal pattern in addition to the simple trend. If we wish to construct a short-term quarterly forecasting model, we surely should take this pattern into account in addition to the steady trend of overall growth.

The most popular way to treat this kind of seasonality is to create a set of dummy variables that indicate in which quarter of the year each observation occurred. Following the treatment of regional location in Chapter 7, the dummy variables *Q1, Q2, Q3,* and *Q4* are defined to be equal to 1 if the observation is in the corresponding quarter and 0 otherwise. These variables are displayed in Table 16.2.

In setting up a time-series regression model involving these seasonal dummies, it is essential that one of them be left out in the specification when the intercept β_0 is also present; it is not important which one of the dummy variables is excluded. Arbitrarily, we leave out the first quarter dummy and specify

$$NPE_i = \beta_0 + \beta_1 T_i + \beta_2 Q2_i + \beta_3 Q3_i + \beta_4 Q4_i + u_i \qquad (16.26)$$

All the observations, including those for first quarters, are used in the estimation.

The results of the estimation are

$$\begin{aligned}\widehat{NPE}_i &= \underset{(0.28)}{12.57} + \underset{(0.025)}{0.108}T_i + \underset{(0.23)}{1.94}Q2_i + \underset{(0.23)}{1.91}Q3_i + \underset{(0.24)}{3.47}Q4_i \\ R^2 &= .977 \qquad SER = 0.281\end{aligned} \qquad (16.27)$$

The coefficient on each dummy variable gives the difference in the predicted value of *NPE* that can be attributed to an observation's occurring in that quarter rather than in the first quarter. Each of the estimated coefficients is positive, indicating that the first quarter of each year tends to have the lowest *NPE*.

Randomness

In addition to the trend and seasonality factors in a time series, there is also a factor that is unique to each time period. This ***randomness*** is specified as the disturbance in regression models such as (16.20) and (16.26).

The simplest case is when the random factor for a given period is not related to that for any other. This corresponds to the regular normal regression assumptions and no special estimation is needed. For forecasts, the random component is assumed to be zero. However, more complex cases are often appropriate in time-series work.

Suppose that the correct model is

$$Y_i = \beta_0 + \beta_1 T_i + u_i \tag{16.28}$$

but that the randomness is described by the first-order autoregressive model

$$u_i = \rho u_{i-1} + \epsilon_i \tag{16.29}$$

This is the situation of autocorrelation, discussed in Chapter 14. If we use a procedure to estimate ρ and then the regression coefficients, we should take this information about ρ into account in making forecasts. Suppose that we wish to forecast Y for period $n + 1$, the next period after the last in our data. The last residual from the estimation is $e_n = Y_n - \hat{Y}_n$, and based on (16.29) we predict the value of the disturbance in period $n + 1$ to be $\hat{\rho}e_n$. (Note that ϵ_{n+1} is predicted to be zero, as u_{n+1} would be if autocorrelation were not present.) Thus, our final prediction for period $n + 1$ is

$$\hat{Y}_{n+1} = \hat{\beta}_0 + \hat{\beta}_1 T_{n+1} + \hat{\rho}e_n \tag{16.30}$$

Similarly, if we were predicting for period $n + 2$, the adjustment for autocorrelation would be $(\hat{\rho})^2 e_n$. Since $|\hat{\rho}| < 1$, this adjustment becomes smaller as the time of forecast moves farther beyond the end of the sample period.

Finally, it may be that what we have classified as randomness includes a regular cyclical variation that is different from the autoregressive model. There are various advanced techniques that can be used to analyze these patterns and make forecasts.

Problems

Section 16.1

16.1 Consider the distributed lag consumption function reported as Equation (16.4). Rewrite this model in difference form, as in Equation (16.12). Suppose that in period i, *DPI* is raised by 10 billion dollars above its actual historical value, but in succeeding years it remains at its actual historical values. What is the impact of this on *CON* in period i and in the four succeeding periods? Graph the results.

* **16.2** Repeat Problem 16.1, except suppose that beginning in period i each value of *DPI* is raised by 10 billion dollars above its actual historical value.

* **16.3** Suppose that Equation (16.3) correctly represents a certain economic process, but a simple regression of Y on current X is estimated. How will the estimate of the slope in the simple regression be related to β_1 in the correct model?

16.4 On one set of axes, display the structure of the distributed lag coefficients for a geometric lag model in which $\lambda = .9$, and for one in which $\lambda = .5$; for simplicity, let $k = 5$.

16.5 Suppose that beginning in year i, *DPI* in each year is raised one billion dollars above its actual historical values. Based on the estimated regression (16.11), compute and graph the impact of this on *CON* for the first five years of the change.

* **16.6** Suppose that in period i, Y_i is less than Y_i^* in the partial adjustment model. If Y_i^* remains at a constant, flat amount for the next five years, what happens to Y_i during those years?

* **16.7** Suppose that in period i, Y_i is less than Y_i^* in the partial adjustment model. If X grows at a constant absolute amount for the next five years, what happens to Y_i during those years?

16.8 Suppose that in period i, X_i is greater than X_i^* in the adaptive expectations model. If X_i remains fixed at a constant, flat amount for the next five years, what happens to Y_i during those years?

* **16.9** Suppose that the Durbin-Watson statistic in Equation (16.11) is $d^* = 1.25$. Taking into account the presence of the lagged dependent variable, would you judge that positive autocorrelation is present?

16.10 Suppose that we wish to estimate the distributed lag model (16.2) with the additional specification that $\beta_2 = (2/3)\beta_1$ and $\beta_3 = (1/3)\beta_1$. Graph the structure of the lag coefficients. Also, use the additional specification to reformulate Equation (16.2) in such a way that the model embodies the specification.

Section 16.2

16.11 Based on regression (16.22), make a simple (unadjusted) prediction of *GNP* for 1985.

16.12 If the standard error of the regression (*SER*) were 0.10, what would be the multiplicative adjustment factor for predicting *GNP* from Equation (16.22)?

★ **16.13** Based on Equation (16.27), determine the estimated coefficients in a seasonally adjusted trend model for *NPE* in which *Q4* is the excluded dummy variable.

16.14 How could we test whether seasonality plays a role in the determination of a time-series variable?

★ **16.15** Based on Equation (16.27), how much higher is predicted *NPE* in the third quarter as compared with the first quarter of the same year?

16.16 Suppose that $\rho^* = .7$ in regression (16.22). Make a simple prediction of *GNP* for 1981, taking autocorrelation into account. (*Hint:* You need Table 2.3.)

Simultaneous-Equation Models

Up to this point, we have viewed each regression equation as a complete model of an economic process. Such a model shows how a dependent variable is determined by a set of explanatory variables, which are taken as fixed by forces outside the process under study, and by a disturbance term. We now turn our attention to models in which more than one equation is necessary to adequately characterize the economic process. In these ***simultaneous-equation models,*** several "dependent" variables are jointly and simultaneously determined by a set of "explanatory" variables and disturbance terms. These models present us with new problems of specification and interpretation as well as new problems of estimation.

17.1 The Nature of the Models

We examine the nature of simultaneous-equation models by considering two examples. The first involves a cross-sectional analysis, and the second is based on time series.

The first model seeks to explain the hours worked and wages received by a relatively homogeneous group of workers. To simplify the economics, we assume that employers do not try to determine a worker's actual skill level or productivity. For concreteness, let us say that we are concerned with married women aged 25–34 who have college degrees with a major in economics. For a variety of reasons there is considerable variation in the wages received and (weekly) hours worked by these women, and our interest is in explaining how wages and hours are determined. From an employer's point of view, a worker who works more hours may be more valuable per hour because she provides continuity and can accept more responsibility. Hence, there is a demand-type relation between the hours an employer demands from a potential worker and the wage to be paid. From the individual woman's point of view, the hours she is willing to work depends on the wage, the other income her family receives, and the number of children in the family. In this context there are two economic behaviors exhibited: demand and supply.

These two patterns of behavior are combined to yield a ***structural model,*** in which each equation describes a single type of economic behavior:

$$\text{(Demand)} \quad HOURS_i = \beta_0 + \beta_1 WAGE_i + u_i \tag{17.1}$$

$$\text{(Supply)} \quad HOURS_i = \gamma_0 + \gamma_1 WAGE_i + \gamma_2 OTHINC_i + \gamma_3 NKIDS_i + v_i \tag{17.2}$$

In this model, *HOURS* is the quantity demanded or supplied. These are observationally equivalent when our data are for women actually in jobs, but sometimes the conceptual distinction is made explicit by using $HOURS^d$ in (17.1) and $HOURS^s$ in (17.2) and then adding a third equation explicitly stating that $HOURS^d = HOURS^s$. For simplicity we have skipped this step. The disturbances u and v are regarded in the usual way. It should be noted that this model is different from standard market models, which deal with aggregate demand and supply: here we examine the behavior of individual employers and employees, and the observations are cross-section data on the characteristics of workers and their jobs.

To analyze this simultaneous-equation model, a new taxonomy of variables is necessary. The variables *HOURS* and *WAGE* are called ***endogenous*** because they are jointly determined inside the system (model). The variables *OTHINC* and *NKIDS* are called ***exogenous*** because they are determined outside the model. The variable *HOURS* appears on the left-hand side of both (17.1) and (17.2), while *WAGE* appears on the right-hand side along with the exogenous variables. This results from a fairly natural specification of the two patterns of behavior. Nonetheless, *HOURS* and *WAGE* have equal stature as endogenous variables. Despite appearances, *HOURS* is no more endogenous than is *WAGE*; both are jointly and simultaneously determined.

To illustrate how the endogenous variables *HOURS* and *WAGE* are determined, we first consider a set of observations that have the same values for *OTHINC* and *NKIDS*. This makes the terms $\gamma_2 OTHINC_i$ and $\gamma_3 NKIDS_i$ in (17.2) effectively constants. In Figure 17.1, the lines labeled "Demand" and "Supply" graph the systematic parts of (17.1) and (17.2), with the disturbances set to zero. Casual economic theory suggests that both slopes are positive, but we have no particular hypothesis with regard to their relative magnitudes.

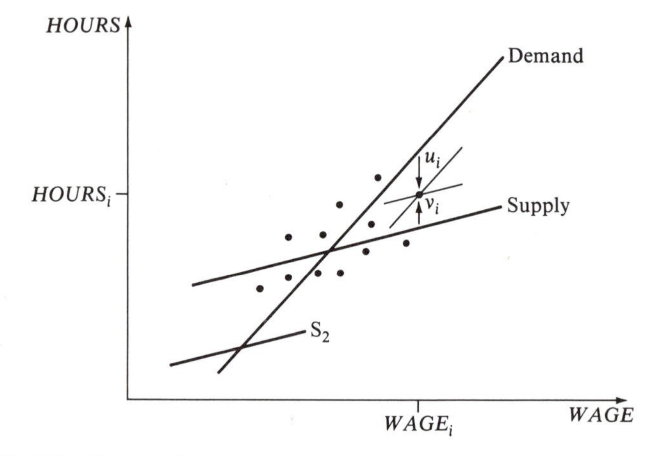

FIGURE 17.1 The lines labeled "Demand" and "Supply" represent the systematic components of Equations (17.1)-(17.2) for a set of observations having the same values for *OTHINC* and *NKIDS*. Suppose that for the ith observation, u_i is negative and v_i is positive. The effective demand and supply relations are represented by the light lines, and their intersection is the *WAGE-HOURS* combination that satisfies both equations. This point is the ith observation. The other observations are generated similarly. The line labeled "S_2" is the systematic part of the supply relation for a set of observations having in common a different pair of values for *OTHINC* and *NKIDS*.

Observations get produced from this process, and they differ from each other because of the randomness of the disturbances. Suppose that for the ith observation u_i is negative and v_i is positive. For this observation, the demand version of the *HOURS-WAGE* relation has a vertical intercept of $\beta_0 + u_i$ and a slope of β_1: it lies below the systematic demand relation by the negative amount u_i. Similarly, for this observation the supply version of the relation lies above its systematic form. Where the two lines intersect, both equations are satisfied; this is the ith observation illustrated in Figure 17.1. Similarly, other observations are produced in a scatter around the intersection of the systematic portions of the demand and supply relations. Given the values for its disturbances u_i and v_i, each observation then satisfies both equations of the model.

To illustrate the full workings of the model (17.1)–(17.2) requires a four-dimensional graph, which is beyond our capacity. However, we can see in Figure 17.1 that for another set of observations having common values of *OTHINC* and *NKIDS*, the supply curve could be graphed as another line (S_2) with the same slope but a different intercept. For example, women with more *OTHINC* and a greater number of *NKIDS* may be willing to work fewer hours at any available wage.

A second example is provided by a small macroeconomic model that is consistent with the simple *IS-LM* formulation of aggregate economic activity. The specification of its structure is

$$CON_i = \beta_0 + \beta_1 GNP_i + \beta_2 RTB_i + \beta_3 CON_{i-1} + u_i \qquad (17.3)$$

$$GNP_i = CON_i + A_i \qquad (17.4)$$

$$RTB_i = \gamma_0 + \gamma_i GNP_i + \gamma_2 M_i + v_i \qquad (17.5)$$

All but one of the variables have been used earlier in this book. *CON* is real consumption, *GNP* is real gross national product, *M* is the real money supply, and *RTB* is the Treasury bill rate. The variable *A* denotes real autonomous expenditures, whose major conceptual components here are investment, government expenditures, and net exports. In terms of our data, *A* is a constructed variable that is defined implicitly by (17.4) as $A_i = GNP_i - CON_i$. We note that (17.3) and (17.5) appear to be single-equation models of specific economic behaviors. By contrast, (17.4) can be viewed as either an accounting identity or an equilibrium condition; in either case, it presents us with no task of econometric estimation. Nonetheless, it is a bona fide element of this simultaneous-equation model.

In economic terms, (17.3) and (17.4) make up the *IS* side of the model. They express a model of equilibrium in the real product market, albeit an unusual one in which investment is autonomous and consumption is interest sensitive. Equation (17.5) is the *LM* side of the economy, combining demand and supply elements in a single-equation model of equilibrium in financial markets.

The lagged value of consumption, CON_{i-1} gives the consumption function a dynamic character, as discussed in Chapter 16. We use *CONLAG* as the name of the regressor that is the one-period lag of *CON*. Whereas *CON* is an endogenous variable, *CONLAG* is not because its value is determined temporally prior to *CON*. In this way, *CONLAG* is similar to the exogenous variables, which are determined logically prior to *CON*. In the taxonomy of simultaneous-equation models, exogenous and lagged endogenous variables together make up a super-classification known as ***predetermined variables.***

There are three endogenous variables in the system (*CON*, *GNP*, and *RTB*), which happen to be the left-hand side variables in the three equations. As we saw in the job market example, however, this is not an essential element in the specification of a simultaneous-equation model. The predetermined variables are *CONLAG*, *A*, and *M*. Taken as a whole, the model is a description of the process by which the economy simultaneously determines *CON*, *GNP*, and *RTB*.

The Reduced Form

A structural model is a set of equations, each of which describes a single type of economic behavior or is like an identity. Equation sets (17.1)-(17.2) and (17.3)-(17.5) are both structural models. We now consider certain algebraic properties and manipulations of these sets of equations in order to gain more insight into their nature.

We look first at the equations of the demand and supply model, (17.1)-(17.2). The spirit of the model is that the values of the exogenous variables are fixed outside the system, the coefficients are parameters (constants), and the values of the disturbances are established by a purely random process. The values of the endogenous variables are determined by all of these. In formal algebraic terms, the model consists of two linear simultaneous equations in two unknowns—the endogenous variables. Once the values of the disturbances are established for a particular observation, the economic process determines the values of the endogenous variables so as to satisfy both equations of the model simultaneously.

By standard algebraic methods, the original set of simultaneous equations can be solved for the unknowns, resulting in

$$WAGE_i = \left(\frac{\gamma_0 - \beta_0}{\beta_1 - \gamma_1}\right) + \left(\frac{\gamma_2}{\beta_1 - \gamma_1}\right)OTHINC_i + \left(\frac{\gamma_3}{\beta_1 - \gamma_1}\right)NKIDS_i + \left(\frac{v_i - u_i}{\beta_1 - \gamma_1}\right) \tag{17.6}$$

and

$$HOURS_i = \left(\frac{\beta_1\gamma_0 - \beta_0\gamma_1}{\beta_1 - \gamma_1}\right) + \left(\frac{\beta_1\gamma_2}{\beta_1 - \gamma_1}\right)OTHINC_i + \left(\frac{\beta_1\gamma_3}{\beta_1 - \gamma_1}\right)NKIDS_i + \left(\frac{\beta_1 v_i - \gamma_1 u_i}{\beta_1 - \gamma_1}\right) \tag{17.7}$$

[A simple solution method here is to solve (17.1) for *WAGE* in terms of *HOURS*, and then substitute (17.2) into that equation and rearrange terms, yielding (17.6). Next, the right side of (17.6) is substituted for *WAGE* in (17.1), which yields (17.7) after rearrangement.] For simplicity, (17.6) and (17.7) are conventionally rewritten as

$$WAGE_i = \pi_{10} + \pi_{11}OTHINC_i + \pi_{12}NKIDS_i + \epsilon_{1i} \tag{17.8}$$

$$HOURS_i = \pi_{20} + \pi_{21}OTHINC_i + \pi_{22}NKIDS_i + \epsilon_{2i} \tag{17.9}$$

where the π and ϵ terms are simply shorthand symbols for the corresponding terms in parentheses in the previous pair of equations. The π coefficients are complex combinations only of β's and γ's, and the ϵ's are combinations of u's, v's, β's, and γ's. Notationally, the first subscript on a π indicates whether it is in the first or second equation, and the second subscript indicates its left-right order in the equation.

In the analytics of simultaneous-equation econometric models, the solutions (17.6)-(17.7) or their simplified reexpressions (17.8)-(17.9) are called the ***reduced form*** of the original structural model (17.1)-(17.2). Each equation of the reduced form specifies how the value of a single endogenous variable is related to the values of the exogenous variables, the disturbances, and the structural coefficients. Loosely speaking, the interdependences among the endogenous variables are solved out. In (17.6)-(17.7), notice that each of the endogenous variables depends on both of the disturbances.

The reduced form of the small macroeconomic model (17.3)-(17.5) is obtained by solving those three linear simultaneous equations for the unknowns *CON*, *GNP*, and *RTB*. The simplified reexpression is

$$CON_i = \pi_{10} + \pi_{11}CONLAG_i + \pi_{12}A_i + \pi_{13}M_i + \epsilon_{1i} \quad (17.10)$$

$$GNP_i = \pi_{20} + \pi_{21}CONLAG_i + \pi_{22}A_i + \pi_{23}M_i + \epsilon_{2i} \quad (17.11)$$

$$RTB_i = \pi_{30} + \pi_{31}CONLAG_i + \pi_{32}A_i + \pi_{33}M_i + \epsilon_{3i} \quad (17.12)$$

To generalize previous statements, each reduced-form equation has a single endogenous variable on the left-hand side and all the predetermined variables on the right-hand side. The laws of algebra lead the right-hand side of each reduced-form equation to be a linear expression involving the whole list of predetermined variables and disturbances. The π coefficients are complex combinations of β's and γ's, and the ϵ's are combinations of β's, γ's, u's, and v's.

If the equations in a structural model are all linear, then for the simultaneous system to be properly specified in an algebraic sense there must be exactly as many equations as there are endogenous variables. This ensures a unique solution for each endogenous variable. If the equations are not all linear, the nature of possible solutions is not so clear. Normally, however, econometric models are specified with the same number of equations and endogenous variables.

The structural and reduced-form versions of a simultaneous-equation model are equally valid formulations of the same economic process. However, the two versions present different problems of estimation and are suited for different applications. For example, if our only purpose is to make predictions of *WAGES* and *HOURS* conditional on specified values of *OTHINC* and *NKIDS*, then after accepting the unestimated structural model (17.1)-(17.2) as theoretically valid we can do all our practical work in terms of the reduced-form version of the model, (17.8)-(17.9). Indeed, if we are interested only in making predictions of the *WAGE*, we can ignore (17.9) and just focus on (17.8). On the other hand, if we want to know how sensitive women's supply of *HOURS* is to changes in *WAGE* or if we want to calculate the impacts on *HOURS* and *WAGE* of the imposition of a payroll tax, we need to know the structural coefficients. The structural model also can be used to make the same kind of predictions that the reduced form produces. In general, it is more useful to estimate the structural model than the reduced form.

Estimation and Simulation

The econometric estimation of the parameters (π's) of the reduced form is relatively straightforward. Each equation stands as a single-equation model and can be estimated by the usual methods. As we have seen, if the disturbances do not meet the assumptions of the normal regression model, ordinary least squares may not be the chosen method.

The estimation of the equations in the structural model is more complicated. In some cases, discussed later, it is possible to estimate the coefficients of each equation by first estimating the reduced form and then solving algebraically to get estimates of the structural coefficients. This is not always fruitful, however, and we are interested in general methods for directly estimating them. Under appropriate circumstances, it is possible simply to use OLS on each of the equations. However, in Section 17.2, we see that even in these circumstances OLS turns out to be biased and inconsistent, and an estimation method having the property of consistency is presented.

An additional problem associated with the specification and estimation of a structural model is that for some reasonable-looking equations, it may be impossible for the coefficients to be estimated meaningfully. These are inappropriate circumstances for applying any estimation model, even OLS. One technical aspect of this, which is known as the problem of "identification," is that it is impossible to obtain consistent estimates of the coefficients. In source and solution this is a problem of specification, not estimation, and we put off its consideration until Section 17.3.

Reserving all these problems for later discussion, we consider here as an example the results of using OLS to estimate the econometric equations of the macroeconomic model (17.3)-(17.5) for the period 1957–1980 ($n = 24$):

$$\begin{aligned} \widehat{CON}_i &= \underset{(6.62)}{-\,29.36} + \underset{(0.050)}{0.335GNP_i} - \underset{(1.08)}{4.23RTB_i} \\ &\quad + \underset{(0.083)}{0.553CONLAG_i} \end{aligned} \tag{17.13}$$

$$GNP_i = CON_i + A_i \tag{17.14}$$

$$\widehat{RTB}_i = \underset{(6.18)}{5.10} + \underset{(0.0019)}{0.0099GNP_i} - \underset{(0.0334)}{0.0455M_i} \tag{17.15}$$

This constitutes an estimated simultaneous-equation model.

One of the reasons for estimating the structural model rather than the reduced form is that it is now possible to look at each of the estimated equations and assess it as a single-equation model of economic behavior. The assessment would depend in part on an examination of the meaning and statistical significance of the coefficients and on a comparison of the results with other empirical work and with theoretical considerations. We do not do this for the example here, but for serious uses of models, this kind of assessment is extremely important.

We proceed to consider the task of making predictions. Once a structural simultaneous-equation model has been estimated, the set of numerically specified equations is sometimes called a ***simulation model*** and the predictions it makes are called ***simulations.*** Algebraically, a prediction or simulation using the model involves solving (17.13)-(17.15) to obtain the values of the endogenous variables that result from a given set of values for the predetermined variables, with disturbances set to zero. When the equations are linear, as they are in this case, the algebraic solution can be based on the same methods that are used for determining the reduced form. In practice, however, most computer programs for simulation use an iterative solution algorithm that leads to the same results.

We can use our estimated model to describe some rudimentary ***policy analysis,*** which usually involves two simulations over a period for which we have data on all the exogenous variables. For example, suppose that we wish to determine the effects of a $10 billion increase in government expenditures beginning in period T. One simulation is made from period T forward, using the historical values of the exogenous variables. A second simulation is the same except that the value of A is set at $10 billion above what it actually was in each period. The effects of the policy are studied by comparing the two simulated time paths for each of the endogenous variables.

A second type of policy analysis involves changes in the parameters. For example, suppose that economic theory and some extra calculations lead us to believe that a proposed decrease in federal tax rates is equivalent to raising the marginal propensity to consume out of *GNP* by 0.020 from whatever amount it now is. We can analyze the impact of this policy change on *CON*, *GNP*, and *RTB* by simulating the model twice, both times using the same historical data on the exogenous variables. First we carry out a simulation using the estimated model. In the second simulation we change the coefficient on *GNP* in (17.13) from 0.335 to 0.355 and carry out the simulation again. The differences in simulated values for the time paths of each of the endogenous variables is the basis for an analysis of the impact of the tax-rate change.

The first policy analysis described could be done using an estimated reduced-form model, because all that is involved is making predictions for alternative values or the exogenous variable A. By contrast, the second policy analysis requires an estimated structural model, because it involves a change in one of the structural coefficients—which cannot be isolated in the reduced form.

17.2 Estimation

In this section we focus on the task of estimating a single structural equation in a simultaneous-equation model. By doing this separately for each equation, the whole model can be estimated. We do not discuss some advanced procedures that involve estimating all the equations together. Also, in this section we presume that the equation of interest is appropriately "identified," so that meaningful estimation can be attempted.

Recall that in Chapter 11 it was shown that for the normal regression model, the technique of ordinary least squares yields unbiased estimates of the regression coefficients. In general, unbiasedness is a desirable property for an estimating techique to have. In the normal regression model we are willing to make the assumption that all the explanatory variables are fixed outside of the process under study, and hence that they are independent of the disturbance. Unfortunately, in simultaneous-equation models this assumption is no longer tenable, because the system implies that a right-hand endogenous variable is correlated with the disturbance in the equation. Hence, the unbiasedness of OLS no longer holds. Further, even as the sample size increases the biasedness remains, and consequently OLS is inconsistent. This situation is often referred to as ***simultaneous-equation bias.***

To see the source of this bias, consider the task of estimating the demand relation (17.1). If this were an isolated single-equation model, we might accept the assumption that $WAGE_i$ is fixed (nonprobabilistic) and therefore uncorrelated with u_i. However, in (17.6), which is one of the reduced-form equations that is fully consistent with the structural model (17.1)-(17.2), we see that $WAGE_i$ depends in part on u_i. Therefore, $WAGE_i$ is correlated with u_i, except under extraordinary circumstances. This correlation of $WAGE_i$ and u_i cannot be assumed away if the structural model is accepted, and therefore the conditions for OLS to be unbiased are not present.

It may still be reasonable to use OLS despite the bias, and in large models this is often done. However, other techniques have been developed that yield consistent (but still biased) estimates of the parameters of each equation, although the appropriateness of some of these techniques is critically dependent on the correctness of the complete model's specification.

Two-Stage Least Squares

The most popular technique for obtaining consistent estimates for the parameters in equations of a simultaneous-equation model is known as ***two-stage least squares*** (TSLS). The essence of the technique is to replace each of the endogenous variables on the right-hand side of an equation with a constructed regressor that serves as a proxy for the original variable and that is essentially uncorrelated with the disturbance. Then OLS is applied to the equation. The resulting estimates of the (original) parameters of the equation may be biased, but this bias vanishes as the sample size increases.

The first stage of the technique is the construction of the proxy regressors. Let Y_j be one of the endogenous variables on the right-hand side of the equation, and let the list of predetermined variables in the entire system of equations be $Z_1, Z_2, \ldots, Z_m$ (there are m predetermined variables in the model). Using OLS, we regress Y_j on all of the Z's. Then, for each observation we calculate the fitted value $\hat{Y}_{ji}$ in the usual way, and this set of fitted values makes up our proxy $\hat{Y}_j$. It should be noticed that this procedure amounts to using OLS to estimate the jth reduced-form equation of the model, and calculating its fitted values. Formally, the proxy values are determined as

$$\hat{Y}_{ji} = \hat{\pi}_{j0} + \hat{\pi}_{j1}Z_{1i} + \hat{\pi}_{j2}Z_{2i} + \cdots + \hat{\pi}_{jm}Z_{mi} \tag{17.16}$$

In this notation i is the index for the observation number, with $i = 1, \ldots, n$ as usual. The index j identifies the endogenous variables: if there are three endogenous variables, then $j = 1$, 2, or 3, and there are three equations in the reduced form (as well as in the structural model). Equation (17.16) represents the estimated version of the jth reduced-form equation, and it is the fitted values of this equation that constitute the proxy for Y_j. This procedure is repeated separately for each of the endogenous variables on the right-hand side of the structural equation that we are seeking to estimate.

Regarding (17.16) we see that $\hat{Y}_j$, the proxy for Y_j, is a linear combination of all the predetermined variables in the system, Z_1 through Z_m. By definition, the predetermined variables are not related to any of the disturbances in the system of structural equations. However, the $\hat{\pi}$'s *are* related to the disturbances of the system, and therefore $\hat{Y}_j$ is correlated with the disturbances. In large samples, however, this correlation vanishes. Hence when $\hat{Y}_j$ is used as a proxy for Y_j in estimating a structural equation, the regressor $\hat{Y}_j$ is asymptotically uncorrelated with the disturbance of that equation.

The second stage of the TSLS technique is simply the OLS estimation of the modified structural equation, in which any endogenous variables that appear on the right-hand side are replaced by their proxy regressors. Whatever predetermined variables originally appeared in the regression remain there. The OLS coefficient estimates in this modified equation are consistent estimates of the parameters in the original equation. In finite-sized samples, which of course are the realm of practical applications, these estimates generally still are biased. The standard errors calculated in the regular way by OLS for this second stage are not correct, since they do not take into account all the information about the estimating procedure.

In practice, many econometric computer programs exist for making TSLS estimates in one fell swoop. From the researcher's point of view it is no more difficult to use TSLS than OLS. It should be noted that with TSLS, as well as with OLS in this context, hypothesis tests involving the estimated coefficients and their standard errors can be used only in an indicative way.

For example, the two-stage least squares estimates (1957–1980) of the small macro model (17.3)-(17.5) are

$$\begin{aligned} \widehat{CON}_i = & -\underset{(11.74)}{18.24} + \underset{(0.064)}{0.248}GNP_i - \underset{(3.05)}{2.57}RTB_i \\ & + \underset{(0.100)}{0.668}CONLAG_i \end{aligned} \tag{17.17}$$

$$GNP_i = CON_i + A_i \tag{17.18}$$

$$\widehat{RTB}_i = \underset{(6.14)}{6.20} + \underset{(0.0019)}{0.0104}GNP_i - \underset{(0.0335)}{0.0525}M_i \tag{17.19}$$

The OLS and TSLS estimates of the interest-rate equation are quite similar, but the estimates of the consumption equation are more noticeably different.

There is no hard and fast rule for determining whether the OLS or TSLS estimates are better. When multicollinearity is strongly present, as in the consumption equation, the bias of both methods tends to be accentuated. It has been shown that under some conditions OLS can be superior to TSLS, especially if sampling variability as well as bias is taken into account. However, the weight of the evidence is that two-stage least squares is the preferable method, and it should be used when possible. Sometimes researchers report both OLS and TSLS estimates of the same simultaneous-equation model so that the reader can assess the differences.

17.3 Identification

As noted earlier, the task of ***identification*** is to determine whether the coefficients of a particular equation can be estimated meaningfully. This is a subtle and difficult problem, and we treat only the easy-to-handle aspects of it.

To begin with an example, we revise the simple job market model by excluding *OTHINC* and *NKIDS* from having an effect on *HOURS* supplied. Thus the structural model is

$$\text{(Demand)} \quad HOURS_i = \beta_0 + \beta_1 WAGE_i + u_i \tag{17.20}$$

$$\text{(Supply)} \quad HOURS_i = \gamma_0 + \gamma_1 WAGE_i + v_i \tag{17.21}$$

We note that there are two endogenous variables and no exogenous variables. The systematic part of each of these is graphed in Figure 17.2a, which is similar to Figure 17.1.

Figure 17.2b shows only the set of observations produced by this economic

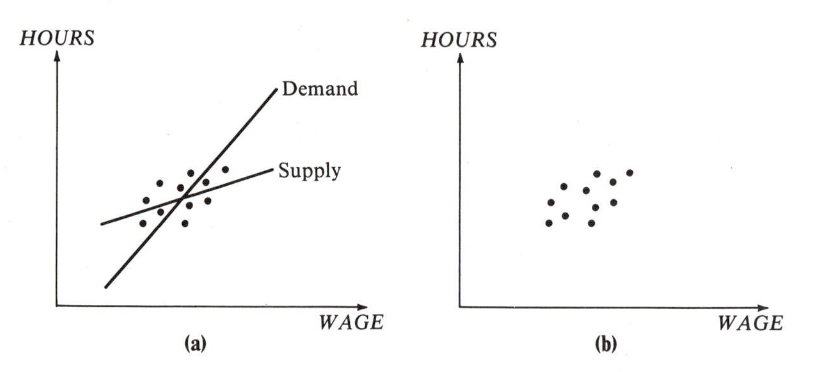

FIGURE 17.2 The model given by Equations (17.20)-(17.21) contains no exogenous variables. In (a), the lines represent the systematic components of the Demand and Supply equations. All the observations on *WAGE* and *HOURS* are produced by the process described for Figure 17.1. The lines are absent in (b), and there is no basis in the data for distinguishing separate demand and supply curves. This reflects the fact that both equations of the model are unidentified.

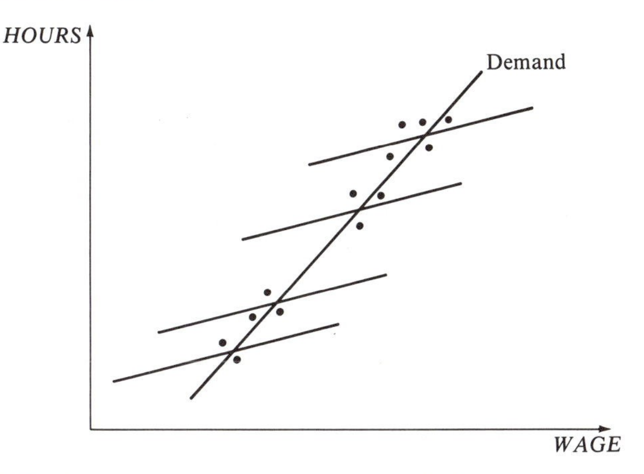

FIGURE 17.3 The model given by Equations (17.22)-(17.23) includes one exogenous variable for supply but none for demand. Observations are generated around various parts of the demand curve as the supply curve effectively shifts. The demand equation is now identified, and its coefficients can be meaningfully estimated.

process. Our experience suggests that we can estimate a regression of *HOURS* on *WAGE*, but we have no basis for saying whether it is an estimate of the demand curve, the supply curve, or something else. In this sense, we have no way of *meaningfully* estimating the demand coefficients or the supply coefficients: they are ***unidentified.***

Suppose instead that the correct model of supply includes the other income received by the family, but not the number of children. The full model is

$$\text{(Demand)} \quad HOURS_i = \beta_0 + \beta_1 WAGE_i + u_i \qquad (17.22)$$

$$\text{(Supply)} \quad HOURS_i = \gamma_0 + \gamma_1 WAGE_i + \gamma_2 OTHINC_i + v_i \qquad (17.23)$$

Figure 17.3 illustrates the systematic part of the demand and supply relations. As before, the systematic demand relation is the same for all observations. Now, however, the *HOURS-WAGE* supply relation differs among the observations because they have different values of *OTHINC*. The effective shifting of the supply curve serves to trace out the demand curve, although the disturbances make this imperfect. A set of data produced by this process provides the basis for estimating the demand version of the *HOURS-WAGE* relation, subject to the usual difficulties of estimation. The demand curve is now ***identified,*** although the supply curve remains unidentified.

What is special about the demand equation (17.22) that makes it identified, whereas the supply equation (17.23) is unidentified? Comparing the two, we see that the exogenous variable *OTHINC* is specifically excluded from the demand relation. The exclusion of exogenous variables is a key to identification.

The Order Condition

The conditions under which an equation within a simultaneous-equation model is identified or not are derived from a matrix algebra approach to simultaneous equations. We consider only one condition, which serves as a useful rule although it does not strictly guarantee the result.

As a preliminary, we recognize that in a complete structural model there are a certain number of endogenous variables and a certain number of predetermined (exogenous and lagged endogenous) ones. In a particular equation there must be present at least one endogenous variable—the one on the left-hand side of the equation. Usually, some endogenous variables are also present on the right-hand side, but not all the endogenous variables in the system need be in the equation; some may be excluded. Similarly, some of the predetermined variables in the system may be included in the equation, on the right-hand side, whereas some may be excluded.

The ***order condition*** for identification states simply that *if an equation is identified, then the number of predetermined variables excluded from the equation must be greater than or equal to the number of endogenous variables included on the right-hand side.*

If the order condition is satisfied with a "greater than" comparison, the equation is ***overidentified,*** whereas if "equal to" holds, the equation is ***just identified.*** Both cases satisfactorily meet the condition for identification of the equation. Technically, the order condition is necessary but not sufficient for identification. Although its satisfaction does not guarantee identification, it is widely used in practice.

We can apply this rule to the job market models. In all of these, there is one endogenous variable (*WAGE*) included on the right-hand side of each equation. In the simplest model (17.20)-(17.21), illustrated in Figure 17.2a, there are no exogenous variables in the system, so there are zero exogenous variables excluded from each equation. Since zero is not greater than or equal to 1, neither equation is identified. In the second model (17.22)-(17.23), illustrated in Figure 17.3, there is one exogenous variable (*OTHINC*) in the entire system. In the demand equation this one exogenous variable excluded; since this just equals the number of endogenous variables included on the right-hand side, the equation is just identified. In the supply equation zero exogenous variables are excluded; the equation is unidentified. In the original job market model (17.1)-(17.2), there are two exogenous variables in the system. Both of these are excluded from the demand equation, and since 2 is greater than 1, the equation is overidentified. The supply equation is unidentified.

Interpretation

What does it mean to say that an equation cannot be estimated meaningfully if it is not identified? Consider the unidentified supply equation (17.23) in the second job market model. Certainly, some estimation *is* possible; given the data, we can use OLS to run a multiple regression of *HOURS* on *WAGE* and *OTHINC*. The computer will accept the job and produce results. The crux of the identification analysis is that these estimated coefficients do not bear a determinate relation to γ_0, γ_1, and γ_2. Consistent estimation is impossible, with OLS or any other method. If we try to use TSLS on this unidentified equation, the technique breaks down. The reduced form proxy regressor for *WAGE* is simply a linear combination of *OTHINC*, which is already in the equation. This situation of perfect multicollinearity violates a basic assumption of the technique and might cause the computer to stop or produce unpredictable results. Similarly, in any unidentified equation, OLS will "work" but produce meaningless results. The TSLS technique will break down because there will be too much overlap between the proxy regressors and the exogenous (predetermined) variables that are included in the equation.

Some additional insight into identification comes from looking at an estimating technique known as ***indirect least squares,*** ILS. Consider (17.8)-(17.9), which represent the simplified reexpression of the reduced form (17.6)-(17.7) of the job market model. The ratio π_{21}/π_{11} equals β_1, as may be seen by carefully inspecting the full form of the coefficients. Similarly, the ratio π_{22}/π_{12} exactly equals β_1 also. Thus, it seems that estimates of the β's can be obtained from estimates of the π's.

The π's of the reduced-form equations can be estimated by ordinary least squares, and the ratios $\hat{\pi}_{21}/\hat{\pi}_{11}$ and $\hat{\pi}_{22}/\hat{\pi}_{12}$ both are ILS estimators of β_1. Somewhat surprisingly, the two ILS estimators do not yield the same estimate of β_1. This situation of having more than one ILS estimator for a particular coefficient in a structural equation is characteristic of that equation's being overidentified. If an equation is just identified, there is just one ILS estimator for each of its coefficients. And if an equation is unidentified, it is impossible to form ILS estimators for all the coefficients. For example, there is no ratio or combination of the π's in (17.8)-(17.9) that yields γ_1, and the supply equation (17.2) is unidentified. In short, we can say that whether or not an equation is identified depends on whether or not the ILS estimators can be formed.

Indirect least squares estimators are biased but consistent. If an equation is just identified, ILS leads to exactly the same estimates as TSLS. If an equation is overidentified, ILS offers multiple estimates with no basis for choosing among them, whereas TSLS gives a single set of estimates. Hence, TSLS is usually preferred in practice.

Finally, we consider what can be done if an equation is unidentified. If the model is fully and correctly specified already, there is nothing to do. In most cases, however, the models we construct are considered to be only reasonable abstractions from the complexities of reality. Thus, respecifying the model can lead to another reasonable specification that may allow estimation of a previously unidentified equation. For example, if employers take a person's measured skill or productivity into account in setting their *HOURS-WAGE* relation, we would want to respecify the demand equation (17.1) to include this new exogenous variable. Presuming that skill or productivity does not affect the supply decision, the supply equation would now be identified, and it could be estimated.

Problems

Section 17.1

⋆ **17.1** Consider a structural model that is identical to equation set (17.1)-(17.2) except that *NKIDS* is not included. Solve for the reduced form.

17.2 Consider a structural model that is identical to equation set (17.1)-(17.2) except that *OTHINC* and *NKIDS* are not included. Solve for the reduced form.

⋆ **17.3** Consider the following structural model:

$$Y_i = \beta_0 + \beta_1 X_i + u_i$$

$$X_i = \gamma_0 + \gamma_1 Z_i + v_i$$

in which Y and X are considered endogenous. Analyze the logical structure of this ***recursive model.***

17.4 Consider Equations (17.3) and (17.4) to be a two-equation model. Taking *CON* and *GNP* to be endogenous, solve for the reduced form.

⋆ **17.5** Suppose that the government decides to assure full employment by letting A adjust to whatever is necessary to fix *GNP* at a desired level. Taking Equations (17.3) and (17.4) to be a two-equation model, this makes *GNP* exogenous and A endogenous. Solve for the reduced form.

17.6 Suppose that Equation (17.13) is viewed as a single-equation model. Will the predicted value of *CON* for a given year (using the standard single-equation method) be the same as the simulated value of *CON* for the same year (using the systems method)?

Section 17.2

17.7 Consider a single-equation simple regression model. Draw a figure showing what the generated data look like if the disturbance is positively correlated with the explanatory variable. Sketch in the (hypothetical) fitted regression line.

17.8 Using the calculated standard errors to be indicators of estimation precision, would you judge the TSLS estimates of the macro model to be substantially different from the OLS estimates?

⋆ **17.9** Does OLS lead to simultaneous-equation bias for the equations in the recursive model discussed in Problem 17.3?

★ **17.10** Why are the $\hat{\pi}$'s in the first stage of TSLS related to the disturbances of the system? (*Hint:* The $\hat{\pi}$'s are OLS estimators.)

Section 17.3

★ **17.11** Assess the identification of Equations (17.3) and (17.5) in the macroeconomic model. [Equation (17.4) needs no estimation, and identification is not an issue.]

17.12 Consider the two-equation model of Problem 17.4. Is the consumption equation identified?

17.13 An alternative form of the order condition is: ''If an equation is identified, the number of excluded variables (counting endogenous and predetermined together) must be greater than or equal to the number of *other* equations in the model.'' Show that this is equivalent to the statement in the text.

★ **17.14** Consider the reduced-form model of Problem 17.1. How can ILS be used to estimate β_1 in the demand equation?

17.15 How does adding ''productivity'' as an exogenous variable to the demand equation (17.22) affect the identification of the demand and supply equations?

Topics in Statistics

Inference for the Mean and Variance

In this chapter we develop and apply methods of statistical inference for the mean and variance of a random variable. The analysis parallels that in Chapters 11 through 13, which present a full discussion of inference in regression. The treatment here extends the earlier ideas and is much more brief.

To set the stage, suppose that there is a random variable having a known probability distribution $p(X)$ with a mean μ_X and variance σ_X^2. If we draw a sample of n observations from this random variable, we expect that the relative frequency distribution of the observed values will resemble the probability distribution $p(X)$ and that the calculated values of the sample data statistics $\overline{X}$ and S_X^2 will be similar to μ_X and σ_X^2. In large samples, we expect that the resemblance and similarities will tend to be great, whereas in small samples they will be less so.

The relations between the sample statistics and the corresponding parameters of the random variable are the subject of sampling theory. In turn, sampling theory is the basis for statistical inference.

18.1 The Sampling Distribution of $\overline{X}$

We focus first on characterizing the sampling variability that is associated with the mean, $\overline{X}$, in a sample taken from a given random variable. Let the random variable X have a known probability distribution $p(X)$ with mean μ_X and standard

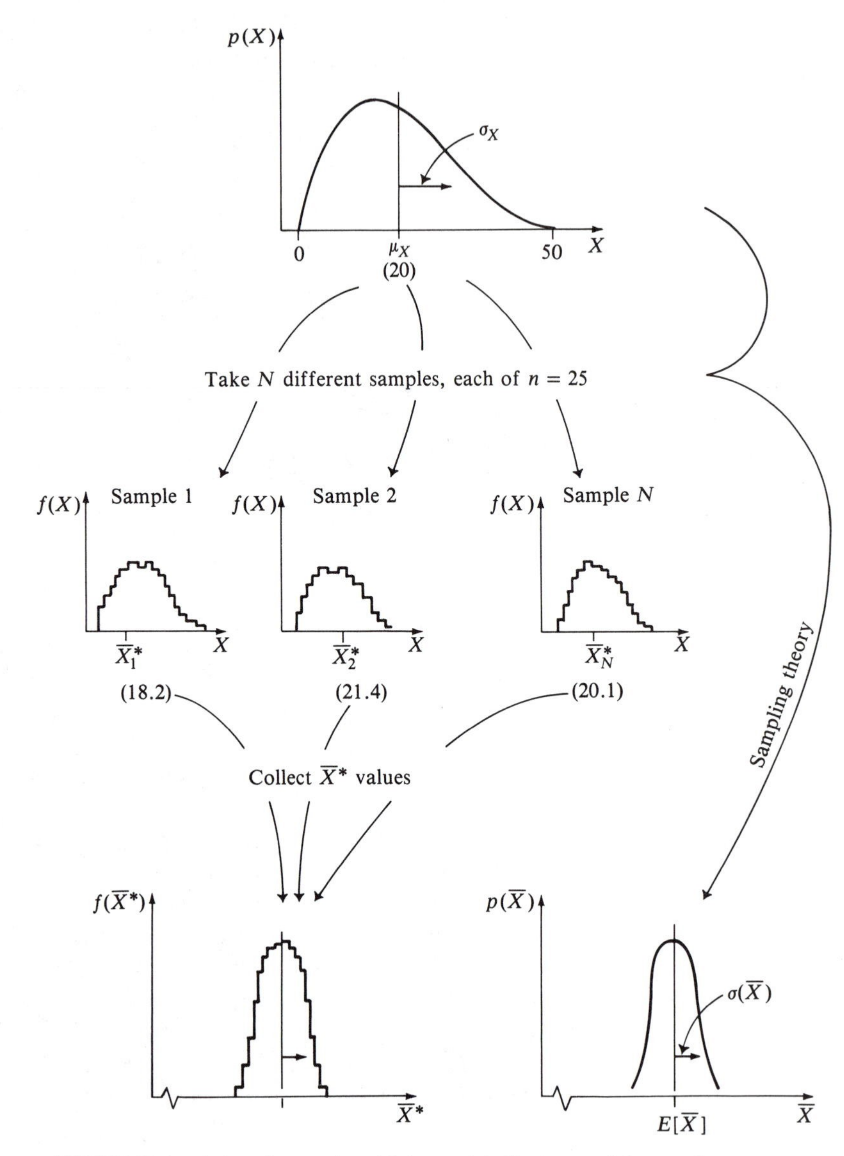

FIGURE 18.1 A thought experiment helps explain the nature of the sampling distribution of $\overline{X}$. We start with a continuous random variable, X, whose probability distribution is shown at the top of the figure. Next, we take N different samples of size n, and in each sample we calculate the mean. These sample means are collected, and their frequency distribution is constructed. Theoretically, as the number of samples (N) increases to infinity, this frequency distribution reaches a limiting form. That limiting form is the sampling distribution of $\overline{X}$, which is the probability distribution that governs the likelihood of different values occurring for $\overline{X}$ in a sample of size n drawn from the original X.

deviation σ_X. As in Chapter 11, we conduct a thought experiment to develop the theory; this is illustrated in Figure 18.1. Suppose that we draw N different samples each containing n observations, the same number in each case. In each sample, we calculate the value of the mean

$$\bar{X} = \frac{1}{n} \sum_{i=1}^{n} X_i \tag{18.1}$$

and across the different samples these are denoted by $\bar{X}_1^*, \bar{X}_2^*, \ldots, \bar{X}_N^*$. (The asterisks indicate actually computed sample values.) Now, the N different $\bar{X}^*$ values can be collected and treated like data themselves: we can organize and determine the relative frequency distribution of the $\bar{X}^*$ values, $f(\bar{X}^*)$, and we can calculate the mean and the standard deviation of these values.

For example, suppose that the possibly occurring values of X range from 0 to 50, that $\mu_X = 20$, and that $\sigma_X = 10$. This specification corresponds to the top panel of Figure 18.1, which shows $p(X)$, the probability distribution of X, to be slightly skewed to the right. Suppose that the sample size (n) is 25, which statisticians consider large but not very large. And suppose that the number of samples (N), only three of which are represented in the figure, is quite large. The relative frequency distribution for each of the N samples bears a close resemblance to the probability distribution, $p(X)$, and therefore they all resemble each other. However, the differences among them are noticeable. For each of the samples, the mean is calculated and the values shown are $\bar{X}_1^* = 18.2$, $\bar{X}_2^* = 21.4$, and $\bar{X}_N^* = 20.1$. Finally, the different $\bar{X}^*$ values are collected and their relative frequency histogram is given in the bottom of the figure.

The frequency distribution of the samples' $\bar{X}$ values, $f(\bar{X}^*)$, is derived from samples that are drawn from the probability distribution of the random variable. Hence, we should expect that knowledge of $p(X)$ would give us information about the characteristics of $f(\bar{X}^*)$. Indeed, from $p(X)$ one can derive mathematically some knowledge regarding a new probability distribution, denoted by $p(\bar{X})$, which is a theoretical analog of the $f(\bar{X}^*)$ constructed from samples in our thought experiment. This probability distribution $p(\bar{X})$ is called the ***sampling distribution*** of $\bar{X}$. In theory, the sampling distribution $p(\bar{X})$ is the limiting form of the frequency distribution $f(\bar{X}^*)$ as the number of samples (N) increases to infinity. In other words, the sampling distribution of $\bar{X}$ is the probability that governs the relative likelihood that different values will occur as the actual $\bar{X}$ in a sample that we draw.

Further mathematical analysis establishes the characteristics of the sampling distribution $p(\bar{X})$. First, if the form of $p(X)$, which is the distribution of the original random variable X, is normal, the form of $p(\bar{X})$ is also normal. Second, if the form of $p(X)$ is not normal, the form of $p(\bar{X})$ is approximately normal anyway if the sample size n is large. This result, which is rather marvelous, is known as the ***Central Limit Theorem.*** It holds true, as an approximation, when X is discrete as well as when it is continuous. The Central Limit Theorem is one of the reasons why the family of normal distributions is so important: other distributions lead to it in the context of sampling. How big does n have to be to make the approximation acceptable? Usually, 15 or 20 is big enough. Hence, for most practical applications we can treat the sampling distribution of $\overline{X}$ as though it were normal. As will be seen, this makes certain important probability calculations rather easy to handle.

The mean and variance of the sampling distribution can be derived by applying results from the appendix to Chapter 9. We now focus on one sample of n observations, rather than the N samples of the thought experiment. The values in the sample (i.e., $X_1, X_2, \ldots, X_n$), are usually thought of as being successive draws from the probability distribution $p(X)$ of the single random variable X. However, we also can think of them as being a set of single draws from n different independent random variables having probability distributions $p(X_1)$, $p(X_2), \ldots, p(X_j), \ldots, p(X_n)$, that are all identical to $p(X)$, with $\mu_j = \mu_X$ and $\sigma_j^2 = \sigma_X^2$. Now, consider defining a new variable named $\overline{X}$ in terms of the outcomes of the n random variables $X_1, X_2, \ldots, X_n$:

$$\overline{X} = \left(\frac{1}{n}\right)X_1 + \left(\frac{1}{n}\right)X_2 + \cdots + \left(\frac{1}{n}\right)X_n = \sum_{j=1}^{n}\left(\frac{1}{n}\right)X_j \tag{18.2}$$

Clearly, $\overline{X}$ is a linear combination of the n variables, and since they are independent (9.30) and (9.31) imply that

$$E[\overline{X}] = \mu_X \tag{18.3}$$

and

$$\sigma^2(\overline{X}) = \frac{\sigma_X^2}{n} \tag{18.4}$$

and thus

$$\sigma(\overline{X}) = \frac{\sigma_X}{\sqrt{n}} \tag{18.5}$$

Why does it make sense to think of the n successive outcomes of a single random variable X as being equivalent to the collection of single draws from each of n independent identically distributed random variables? The nature of a single random variable is that what occurs on one outcome is not affected at all by what occurred on any previous outcome, and this is the key to independence also.

Equation (18.3) shows that the mean of the sampling distribution for $\overline{X}$, which is denoted by $E[\overline{X}]$, is exactly equal to the mean of the underlying distribution of X. In other words, the expected value of the $\overline{X}$ calculated for a sample drawn from a random variable with a mean μ_X is exactly equal to that μ_X. This does not say, of course, that in taking one sample from X the $\overline{X}$ value will be equal to μ_X, but it says that on average it will be.

Since $\sigma(\overline{X})$ is greater than zero, we realize that in any sample there will be a difference between the value of $\overline{X}$ and the value μ_X. Since $\overline{X}$ is used to estimate the unknown μ_X, this difference is known as the ***estimation error,*** or ***sampling error:***

$$\text{estimation error} = \overline{X} - \mu_X \tag{18.6}$$

The estimation error may be positive or negative, but on average it is zero [see (18.3)].

The typical (i.e., anticipated) magnitude of the estimation error (neglecting sign), is given by the ***standard error*** of $\overline{X}$; this is equal to the standard deviation of its sampling distribution, $\sigma(\overline{X})$. As shown by (18.5) , $\sigma(\overline{X})$ is exactly equal to $\sigma_X/\sqrt{n}$. As n gets larger, the standard deviation of the sampling distribution gets smaller. Because n enters the expression through $\sqrt{n}$, this relation is not proportional; in order to cut the standard error, $\sigma(\overline{X})$, in half, the sample size (n) must be quadrupled. This is illustrated in Figure 18.2, which shows $p(\overline{X})$ for samples of size 10 and 40 from the same random variable X.

In summary, for most practical purposes, the sampling (probability) distribution of the mean

$$\overline{X} = \frac{1}{n}\sum_{i=1}^{n} X_i \tag{18.7}$$

in a sample of size n is normal, with

$$E[\overline{X}] = \mu_X \tag{18.8}$$

and

$$\sigma(\overline{X}) = \frac{\sigma_X}{\sqrt{n}} \tag{18.9}$$

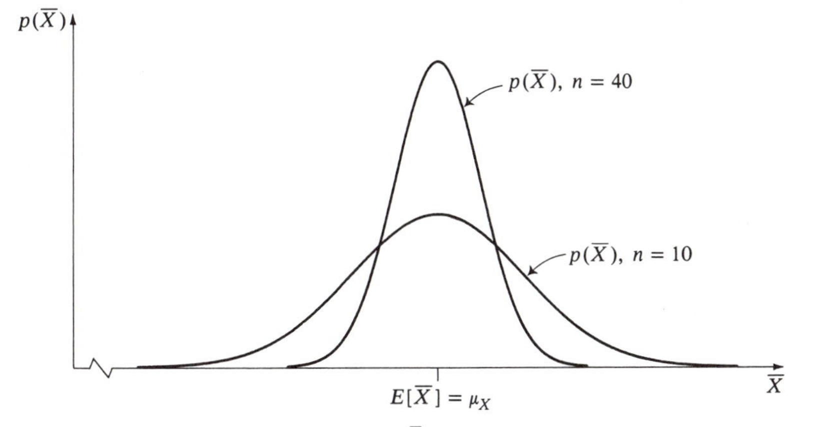

FIGURE 18.2 The sampling distribution of $\overline{X}$ is exactly normal if $p(X)$ is normal. When $p(X)$ is not normal, the sampling distribution of $\overline{X}$ is approximately normal if n is large, according to the Central Limit Theorem. The mean of $p(\overline{X})$ is equal to μ_X, and the standard error is equal to $\sigma_X/\sqrt{n}$. The sampling distributions of $\overline{X}$ shown here are for samples of size 10 and 40 drawn from the same random variable. When the sample size is quadrupled, the standard error is cut in half.

For example, we considered earlier a random variable X that ranges from 0 to 50 and that has a probability distribution $p(X)$ with a mean $\mu_X = 20$ and a standard deviation $\sigma_X = 10$. Suppose that we consider taking one sample of size $n = 25$ and are interested in the $\overline{X}$ value that we will get. Even before all the theory, we might have guessed that it would be about 20, but this is rather imprecise. Sampling theory tells us that the probability distribution of $\overline{X}$ is approximately normal with a mean of $E[\overline{X}] = 20$ and a standard deviation equal to $\sigma(\overline{X}) = 10/\sqrt{25} = 2$.

Thus the theory confirms our intuitive guess, but it also allows us to make some probability calculations. Knowing that $\mu_X = 20$ and $\sigma_X = 10$, we might wonder, for example, what the probability is that $\overline{X}$ in a sample of 25 will be within one unit of the true mean. Formally, we seek to determine Pr $(19 \leq \overline{X} \leq 21)$. This notation highlights the fact that in thinking of the sampling distribution of $\overline{X}$, the $\overline{X}$ itself is the random variable being described. Now, since $\overline{X}$ is taken to be normal with $E[\overline{X}] = 20$ and $\sigma(\overline{X}) = 2$, we can solve this probability problem as we do with any one involving a normal distribution: we solve the equivalent probability problem in terms of Z, the standard normal distribution. The $\overline{X}$ value 19 corresponds to

$$Z_k = \frac{\overline{X}_k - E[\overline{X}]}{\sigma(\overline{X})} = \frac{19 - 20}{2} = -0.5 \tag{18.10}$$

and 21 corresponds to $Z_k = (21 - 20)/2 = 0.5$. Hence we want to determine Pr $(-0.5 \leq Z \leq 0.5)$, because this probability amount is the same as Pr $(19 \leq \overline{X} \leq 21)$. Table A.1 shows us that $\Pr(Z \geq 0.5) = .3085$; after drawing a diagram, recognizing symmetry, and doing a little arithmetic, we find Pr $(-0.5 \leq Z \leq 0.5)$ to be .3830. Hence the probability that $\overline{X}$ in a sample of 25 will be within one unit of the true mean is 38.3 percent in this case.

Continuing the same example, what is the chance that the estimation error for μ_X will be as large as 3 units? (Note that "as large as 3" means "3 or larger" and that the context implies that we are concerned with negative as well as positive errors.) In other words, we seek to determine $\Pr(|\overline{X} - \mu_X| \geq 3)$. After drawing a diagram and noting that $\mu_X = E[\overline{X}] = 20$, we can restate the event $|\overline{X} - \mu_X| \geq 3$ as "$\overline{X} \leq 17$ or $\overline{X} \geq 23$." Thus we seek to determine Pr $(X \leq 17) + \Pr(\overline{X} \geq 23)$. Working with the normal distribution, we find the Z values to be -1.5 and 1.5, respectively, so the total probability is .1336. Thus the chance of having an estimation error as large as 3 is about 13.36 percent in this case.

18.2 Inference Regarding μ_X

When X has a normal distribution, the procedures for carrying out hypothesis tests and for making interval estimates for μ_X are basically the same as those presented in Chapters 12 and 13 for regression coefficients. These are reviewed here, and appropriate differences are noted. If the distribution of X is not normal, these procedures can be applied as good approximations when the sample size is large.

We start from the sampling theory summarized in (18.7)-(18.9). In practical applications we never know the standard error $\sigma(\overline{X})$ because it depends on σ_X, the standard deviation of $p(X)$. Since S_X is used widely as an estimator of σ_X, we use it here to help estimate $\sigma(\overline{X})$. The resulting estimator of the standard error of $\overline{X}$ is given by

$$s(\overline{X}) = \frac{S_X}{\sqrt{n}} \tag{18.11}$$

Now, since the sampling distribution $p(\overline{X})$ is normal with a mean of μ_X and a standard deviation estimated by $s(\overline{X})$, the variable t defined by

$$t = \frac{\overline{X} - \mu_X}{s(\overline{X})} \tag{18.12}$$

can be expected to have a probability distribution that is similar to the standard normal. In fact, it turns out that this t has a t distribution with $n - 1$ degrees of freedom.

Since $\overline{X} - \mu_X$, which is the numerator in (18.12), is the estimation error involved in estimating μ_X, t can be interpreted as a measure of the estimation error relative to the typical estimation error. This sampling statistic, t, is the key to carrying out hypothesis tests and making interval estimates.

Hypothesis Tests

Hypothesis testing regarding μ_X can be carried out using the same five steps outlined in Chapter 12:

1. State the hypotheses clearly.
2. Choose the level of significance, α.
3. Construct the decision rule.
4. Determine the value of the test statistic, t^*.
5. State the conclusion of the test.

For example, suppose that we have a random variable X with a probability distribution that we believe is identical to the one used in Figure 18.1 and earlier examples in this chapter. We believe that $\mu_X = 20$. To test this belief we gather a sample of 36 observations on X.

The hypotheses are statements about an unknown parameter, here μ_X. The null hypothesis must be specific, stating that μ_X is equal to a particular value. The alternative hypothesis must be vague about the value of μ_X; its particular form determines whether the test is one-tailed or two-tailed. In our example we have no reason to believe that μ_X would be greater than 20 rather than less than 20, so the appropriate hypotheses are:

$$\begin{aligned} H_0&: \mu_X = 20 \\ H_1&: \mu_X \neq 20 \end{aligned} \tag{18.13}$$

The goal of the test procedure is to determine if the data contradict the null hypothesis.

The level of significance, α, is chosen by the researcher, and it is usually taken to be a conventional amount such as 5 percent. The chosen α is used in the specification of the decision rule, and it is the probability that the null hypothesis will be mistakenly rejected (a Type I error).

The decision rule for a two-tailed test, which is called for by the hypotheses (18.13), is of the form:

$$\text{Reject } H_0 \text{ if } |t^*| \geq t^c \tag{18.14}$$

where

$$\Pr(|t| \geq t^c) = \alpha \tag{18.15}$$

We recognize that t defined by (18.12) can take on any value from $-\infty$ to ∞. The logic of the decision rule is that the null hypothesis ought to be rejected if the estimation error is so large (in absolute value) relative to the standard error as to be sufficiently unlikely to have occurred if H_0 were true. The measure of "sufficiently unlikely" is the probability amount α, the level of significance. Given α, the critical values $-t^c$ and t^c that bound the range of "so large . . . as to be sufficiently unlikely" are determined from (18.15). This range of values is the critical region, and it is indicated by the range of t values under the shaded

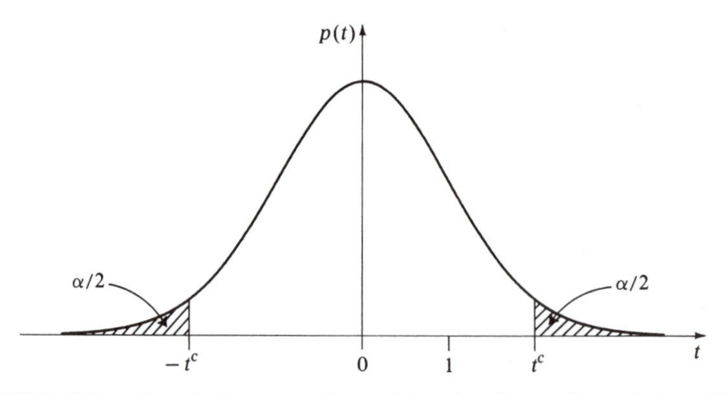

FIGURE 18.3 A hypothesis test regarding μ_X is based on the sampling statistic t, which is equal to $(\bar{X} - \mu_X)/s(\bar{X})$. This t has a t distribution with $n - 1$ degrees of freedom. In a two-tailed test, the decision rule is to reject the null hypothesis if $|t^*| \geq t^c$. The critical region corresponds to the shaded areas here. The critical values t^c and $-t^c$ each bound $\alpha/2$ probability in a tail, where α is the level of significance.

areas in Figure 18.3. In our example, with $\alpha = .05$ and $n = 36$ (so df $= 35$), we see from Table A.3 that $t^c = 2.03$ approximately.

Only now, in the fourth step of the test, do we look at the findings of the sample. Suppose that $\bar{X}^* = 17.4$ and $S_X^* = 9$. The value of the test statistic is

$$t^* = \frac{\bar{X}^* - \mu_X}{s^*(\bar{X})} = \frac{17.4 - 20}{9/\sqrt{36}} = -1.73 \qquad (18.16)$$

That is, presuming that $\mu_X = 20$, the negative estimation error associated with finding $\bar{X}^* = 17.4$ amounts to 1.73 standard errors.

The conclusion of the test is to not reject H_1 because t^* is not in the critical region, and we say that the null hypothesis that $\mu_X = 20$ is accepted. Also, we say that the $\bar{X}^*$ value is not significantly different from 20. However, from Table A.2 we see that the chance of getting a negative sampling error as large as 1.75 standard errors is only about 4.5 percent. Therefore, finding $t^* = -1.73$ should make us somewhat suspicious of the validity of the null hypothesis even though it has not been rejected formally.

Interval Estimates

The procedure for constructing a confidence interval for estimating μ_X follows along the same lines as Chapter 13. The level of confidence, $1 - \alpha$, is chosen by the researcher, with .95 being the most common amount.

It can be shown that the interval

$$\bar{X}^* \pm h, \quad \text{where} \quad h = t^c s^*(\bar{X}) \qquad (18.17)$$

in which t^c is defined by

$$\Pr(|t| \geq t^c) = \alpha \qquad (18.18)$$

can be used to construct an interval estimate of an unknown μ_X at the $1 - \alpha$ level of confidence.

For example, for the sample data used in the hypothesis test, we determine $t^c \approx 2.03$ from Table A.3 for a 95 percent confidence interval. The interval is

$$17.4 \pm (2.03)(9/\sqrt{36}), \quad \text{or} \quad 17.4 \pm 3.05 \qquad (18.19)$$

which ranges from 14.35 to 20.45. We notice that the earlier null hypothesis value $\mu_X = 20$ lies within this interval. Indeed, it can be shown that there is a perfect correspondence between the findings of a $1 - \alpha$ confidence interval and a two-tailed test using an α level of significance: if the value of μ_X under the null hypothesis lies within the confidence interval, H_0 is not rejected, and if μ_X lies outside the interval, H_0 is rejected.

18.3 The Sampling Distribution of S_X^2

We turn now to characterizing the sampling variability that is associated with the variance, S_X^2, to be calculated in a sample from a given random variable. The variance of data in a sample is defined as

$$S_X^2 = \frac{\sum_{i=1}^{n} (X_i - \overline{X})^2}{n - 1} \tag{18.20}$$

Following the same logic as in the earlier thought experiment, we realize that the variance calculated in a sample is a random variable, S_X^2, with some sampling (probability) distribution denoted by $p(S_X^2)$. It can be shown that the mean of this distribution is

$$E[S_X^2] = \sigma_X^2 \tag{18.21}$$

which means that S_X^2 is an unbiased estimator of σ_X^2. This property would not hold true if the variance were defined with n rather than $n - 1$ in the denominator (i.e., if it were the mean squared deviation). Putting this in reverse, we see that the reason for defining S_X^2 with $n - 1$ in the denominator is to make this statistic an unbiased estimator of σ_X^2, the variance of the underlying random variable.

The other characteristics of the sampling distribution $p(S_X^2)$ depend crucially on the shape of $p(X)$. There is nothing like the Central Limit Theorem to point to a limiting form, except in special cases.

Consider the case of a random variable X having a normal distribution with mean μ_X and variance σ_X^2. It can be shown that the statistic

$$\chi^2 = \frac{(n - 1)S_X^2}{\sigma_X^2} = \frac{\sum (X_i - \overline{X})^2}{\sigma_X^2} = \sum_{i=1}^{n} \left(\frac{X_i - \overline{X}}{\sigma_X}\right)^2 \tag{18.22}$$

is a new random variable having a chi-square probability distribution with $n - 1$ degrees of freedom. To understand why, note that if $\overline{X}$ were replaced with μ_X in the rightmost expression, each term in parentheses would be a transformed variable having a standard normal distribution. Squaring and summing these would yield a χ^2 having a chi-square distribution with n degrees of freedom (see the appendix to Chapter 10). The true χ^2 defined by (18.22) has $n - 1$ degrees of freedom, the lost "1" being due to the algebra of isolating $\overline{X}$ and seemingly replacing it with μ_X.

18.4 Inference Regarding σ_X^2

When X has a normal distribution, hypothesis tests regarding the value of σ_X^2 can be carried out using χ^2 as the test statistic. If the null hypothesis H_0: $\sigma_X^2 = \sigma_\circ^2$ is true, then

$$\chi^2 = \frac{(n-1)S_X^2}{\sigma_\circ^2} \tag{18.23}$$

has a chi-square distribution with $n - 1$ degrees of freedom. Table A.4 permits us to find the critical values of χ^2 that bound selected areas in the left side and in the right-hand tail of the distribution, as appropriate.

For example, again using the sample explored earlier (with $n = 36$), suppose we believe that $\sigma_X^2 = 100$ (i.e., $\sigma_X = 10$). The hypotheses for testing this belief are:

$$H_0\colon \sigma_X^2 = 100$$
$$H_1\colon \sigma_X^2 \neq 100 \tag{18.24}$$

Choosing the .05 level of significance, the left-side critical value is $(\chi^2)_l^c =$ 20.56 and the right-hand critical value is $(\chi^2)_r^c = 53.22$. The decision rule is:

$$\text{Reject } H_0 \text{ if } (\chi^2)^* \leq (\chi^2)_l^c \text{ or } (\chi^2)^* \geq (\chi^2)_r^c \tag{18.25}$$

The sampling distribution of χ^2 if the null hypothesis is true is illustrated in Figure 18.4, and the critical region is indicated. Based on the sample in which $S_X^{2*} = 81$, the value of the test statistic is

$$(\chi^2)^* = \frac{(n-1)S_X^{2*}}{\sigma_\circ^2} = \frac{(35)(81)}{100} = 28.35 \tag{18.26}$$

Since $(\chi^2)^*$ is not in the critical region, we do not reject the null hypothesis. Note, however, that this test is properly applied only if it assumed that $p(X)$ is normal, which was not part of the original example.

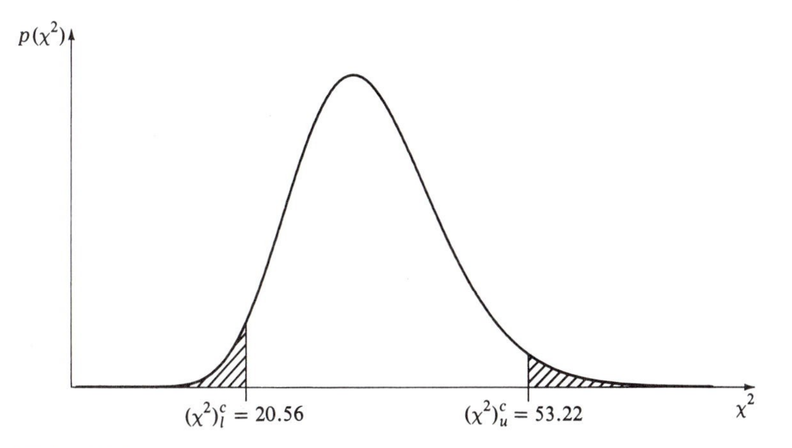

FIGURE 18.4 A hypothesis test regarding σ_X^2 is based on the sampling distribution of the χ^2 test statistic, which is defined in Equation (18.23). This has a chi-square distribution with $n - 1$ degrees of freedom. In a two-tailed test, the decision rule is to reject the null hypothesis if $(\chi^2)^* \leq (\chi^2)_l^c$ or $(\chi^2)^* \geq (\chi^2)_r^c$. As illustrated here, the critical value $(\chi^2)_l^c$ bounds $\alpha/2$ probability in the left side of the distribution, and $(\chi^2)_r^c$ bounds $\alpha/2$ in the right-hand tail; α is the level of significance.

Finally, confidence intervals can be constructed for estimating σ_X^2, based on the sample value S_X^2. Since the sampling distribution of S_X^2 is not symmetric, we are confronted with a new problem in constructing the interval. One approach leads to specifying the following

$$\text{confidence interval for } \sigma_X^2\text{: } \frac{(n-1)S_X^{2*}}{(\chi^2)_r^c} \text{ to } \frac{(n-1)S_X^{2*}}{(\chi^2)_l^c} \tag{18.27}$$

where $(\chi^2)_r^c$ bounds $\alpha/2$ probability in the right-hand tail and $(\chi^2)_l^c$ bounds $\alpha/2$ probability in the left side. For the sample data used above, this 95 percent confidence interval for estimating σ_X^2 ranges from 53.27 to 137.89.

Problems

Section 18.1

18.1 Suppose that X is a random variable having a normal distribution with a mean of 0 and a standard deviation of 4.

(a) On one set of axes draw the probability distribution of X and also a hypothetical relative frequency distribution for a sample of $n = 16$ taken from it.

(b) On another set of axes draw the probability distribution of X and also the sampling distribution $p(\overline{X})$ for the mean of a sample of size 16.

⋆ **18.2** Suppose that X is a random variable having a binomial distribution with $n = 100$ and $P = .5$.

(a) If we take a sample of size 16, determine the mean and standard deviation of the sampling distribution $p(\overline{X})$.

(b) Determine approximately the probability that $\overline{X}$ will be less than 48.

18.3 Consider the frequency distribution $f(\overline{X}^*)$ in Figure 18.1.

(a) What happens to it as N increases?

(b) What happens to it as n increases?

18.4 Suppose that X is a random variable with an unknown mean and a standard deviation equal to 150.

(a) How large a sample must be taken in order that the typical estimation error for μ_X will be 10 or less?

(b) How large a sample must be taken in order that the typical estimation error for μ_X will be 5 or less?

⋆ **18.5** Suppose that X is a discrete random variable whose two possible outcomes are -1 and 1, each occurring with probability 1/2. Determine the sampling distribution of $\overline{X}$ in a sample of size 2: what are its possible outcomes and what are their probabilities?

Section 18.2

18.6 Suppose that X is a random variable. In a sample of size 81 we find $\overline{X}^* = 103$ and $S_X = 15$. Test the hypothesis that $E[X] = 100$, using a 5 percent level of significance.

⋆ **18.7** Continuing Problem 18.6, suppose that the same values for the sample mean and standard deviation were found in a sample of size 400. How would the results of the test be affected?

⋆ **18.8** Determine the 95 percent confidence interval for estimating μ_X in the situation of Problem 18.6.

18.9 What happens to the width of the confidence interval for estimating μ_X if the sample size is quadrupled?

Section 18.3

18.10 Suppose that X has a normal distribution. If we consider taking a sample of size 10 from X, sketch (on separate graphs) the sampling distributions of $\overline{X}$ and of S_X^2.

⋆ **18.11** Suppose that X has a standard normal distribution. In a sample of size 10:

(a) What is the expected value of S_X^2?

(b) Determine (approximately) the probability that S_X^2 will be greater than 2.

Section 18.4

18.12 Suppose that X has a normal distribution. In a sample of size 16 we find $S_X^{2*} = 55$. Does this contradict to the hypothesis that $\sigma_X^2 = 50$?

⋆ **18.13** Construct a confidence interval for estimating σ_X^2 in Problem 18.12.

Chi-Square Tests and Analysis of Variance

This chapter surveys two statistical techniques that are used commonly in fields other than economics. The econometric approach usually involves a specification of the structure of the relationship among variables, and it permits the testing of behavioral hypotheses in that context. Chi-square tests lack this structure, and often are less powerful. By contrast, the analysis of variance specifies a structure, but it turns out that the resulting tests can be carried out easily using regression analysis.

19.1 Chi-Square Tests

In general, ***chi-square tests*** lend themselves to situations in which observations are categorized into a number of groups. This categorization may be on the basis of a single variable or more than one variable. Suppose that n observations are divided into a certain number of categories, with the absolute frequency in the kth (typical) category being denoted by n_k. Now suppose that a theory or belief leads to a prediction that out of the n observations a certain number should be in each category, with the number predicted to be in the kth category being denoted by p_k. The question arises as to whether the observed frequencies, n_k, are sufficiently different from the predicted frequencies, p_k, to cast doubt on the validity of the stated theory or belief. The question is answered by carrying out a hypothesis test.

We seek a test statistic that serves as an overall measure of how much difference there is between the n_k and the p_k. To be useful, the sampling distribution of the test statistic must be known under the condition that the underlying theory is correct. A statistic based simply on $\Sigma\,(n_k - p_k)$ will not serve, because this is always equal to zero. Squaring the differences gets rid of the canceling out of positive and negative values, and it turns out that dividing these squares by p_k standardizes the measure in such a way that the sampling distribution can be determined. Formally, the test statistic is

$$\chi^2 = \sum \frac{(n_k - p_k)^2}{p_k} \tag{19.1}$$

where the summation is understood to include all the defined categories. It turns out that the sampling distribution of χ^2 is approximately chi-square. Generally, the approximation is good if all the p_k values are at least 5.

If the observed frequencies are exactly equal to the predicted frequencies (i.e., if $n_k = p_k$ for all k), then χ^2 equals zero. If the n_k are very different from the p_k, χ^2 is large. Hence large values of the test statistic are used to reject the null hypothesis that the observed frequency distribution arises from a process that leads to the distribution of predicted frequencies. As with all hypothesis tests, we recognize that rejections using this decision rule will sometimes be mistaken. The probability of making this kind of mistake depends on the critical value for the test statistic, and the critical value is chosen so as to make this probability equal to a specified amount—the level of significance.

Test of Goodness of Fit

One major application of the chi-square test is in situations where the predicted values, p_k, are derived from a theory that predicts a certain distribution for the values of a single variable. This ***test of goodness of fit*** examines whether the observed data fit the theoretical distribution, and vice versa. If the number of categories defined in the distribution is r, the χ^2 test statistic has $r - 1$ degrees of freedom. (Given the total number of observations, the number in the rth category is set by the numbers in the other $r - 1$ categories; hence there are $r - 1$ degrees of freedom here.)

For example, suppose we have a theory of organizations that predicts that for every manager in a certain firm, there will be 3 white-collar workers and 6 blue-collar workers. Thus, 10 percent of the total number of employees will be managers, 30 percent will be white-collar workers, and 60 percent will be blue-collar workers. Suppose also that we have complete personnel data for a firm of this type with 200 employees, and that the categorical breakdown shows 25 managers, 65 white-collar workers, and 110 blue-collar workers. Based on the organizational theory, the corresponding predicted frequencies (p_k) are 20, 60, and 120. These observed and predicted frequencies are displayed in Table 19.1.

TABLE 19.1 Observed and Predicted Occupation Frequencies

	Observed n_k	Predicted p_k	$n_k - p_k$	$\frac{(n_k - p_k)^2}{p_k}$
Manager	25	20	5	1.25
White-collar	65	60	5	0.42
Blue-collar	110	120	−10	0.83
				χ^2 = 2.50

It looks as though this firm is a bit overloaded with managers and white-collar workers, possibly contradicting the theory in this case, but we need a hypothesis test to determine whether the differences could be due simply to chance variation.

The null hypothesis is that the distribution of predicted frequencies correctly characterizes the process underlying the generation of the observed data. We carry out the test at the .05 level of significance. With three occupational groups, the number of degrees of freedom is 2 (df $= r - 1$ here), and from Table A.4 we find that the critical value is $(\chi^2)^c = 5.99$. The sampling distribution of χ^2 and the critical region for the test are illustrated in Figure 19.1. Table 19.1 shows the calculations needed to determine the value of the test statistic, $(\chi^2)^* = 2.50$. Since $(\chi^2)^*$ is not greater than $(\chi^2)^c$, we do not reject the null hypothesis. The differences between actual and predicted frequencies are not so large as to be inconsistent with the organizational theory from which the predicted frequencies were determined.

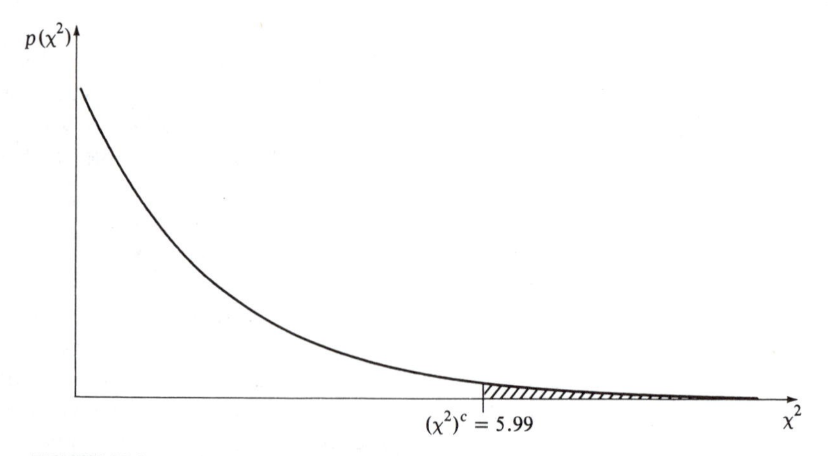

FIGURE 19.1 A chi-square test can be used to judge whether the observed and predicted frequencies in Table 19.1 are consistent. The test statistic, χ^2, has a chi-square distribution with 2 degrees of freedom. The critical value that bounds 5 percent of the probability in the right-hand tail is 5.99, and the critical region consists of the values of χ^2 in the shaded area.

Test of Independence

A second application of the same basic approach leads to what is known as the ***test of independence*** in a contingency table.

Suppose that we still have available all the personnel data from the firm in the previous example, but now we wish to investigate whether an employee's sex plays a role in the process of occupational hiring or assignment. Each employee is an observation in the data, and we have measurements on two variables: sex and occupation. One way to look at the data is given in Table 19.2, which shows the number of employees in each sex-occupation category.

The numbers in parentheses show the frequencies that would be predicted if occupation were independent of sex in the data. These are arrived at by taking as given the proportions of workers of each sex and in each occupation. We can quickly see that 60 percent of the workers are male and 40 percent are female; similarly, 12.5 percent of the workers are managers, 32.5 percent are white-collar workers, and 55 percent are blue-collar workers. Based on these overall (marginal) observed relative frequencies, we would predict that if occupational hiring and assignment were independent of sex, then 60 percent of all managers would be male and 40 percent would be female. Given 25 actual managerial jobs, this prediction translates into 15 male managers and 10 female managers. The sex-specific predictions for the other occupations are made in the same way, by multiplying the total number of jobs of the occupational type by the (marginal) relative frequency of each sex type. It should be noted that exactly the same set of predicted frequencies for the six categories is obtained when the total number of workers of each sex is multiplied by the (marginal) relative frequency of each occupation type.

It appears from the data that women are more likely to be in white-collar jobs than in either managerial or blue-collar jobs, compared with the predicted frequencies. Put differently, managers and blue-collar workers are more likely to be men than is predicted. Do these differences contradict the null hypothesis that sex and occupation are independent, or are they due just to chance fluctuation?

This question is answered by carrying out a hypothesis test using the test statistic χ^2 defined in (19.1). If the null hypothesis is correct, χ^2 has a chi-square distribution with $(r - 1)(c - 1)$ degrees of freedom, where r is the number of rows and c is the number of columns in the main body of the

TABLE 19.2 Sex-Occupation Contingency Table

	Male		Female		Both Sexes
Manager	21	(15)	4	(10)	25
White-collar	27	(39)	38	(26)	65
Blue-collar	72	(66)	38	(44)	110
All occupations	120		80		

contingency table. (The counting of r and c excludes the marginal totals.) The proper number of degrees of freedom can be interpreted as the number of unrestricted comparisons or differences that lead to the calculation of χ^2. In any column, the last entry can be determined from the $r - 1$ entries above it and the marginal frequency below it. Similarly, in any row the last entry can be determined from the $c - 1$ entries to the left of it and the marginal frequency to the right of it. Hence there are $(r - 1)(c - 1)$ unrestricted entries.

Returning to the example, with three rows and two columns there are two degrees of freedom for the relevant chi-square distribution. Large differences between observed and predicted frequencies in the categories lead to large values of χ^2. Hence large values of χ^2 constitute the critical region of values that lead us to reject the null hypothesis that sex and occupation are independent. At the .05 level of significance, Table A.4 gives $(\chi^2)^c = 5.99$. From Table 19.2 we calculate $(\chi^2)^* = 16.6$. Since $(\chi^2)^*$ is greater than $(\chi^2)^c$, we reject the null hypothesis and conclude that sex and occupation are not independent.

It should be noted that in going from the goodness-of-fit test to the test of independence for the same set of data, the nature of the analysis changes. In the first case we explicitly test a theory of occupational organization. In the second test, we take as given both the occupational structure and the relative numbers of men and women, and we test just for their independence.

19.2 Analysis of Variance and Regression

Analysis of variance, commonly referred to as ANOVA, is a statistical technique that seeks to determine whether differences in the values of a variable can be explained by categorization of the observations. This technique was developed for application to experimental data. Most of the results can be obtained through regression, and we approach ANOVA from this point of view.

ANOVA Table for Regression

The traditional famework for analysis of variance gives rise to a special format for organizing some of the results of regression estimation and hypothesis

TABLE 19.3 ANOVA Table for Regression

Source of Variation	Sum of Squares	Degrees of Freedom	Mean Square	F Ratio
Explained (by regression)	$\sum (\hat{Y}_i - \bar{Y})^2$	k	$\dfrac{\sum (\hat{Y}_i - \bar{Y})^2}{k}$	$\dfrac{\sum (\hat{Y}_i - \bar{Y})^2/k}{\sum (Y_i - \hat{Y}_i)^2/(n - k - 1)}$
Unexplained (due to error)	$\sum (Y_i - \hat{Y}_i)^2$	$n - k - 1$	$\dfrac{\sum (Y_i - \hat{Y}_i)^2}{n - k - 1}$	
Total	$\sum (Y_i - \bar{Y})^2$	$n - 1$		

testing. This format, which is known as an ANOVA table, is shown generically for the results of a multiple regression in Table 19.3. Many computer programs print such a table for each estimated regression, replacing the formulas and symbols given here for their calculated values.

The core of the table is the column labeled "Sum of Squares." The three items here represent the identity derived earlier as (5.24), which states that the total variation in Y can be decomposed into the explained variation and the unexplained variation:

$$\sum (Y_i - \overline{Y})^2 = \sum (\hat{Y}_i - \overline{Y})^2 + \sum (Y_i - \hat{Y}_i)^2 \tag{19.2}$$

The coefficient of determination, R^2, can be computed as the ratio of the sum of squares in the first row to that in the third; equivalently, it is 1 minus the ratio of the second to the third:

$$R^2 = \frac{\sum (\hat{Y}_i - \overline{Y})^2}{\sum (Y_i - \overline{Y})^2} = 1 - \frac{\sum (Y_i - \hat{Y}_i)^2}{\sum (Y_i - \overline{Y})^2} \tag{19.3}$$

The assignment of values to the column labeled "Degrees of Freedom" is definitional, and it is made as an intermediate step to later results. The total variation is assigned $n - 1$ degrees of freedom and the explained variation is assigned k, which is the number of regressors. The difference between these, $n - k - 1$, is assigned to the unexplained variation.

The "Mean Square" for each row is that row's sum of squares divided by its degrees of freedom. The mean square due to the regression is referred to as the explained variance. The mean square due to the error is an unbiased estimate of the variance of the disturbance, σ_u^2, and the square root of this item is the standard error of the regression, *SER*. Were we to calculate the mean square for the third row, it would be simply the variance of Y. It should be noted that the explained variance plus the unexplained variance does not equal the (total) variance of Y.

The only item in the "F" ratio column is the ratio of the first mean square to the second, and it is sometimes called the variance ratio. This ratio is the same as the value of the F statistic derived earlier (15.12) to test the null hypothesis that all the regression coefficients except the intercept are equal to zero. SSR_U in the earlier terminology is the same as the unexplained sum of squares here. SSR_R results from a regression on a constant, because all β_j are restricted to be zero. The resulting SSR is identical to the total sum of squares in the ANOVA table here (Table 19.3). Hence $SSR_R - SSR_U$ is equal to the explained sum of squares in the ANOVA table, and thus the F ratios are identical.

One-Way ANOVA

We develop some of the ideas of analysis of variance with examples continuing from earlier in this chapter. Suppose that we have data on the salaries (or annual wage payments) made to the employees of a particular firm. This salary is denoted by Y, with the mean being $\overline{Y}$. Now suppose that the observations are categorized into three occupations: manager, white-collar workers, and blue-collar workers. The salary means within these three groups are $\overline{Y}_1$, $\overline{Y}_2$, and $\overline{Y}_3$, respectively. It seems natural to ask whether observed differences in these means reflect true interoccupational differences or just chance fluctuation.

Underlying the ANOVA technique is a set of assumptions about how salaries are generated. It is assumed that for the jth occupation, observed Y's are outcomes from a normally distributed random variable with mean μ_j and variance σ^2. The means may differ among the occupations, but the variances are assumed to be the same. Our inquiry into the data is to test the null hypothesis

$$H_0\colon \mu_1 = \mu_2 = \mu_3 \tag{19.4}$$

against the alternative that at least one of them is different from the others.

Given data on salary (Y) and the categorization of observations into the three occupations, it is simple to determine the mean Y for all observations ($\overline{Y}$) and the mean Y within each occupation, denoted generically by $\overline{Y}_j$. These occupational means can be used to "predict" or help explain the observed Y values within each occupation. For an observation in the jth occupation, the predicted salary is simply $\overline{Y}_j$ and the error associated with this prediction is $Y_i - \overline{Y}_j$. For each observation,

$$(Y_i - \overline{Y}) = (\overline{Y}_j - \overline{Y}) + (Y_i - \overline{Y}_j) \tag{19.5}$$

is a simple algebraic identity that can be interpreted as a decomposition of the total deviation of Y_i (from $\overline{Y}$) into two parts: the deviation of the group mean $\overline{Y}_j$ from the overall mean $\overline{Y}$, and the error associated with using this group mean to predict Y_i. This equation can be squared to yield another valid equation, and the n squared equations like that (one for each observation) can be added to yield, after extensive algebraic manipulation,

$$\sum (Y_i - \overline{Y})^2 = \sum (\overline{Y}_j - \overline{Y})^2 + \sum (Y_i - \overline{Y}_j)^2 \tag{19.6}$$

This expression, whose evaluation requires careful attention to the correct j and i indices, can be reexpressed as

$$SS_T = SS_E + SS_U \tag{19.7}$$

In this simplified expression, SS_T is the total variation of Y from its mean, and it is called the total sum of squares. SS_E is the portion of SS_T that is derived from terms like $\overline{Y}_j - \overline{Y}$, which help "explain" deviations of Y_i from $\overline{Y}$ on the basis of the occupational mean $\overline{Y}_j$. For this reason, SS_E is called the explained sum of squares. Finally, SS_U is the portion of SS_T that is derived from terms like $Y_i - \overline{Y}_j$, which is "unexplained" even after occupational categorization. This SS_U is called the unexplained sum of squares.

These sums of squares can be arranged as in Table 19.4. Letting r be the number of groups (occupations), we assign $r - 1$ degrees of freedom to the

TABLE 19.4 One-Way ANOVA Table

Source of Variation	Sum of Squares	Degrees of Freedom	Mean Square	*F* Ratio
Explained (by differences in $\overline{Y}_j$)	SS_E	$r - 1$	$\frac{SS_E}{r - 1}$	$\frac{SS_E/(r - 1)}{SS_U/(n - r)}$
Unexplained	SS_U	$n - r$	$\frac{SS_U}{n - r}$	
Total	SS_T	$n - 1$		

explained variation and $n - r$ to the unexplained variation, adding up to $n - 1$ for the total. The mean square in each row is the sum of squares divided by the number of degrees of freedom. The F ratio is the ratio of the first two mean squares, and sometimes is referred to as the variance ratio.

It can be shown that if the null hypothesis (19.4) is true, this variance ratio indeed has an F distribution, with $r - 1$ and $n - r$ degrees of freedom. Large values of the F ratio reflect considerable differences among the $\overline{Y}_j$, and therefore large values of F form the critical region for rejecting the null hypothesis.

We have avoided explaining the detail of this method and hypothesis test because exactly the same results can be obtained using regression analysis with dummy variables. To see this, let the dummy variable *D1* take on the value 1 if the observation is a manager and 0 otherwise, and let the variable *D2* take on the value 1 if the observation is a white-collar worker and 0 otherwise. The regression specification

$$Y_i = \beta_0 + \beta_1 D1_i + \beta_2 D2_i + u_i \tag{19.8}$$

can be estimated by OLS, and an ANOVA table like Table 19.3 could be produced.

A test of the proposition that $E[Y]$ is the same for each occupation is equivalent to testing the null hypothesis that occupation has no effect on Y (i.e., that $\beta_1 = \beta_2 = 0$). This test is carried out using the basic F test of regression, for which the value of the test statistic, F^*, is found in the regression ANOVA table. The validity of this test depends on the correctness of the assumptions underlying the regression model. In particular, the regular assumptions amount to saying that for each occupation the distribution of Y is assumed to be normal, with the same variance.

It turns out that $\hat{\beta}_0 = \overline{Y}_3$, $\hat{\beta}_1 = \overline{Y}_1 - \overline{Y}_3$, and $\hat{\beta}_2 = \overline{Y}_2 - \overline{Y}_3$. Hence the regression can be interpreted as "explaining" the observed values of Y on the basis of the mean value of the salary for the occupation to which each observation belongs. The unexplained variation, $\Sigma\,(Y_i - \hat{Y}_i)^2$, measures the variation in Y that arises within each occupation. That is, within each occupation $\hat{Y} = \overline{Y}_j$ $(j = 1, 2, 3)$, and for each worker therein $Y_i - \hat{Y}_i = Y_i - \overline{Y}_j$. Thus the unexplained variation within each occupation is the sum of squares like

$(Y_i - \overline{Y}_j)^2$, and the unexplained variation in the whole regression is the sum of the three occupations' sums of squares. Thus regression and ANOVA lead to precisely the same decomposition of the total variation of Y, and the F statistics are identical.

Two-Way ANOVA

We extend our example further by recognizing explicitly that salaries may be explained by more factors that just occupation. Specifically, we suppose that the employee's sex is a variable that may help determine salary levels. It is natural to ask whether there are systematic differences in salary by occupation, given that salary is also related to the employee's sex. Symmetrically, one might ask whether there are systematic differences by sex, given occupation.

We approach this problem first through regression. Extending the previous specification (19.8), we let S be a dummy variable taking the value 1 for male workers and 0 for female workers. The resulting model is

$$Y_i = \beta_0 + \beta_1 D1_i + \beta_2 D2_i + \beta_3 S_i + u_i \tag{19.9}$$

In its theoretical specification, this model states that the expected salary within each group is as given in Table 19.5.

The null hypothesis that occupation has no effect on salary is H_0: $\beta_1 = \beta_2 = 0$, and this is tested by using a regression F test. The unconstrained model is simply (19.9), and the constrained model is

$$Y_i = \beta_0 + \beta_3 S_i + u_i \tag{19.10}$$

The null hypothesis that there is no difference by sex is H_0: $\beta_3 = 0$. This also can be tested with an F test in regression, but the same result is achieved simply by using the basic significance test (a t test) on β_2 in (19.9).

The analysis of variance treatment of the same questions is based on equivalent assumptions and leads to identical F tests. In the special case when the two factors are independent, as occurs in some experimental designs, there can be a decomposition of the total sum of squares into two explained sums of squares (one for each factor) and an unexplained sum of squares. The F tests are then based on the appropriate variance ratios. However, when the factors

TABLE 19.5 $E(Y)$ for a Regression Model

	Male	Female
Manager	$\beta_0 + \beta_1 + \beta_3$	$\beta_0 + \beta_1$
White-collar	$\beta_0 + \beta_2 + \beta_3$	$\beta_0 + \beta_2$
Blue-collar	$\beta_0 + \beta_3$	β_0

are not independent this decomposition is not so simple and the tests are more easily handled through regression.

Problems

Section 19.1

* **19.1** In the example leading to Table 19.1, suppose that there are 600 workers in the firm instead of 200 and that the actual proportionate breakdown by occupation is the same as in the example. Test to see if the data contradict the organizational theory.

19.2 Using the 120 observations on men in Table 19.2, test to see if the theory leading to the test in Table 19.1 is supported.

19.3 Suppose that 60 rolls of a die result in 15 occurrences of an ace (one spot) and 9 occurrences of each of the other possible outcomes. Test whether this is a fair die.

* **19.4** Suppose that 100 observations occur as follows: 10 less than -1, 45 between -1 and 0, 30 between 0 and 1, and 15 greater than 1. Use a test of goodness of fit to determine if these observations might have been taken from a standard normal random variable.

19.5 The following table shows the racial and poverty status breakdown among 900 people. Is being poor related to race?

	Blacks	Whites
Poor	22	113
Not poor	78	687

* **19.6** The following table shows the educational attainment and earnings status of 600 people. Is income related to education?

	Grade School	High School	College
Less than \$10,000	70	90	40
\$10,000–\$20,000	20	160	70
More than \$20,000	10	50	90

Section 19.2

19.7 Given the specification of a regression, the values of R^2 and SSR from its estimation, and the value of n, explain how an ANOVA table can be created from this information.

For the next four problems, consider the following three regressions, which are hypothetical estimates of equations (19.8)–(19.10) with $n = 100$:

$$\hat{Y}_i = 12.0 + 18.0D1_i + 3.0D2_i$$
$$R^2 = .350 \qquad SSR = 920.0$$

$$\hat{Y}_i = 13.0 + 16.0D1_i + 4.0D2_i - 3.0S_i$$
$$R^2 = 3.70 \qquad SSR = 891.7$$

$$\hat{Y}_i = 19.0 - 7.0S_i$$
$$R^2 = .250 \qquad SSR = 1061.5$$

19.8 In a one-way ANOVA for salaries, are there significant differences by sex?

★ **19.9** In a one-way ANOVA for salaries, are there significant differences by occupation?

★ **19.10** In a two-way ANOVA for salaries, are there significant differences by sex?

19.11 In a two-way ANOVA for salaries, are there significant differences by occupation?

Statistical Tables

Use of Table A.1: The Standard Normal Distribution

The main body of this table gives values of Pr $(Z \geq Z^*)$, where Z is a random variable having a standard normal distribution (see Section 10.1). Each Z^* is the sum of the values of the row and column labels. In the following figure, Z^* is a specific value of Z, and Pr $(Z \geq Z^*)$ is given by the shaded area.

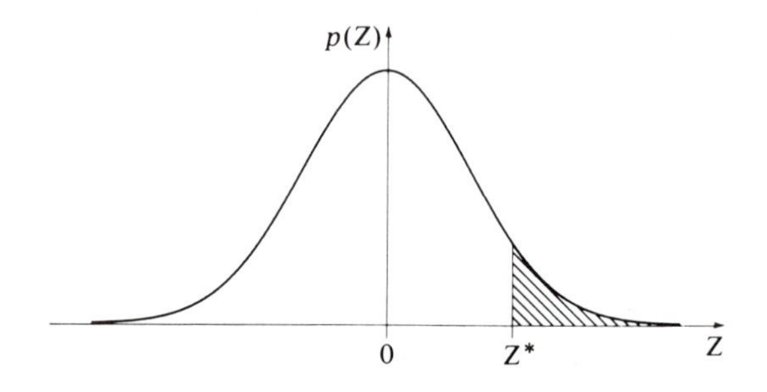

This table is used to answer two types of probability problems. First,

$$\text{Find } \alpha \text{ such that Pr } (Z \geq Z^*) = \alpha$$

In this case the Z^* value is given, and the table is used to determine the corresponding probability, which is denoted by α. Second.

$$\text{Find } Z^c \text{ such that Pr } (Z \geq Z^c) = \alpha$$

In this case the probability amount (α) is given, and the table is used to determine the critical value of Z (denoted generally by Z^c, but corresponding to Z^* in the table) that bounds exactly α probability in the right-hand tail. In both cases, interpolation can be used if greater precision is desired.

The lower section of the table gives several probability statements that are of special interest for statistical inference.

For example, given $Z^* = 1.53$ we look in the row labeled 1.50 and the column labeled .03; the table entry gives Pr $(Z \geq 1.53) = .063$. If we are given that Pr $(Z \geq Z^c) = .25$, we find that Z^c lies between 0.67 and 0.68.

TABLE A.1 Standard Normal Distribution

	.00	.01	.02	.03	.04	.05	.06	.07	.08	.09
.00	.500	.496	.492	.488	.484	.480	.476	.472	.468	.464
.10	.460	.456	.452	.448	.444	.440	.436	.433	.429	.425
.20	.421	.417	.413	.409	.405	.401	.397	.394	.390	.386
.30	.382	.378	.374	.371	.367	.363	.359	.356	.352	.348
.40	.345	.341	.337	.334	.330	.326	.323	.319	.316	.312
.50	.309	.305	.302	.298	.295	.291	.288	.284	.281	.278
.60	.274	.271	.268	.264	.261	.258	.255	.251	.248	.245
.70	.242	.239	.236	.233	.230	.227	.224	.221	.218	.215
.80	.212	.209	.206	.203	.200	.198	.195	.192	.189	.187
.90	.184	.181	.179	.176	.174	.171	.169	.166	.164	.161
1.00	.159	.156	.154	.152	.149	.147	.145	.142	.140	.138
1.10	.136	.133	.131	.129	.127	.125	.123	.121	.119	.117
1.20	.115	.113	.111	.109	.107	.106	.104	.102	.100	.099
1.30	.097	.095	.093	.092	.090	.089	.087	.085	.084	.082
1.40	.081	.079	.078	.076	.075	.074	.072	.071	.069	.068
1.50	.067	.066	.064	.063	.062	.061	.059	.058	.057	.056
1.60	.055	.054	.053	.052	.051	.049	.048	.047	.046	.046
1.70	.045	.044	.043	.042	.041	.040	.039	.038	.038	.037
1.80	.036	.035	.034	.034	.033	.032	.031	.031	.030	.029
1.90	.029	.028	.027	.027	.026	.026	.025	.024	.024	.023
2.00	.023	.022	.022	.021	.021	.020	.020	.019	.019	.018
2.10	.018	.017	.017	.017	.016	.016	.015	.015	.015	.014
2.20	.014	.014	.013	.013	.013	.012	.012	.012	.011	.011
2.30	.011	.010	.010	.010	.010	.009	.009	.009	.009	.008
2.40	.008	.008	.008	.008	.007	.007	.007	.007	.007	.006
2.50	.006	.006	.006	.006	.006	.005	.005	.005	.005	.005
2.60	.005	.005	.004	.004	.004	.004	.004	.004	.004	.004
2.70	.003	.003	.003	.003	.003	.003	.003	.003	.003	.003
2.80	.003	.002	.002	.002	.002	.002	.002	.002	.002	.002
2.90	.002	.002	.002	.002	.002	.002	.002	.001	.001	.001
3.00	.001	.001	.001	.001	.001	.001	.001	.001	.001	.001

$\Pr(Z \geq 1.282) = .10$
$\Pr(Z \geq 1.645) = .05$
$\Pr(Z \geq 1.960) = .025$
$\Pr(Z \geq 2.326) = .01$
$\Pr(Z \geq 2.576) = .005$

Note: Table entry gives $\Pr(Z \geq Z^*)$, where Z^* is the sum of the values of the row and column labels.

Source: Computed using Fortran subroutines from the IMSL Library.

Use of Table A.2: The t Distribution

This table gives selected right-hand tail probabilities for a random variable having a t distribution with df degrees of freedom (see Section 10.2). Each entry in the table is the answer to a problem of the form:

$$\text{Find } \alpha \text{ such that } \Pr(t \geq t^*) = \alpha$$

In hypothesis testing, this α is known as the P-value. In the following figure, t^* is a specific value of the random variable t and the probability α is given by the shaded area. In the table, various t^* values are given in the column headings, and the corresponding α is found in the row labeled by df.

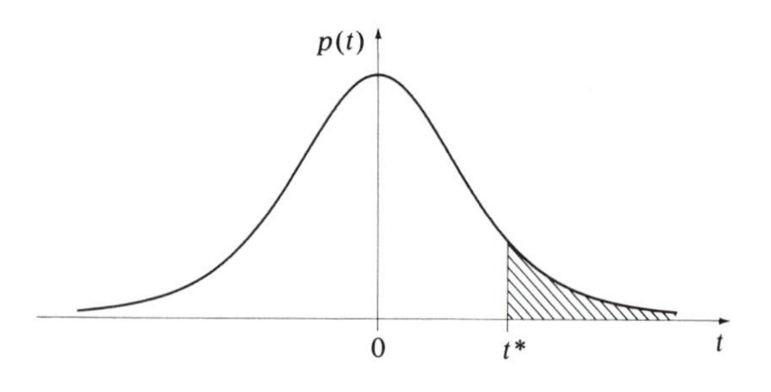

For example, if df $=$ 15, we find that $\Pr(t \geq 1.50) = .077$ and $\Pr(t \geq 2.50) = .012$.

TABLE A.2 *P*-Values for the *t* Distribution

t^* =	0.50	1.00	1.25	1.50	1.75	2.00	2.25	2.50	3.00
df = 1	.352	.250	.215	.187	.165	.148	.133	.121	.102
2	.333	.211	.169	.136	.111	.092	.077	.065	.048
3	.326	.196	.150	.115	.089	.070	.055	.044	.029
4	.322	.187	.140	.104	.078	.058	.044	.033	.020
5	.319	.182	.133	.097	.070	.051	.037	.027	.015
6	.317	.178	.129	.092	.065	.046	.033	.023	.012
7	.316	.175	.126	.089	.062	.043	.030	.020	.010
8	.315	.173	.123	.086	.059	.040	.027	.018	.009
9	.315	.172	.121	.084	.057	.038	.026	.017	.007
10	.314	.170	.120	.082	.055	.037	.024	.016	.007
11	.313	.169	.119	.081	.054	.035	.023	.015	.006
12	.313	.169	.118	.080	.053	.034	.022	.014	.006
13	.313	.168	.117	.079	.052	.033	.021	.013	.005
14	.312	.167	.116	.078	.051	.033	.021	.013	.005
15	.312	.167	.115	.077	.050	.032	.020	.012	.004
16	.312	.166	.115	.077	.050	.031	.019	.012	.004
17	.312	.166	.114	.076	.049	.031	.019	.011	.004
18	.312	.165	.114	.075	.049	.030	.019	.011	.004
19	.311	.165	.113	.075	.048	.030	.018	.011	.004
20	.311	.165	.113	.075	.048	.030	.018	.011	.004
21	.311	.164	.113	.074	.047	.029	.018	.010	.003
22	.311	.164	.112	.074	.047	.029	.017	.010	.003
23	.311	.164	.112	.074	.047	.029	.017	.010	.003
24	.311	.164	.112	.073	.046	.028	.017	.010	.003
25	.311	.163	.111	.073	.046	.028	.017	.010	.003
26	.311	.163	.111	.073	.046	.028	.017	.010	.003
27	.311	.163	.111	.073	.046	.028	.016	.009	.003
28	.310	.163	.111	.072	.046	.028	.016	.009	.003
29	.310	.163	.111	.072	.045	.027	.016	.009	.003
30	.310	.163	.110	.072	.045	.027	.016	.009	.003
40	.310	.162	.109	.071	.044	.026	.015	.008	.002
50	.310	.161	.109	.070	.043	.025	.014	.008	.002
60	.309	.161	.108	.069	.043	.025	.014	.008	.002
70	.309	.160	.108	.069	.042	.025	.014	.007	.002
80	.309	.160	.107	.069	.042	.024	.014	.007	.002
90	.309	.160	.107	.069	.042	.024	.013	.007	.002
100	.309	.160	.107	.068	.042	.024	.013	.007	.002
125	.309	.160	.107	.068	.041	.024	.013	.007	.002
150	.309	.159	.107	.068	.041	.024	.013	.007	.002
200	.309	.159	.106	.068	.041	.023	.013	.007	.002
∞	.309	.159	.106	.067	.040	.023	.012	.006	.001

Note: Table entry gives Pr ($t \geq t^*$) for t^* in column heading.
Source: Computed using Fortran subroutines from the IMSL Library.

Use of Table A.3: The t Distribution

This table gives critical values for a random variable having a t distribution with df degrees of freedom (see Section 10.2). The table is arranged to answer problems of two forms. First:

$$\text{Find } t^c \text{ such that } \Pr(t \geq t^c) = \alpha$$

Various one-tailed α values are given in the column headings, and the corresponding critical values, t^c, are found in the row for a particular number of degrees of freedom. In the following figure, the shaded area corresponds to the given α, and t^c is the t value that bounds α probability in the right-hand tail of the distribution.

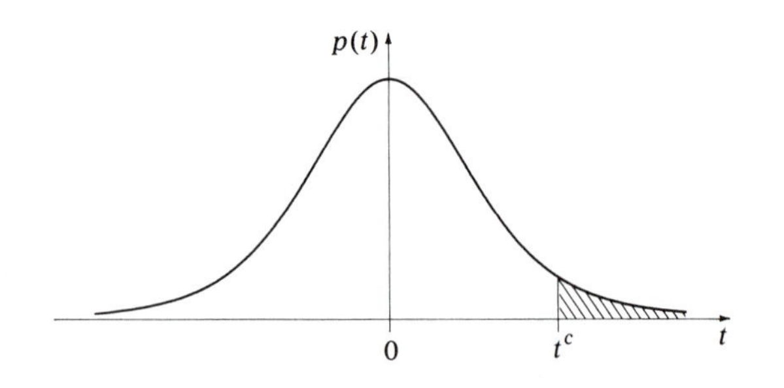

The second problem is of the form:

$$\text{Find } t^c \text{ such that } \Pr(|t| \geq t^c) = \alpha$$

The probability amount α is split between the two tails of the distribution, each containing $\alpha/2$. Various two-tailed α values are given in the column headings, and the corresponding positive critical value, t^c, is given in the appropriate row (the other critical value is $-t^c$). In the following figure the two shaded areas together correspond to the given α, and t^c is the t value that bounds $\alpha/2$ probability in the right-hand tail.

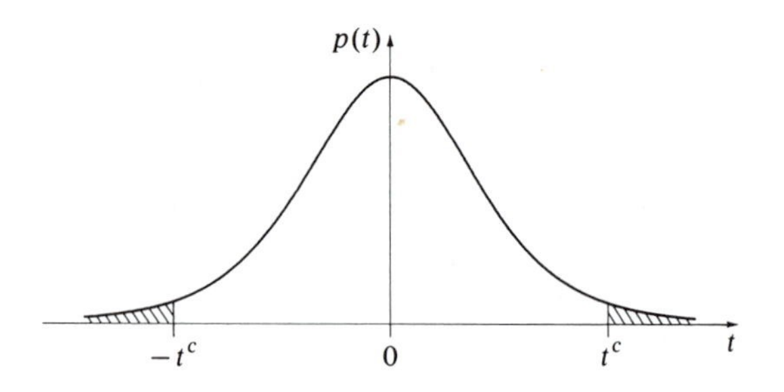

For example, if df $= 15$ and $\alpha = .05$, $t^c = 1.753$ for a one-tailed test, and $t^c = 2.131$ for a two-tailed test. Note that t^c for a one-tailed test with $\alpha = .05$ is the same as t^c for a two-tailed test with $\alpha = .10$.

TABLE A.3 Critical Values for the t Distribution

One-tailed α = .10 Two-tailed α = .20	.05 .10	.025 .05	.01 .02	.005 .01
df = 1 3.078	6.314	12.706	31.821	63.657
2 1.886	2.920	4.303	6.965	9.925
3 1.638	2.353	3.182	4.541	5.841
4 1.533	2.132	2.776	3.747	4.604
5 1.476	2.015	2.571	3.365	4.032
6 1.440	1.943	2.447	3.143	3.707
7 1.415	1.895	2.365	2.998	3.499
8 1.397	1.860	2.306	2.896	3.355
9 1.383	1.833	2.262	2.821	3.250
10 1.372	1.812	2.228	2.764	3.169
11 1.363	1.796	2.201	2.718	3.106
12 1.356	1.782	2.179	2.681	3.055
13 1.350	1.771	2.160	2.650	3.012
14 1.345	1.761	2.145	2.624	2.977
15 1.341	1.753	2.131	2.602	2.947
16 1.337	1.746	2.120	2.583	2.921
17 1.333	1.740	2.110	2.567	2.898
18 1.330	1.734	2.101	2.552	2.878
19 1.328	1.729	2.093	2.539	2.861
20 1.325	1.725	2.086	2.528	2.845
21 1.323	1.721	2.080	2.518	2.831
22 1.321	1.717	2.074	2.508	2.819
23 1.319	1.714	2.069	2.500	2.807
24 1.318	1.711	2.064	2.492	2.797
25 1.316	1.708	2.060	2.485	2.787
26 1.315	1.706	2.056	2.479	2.779
27 1.314	1.703	2.052	2.473	2.771
28 1.313	1.701	2.048	2.467	2.763
29 1.311	1.699	2.045	2.462	2.756
30 1.310	1.697	2.042	2.457	2.750
40 1.303	1.684	2.021	2.423	2.704
50 1.299	1.676	2.009	2.403	2.678
60 1.296	1.671	2.000	2.390	2.660
70 1.294	1.667	1.994	2.381	2.648
80 1.292	1.664	1.990	2.374	2.639
90 1.291	1.662	1.987	2.368	2.632
100 1.290	1.660	1.984	2.364	2.626
125 1.288	1.657	1.979	2.357	2.616
150 1.287	1.655	1.976	2.351	2.609
200 1.286	1.653	1.972	2.345	2.601
∞ 1.282	1.645	1.960	2.326	2.576

Note: Table entry gives t^c corresponding to Pr $(t \geq t^c) = \alpha$ for one-tailed tests and Pr $(|t| \geq t^c) = \alpha$ for two-tailed tests.

Source: Computed using Fortran subroutines from the IMSL Library.

Use of Table A.4: The Chi-Square Distribution

This table gives critical values for a random variable having a chi-square distribution with df degrees of freedom (see the appendix to Chapter 10). Each entry in the table is the answer to a problem of the form:

$$\text{Find } (\chi^2)^c \text{ such that Pr } (\chi^2 \geq (\chi^2)^c) = \alpha$$

Various values of α are given in the column headings, and the corresponding $(\chi^2)^c$ is found in the row labeled by df. Critical values for areas in the left side of the distribution can be found by subtraction.

For example, if df = 35, $(\chi^2)^c = 53.22$ for $\alpha = .025$ and $(\chi^2)^c = 20.56$ for $\alpha = .975$. Thus $(\chi^2)^c = 53.22$ bounds .025 probability in the right-hand tail and $(\chi^2)^c = 20.56$ bounds .025 probability in the left side.

Use of Tables A.5 and A.6: The *F* Distribution

These tables give critical values for a random variable having an F distribution with n degrees of freedom in the numerator and d degrees of freedom in the denominator (see the appendix to Chapter 10). Each entry in the tables is the answer to a problem of the form:

$$\text{Find } F^c \text{ such that Pr } (F_{n,d} \geq F^c) = \alpha$$

Table A.5 contains the F^c values for the .05 level of significance, and Table A.6 contains the F^c values for $\alpha = .01$. The entry for a particular distribution is in the column headed by n and the row labeled by d.

For example, if $n = 4$ and $d = 50$, $F^c = 2.56$ for $\alpha = .05$ and $F^c = 3.72$ for $\alpha = .01$.

Use of Tables A.7 and A.8: The Durbin–Watson Statistic

These tables give the critical values (significance points) of the lower and upper bounds, d_l and d_u, for the Durbin–Watson statistic (see Section 14.1). These bounds depend on the number of observations (n) and the number of regressors (k). Both tables are appropriate for use at the .05 level of significance: Table A.7 is for one-tailed tests, and Table A.8 is for two-tailed tests. (Table A.7 is also for two-tailed tests with $\alpha = .10$, and Table A.8 is also for one-tailed tests with $\alpha = .025$.)

For example, suppose that $n = 25$, $k = 3$, and $\alpha = .05$. For a one-tailed test, $d_l = 1.12$ and $d_u = 1.66$; if d^* falls between these values, the result of the test is indeterminate. For a two-tailed test, $d_l = 1.02$ and $d_u = 1.54$; on the right side of the distribution (see Figure 14.4) the range of indeterminacy extends from 2.46 to 2.98.

TABLE A.4 Critical Values for the Chi-Square Distribution

α =	.990	.975	.950	.900	.100	.050	.025	.010
df = 1	—	—	—	0.02	2.71	3.84	5.02	6.64
2	0.02	0.05	0.10	0.21	4.60	5.99	7.38	9.22
3	0.11	0.22	0.35	0.58	6.25	7.82	9.36	11.32
4	0.30	0.48	0.71	1.06	7.78	9.49	11.15	13.28
5	0.55	0.83	1.15	1.61	9.24	11.07	12.84	15.09
6	0.87	1.24	1.63	2.20	10.65	12.60	14.46	16.81
7	1.24	1.69	2.17	2.83	12.02	14.07	16.02	18.47
8	1.64	2.18	2.73	3.49	13.36	15.51	17.55	20.08
9	2.09	2.70	3.32	4.17	14.69	16.93	19.03	21.65
10	2.55	3.24	3.94	4.86	15.99	18.31	20.50	23.19
11	3.05	3.81	4.57	5.58	17.28	19.68	21.93	24.75
12	3.57	4.40	5.22	6.30	18.55	21.03	23.35	26.25
13	4.10	5.01	5.89	7.04	19.81	22.37	24.75	27.72
14	4.65	5.62	6.57	7.79	21.07	23.69	26.13	29.17
15	5.23	6.26	7.26	8.55	22.31	25.00	27.50	30.61
16	5.81	6.90	7.96	9.31	23.55	26.30	28.86	32.03
17	6.40	7.56	8.67	10.08	24.77	27.59	30.20	33.44
18	7.00	8.23	9.39	10.86	25.99	28.88	31.54	34.83
19	7.63	8.90	10.11	11.65	27.21	30.15	32.87	36.22
20	8.25	9.59	10.85	12.44	28.42	31.42	34.18	37.59
21	8.89	10.28	11.59	13.24	29.62	32.68	35.49	38.96
22	9.53	10.98	12.34	14.04	30.82	33.93	36.79	40.31
23	10.19	11.69	13.09	14.85	32.01	35.18	38.09	41.66
24	10.85	12.40	13.84	15.66	33.20	36.42	39.38	43.00
25	11.51	13.11	14.61	16.47	34.38	37.66	40.66	44.34
26	12.19	13.84	15.38	17.29	35.57	38.89	41.94	45.66
27	12.87	14.57	16.15	18.11	36.74	40.12	43.21	46.99
28	13.55	15.30	16.92	18.94	37.92	41.34	44.47	48.30
29	14.24	16.04	17.70	19.77	39.09	42.56	45.74	49.61
30	14.94	16.78	18.49	20.60	40.26	43.78	46.99	50.91
35	18.49	20.56	22.46	24.79	46.06	49.81	53.22	57.36
40	22.14	24.42	26.51	29.06	51.80	55.75	59.34	63.71
45	25.88	28.36	30.61	33.36	57.50	61.65	65.41	69.98
50	29.68	32.35	34.76	37.69	63.16	67.50	71.42	76.17
55	33.55	36.39	38.96	42.06	68.79	73.31	77.38	82.31
60	37.46	40.47	43.19	46.46	74.39	79.08	83.30	88.40
65	41.42	44.60	47.45	50.89	79.97	84.82	89.18	94.44
70	45.42	48.75	51.74	55.33	85.52	90.53	95.03	100.44
75	49.46	52.94	56.05	59.80	91.06	96.21	100.84	106.41
80	53.52	57.15	60.39	64.28	96.57	101.88	106.63	112.34
85	57.62	61.38	64.75	68.78	102.07	107.52	112.40	118.25
90	61.74	65.64	69.13	73.29	107.56	113.14	118.14	124.13
95	65.88	69.92	73.52	77.82	113.03	118.75	123.86	129.99
100	70.05	74.22	77.93	82.36	118.49	124.34	129.56	135.82

Note: Table entry gives $(\chi^2)^c$ corresponding to $\Pr(\chi^2 \geq (\chi^2)^c) = \alpha$.

Source: Computed using Fortran subroutines from the IMSL Library. Some values differ slightly from those in other published tables.

TABLE A.5 Critical Values for the F Distribution ($\alpha = .05$)

	$n = 1$	2	3	4	5	6	8	10	15
$d = 1$	161.4	199.5	215.7	224.6	230.2	234.0	238.9	241.9	245.9
2	18.51	19.00	19.16	19.25	19.30	19.33	19.37	19.40	19.43
3	10.13	9.55	9.28	9.12	9.01	8.94	8.85	8.79	8.70
4	7.71	6.94	6.59	6.39	6.26	6.16	6.04	5.96	5.86
5	6.61	5.79	5.41	5.19	5.05	4.95	4.82	4.74	4.62
6	5.99	5.14	4.76	4.53	4.39	4.28	4.15	4.06	3.94
7	5.59	4.74	4.35	4.12	3.97	3.87	3.73	3.64	3.51
8	5.32	4.46	4.07	3.84	3.69	3.58	3.44	3.35	3.22
9	5.12	4.26	3.86	3.63	3.48	3.37	3.23	3.14	3.01
10	4.96	4.10	3.71	3.48	3.33	3.22	3.07	2.98	2.85
11	4.84	3.98	3.59	3.36	3.20	3.09	2.95	2.85	2.72
12	4.75	3.89	3.49	3.26	3.11	3.00	2.85	2.75	2.62
13	4.67	3.81	3.41	3.18	3.03	2.92	2.77	2.67	2.53
14	4.60	3.74	3.34	3.11	2.96	2.85	2.70	2.60	2.46
15	4.54	3.68	3.29	3.06	2.90	2.79	2.64	2.54	2.40
16	4.49	3.63	3.24	3.01	2.85	2.74	2.59	2.49	2.35
17	4.45	3.59	3.20	2.96	2.81	2.70	2.55	2.45	2.31
18	4.41	3.55	3.16	2.93	2.77	2.66	2.51	2.41	2.27
19	4.38	3.52	3.13	2.90	2.74	2.63	2.48	2.38	2.23
20	4.35	3.49	3.10	2.87	2.71	2.60	2.45	2.35	2.20
21	4.32	3.47	3.07	2.84	2.68	2.57	2.42	2.32	2.18
22	4.30	3.44	3.05	2.82	2.66	2.55	2.40	2.30	2.15
23	4.28	3.42	3.03	2.80	2.64	2.53	2.37	2.27	2.13
24	4.26	3.40	3.01	2.78	2.62	2.51	2.36	2.25	2.11
25	4.24	3.39	2.99	2.76	2.60	2.49	2.34	2.24	2.09
26	4.23	3.37	2.98	2.74	2.59	2.47	2.32	2.22	2.07
27	4.21	3.35	2.96	2.73	2.57	2.46	2.31	2.20	2.06
28	4.20	3.34	2.95	2.71	2.56	2.45	2.29	2.19	2.04
29	4.18	3.33	2.93	2.70	2.55	2.43	2.28	2.18	2.03
30	4.17	3.32	2.92	2.69	2.53	2.42	2.27	2.16	2.01
40	4.08	3.23	2.84	2.61	2.45	2.34	2.18	2.08	1.92
50	4.03	3.18	2.79	2.56	2.40	2.29	2.13	2.03	1.87
60	4.00	3.15	2.76	2.53	2.37	2.25	2.10	1.99	1.84
70	3.98	3.13	2.74	2.50	2.35	2.23	2.07	1.97	1.81
80	3.96	3.11	2.72	2.49	2.33	2.21	2.06	1.95	1.79
90	3.95	3.10	2.71	2.47	2.32	2.20	2.04	1.94	1.78
100	3.94	3.09	2.70	2.46	2.31	2.19	2.03	1.93	1.77
125	3.92	3.07	2.68	2.44	2.29	2.17	2.01	1.91	1.75
150	3.90	3.06	2.66	2.43	2.27	2.16	2.00	1.89	1.73
200	3.89	3.04	2.65	2.42	2.26	2.14	1.98	1.88	1.72
∞	3.84	3.00	2.60	2.37	2.21	2.10	1.94	1.83	1.67

Note: Table entry gives F^c corresponding to Pr $(F_{n,d} \geq F^c) = .05$.
Source: Computed using Fortran subroutines from the IMSL Library.

TABLE A.6 Critical Values for the F Distribution (α = .01)

	n = 1	2	3	4	5	6	8	10	15
d = 1	4052.	4999.	5403.	5625.	5764.	5859.	5981.	6056.	6157.
2	98.50	99.00	99.17	99.25	99.30	99.33	99.37	99.40	99.43
3	34.12	30.82	29.46	28.71	28.24	27.91	27.49	27.23	26.87
4	21.20	18.00	16.69	15.98	15.52	15.21	14.80	14.55	14.20
5	16.26	13.27	12.06	11.39	10.97	10.67	10.29	10.05	9.72
6	13.75	10.92	9.78	9.15	8.75	8.47	8.10	7.87	7.56
7	12.25	9.55	8.45	7.85	7.46	7.19	6.84	6.62	6.31
8	11.26	8.65	7.59	7.01	6.63	6.37	6.03	5.81	5.52
9	10.56	8.02	6.99	6.42	6.06	5.80	5.47	5.26	4.96
10	10.04	7.56	6.55	5.99	5.64	5.39	5.06	4.85	4.56
11	9.65	7.21	6.22	5.67	5.32	5.07	4.74	4.54	4.25
12	9.33	6.93	5.95	5.41	5.06	4.82	4.50	4.30	4.01
13	9.07	6.70	5.74	5.21	4.86	4.62	4.30	4.10	3.82
14	8.86	6.51	5.56	5.04	4.69	4.46	4.14	3.94	3.66
15	8.68	6.36	5.42	4.89	4.56	4.32	4.00	3.80	3.52
16	8.53	6.23	5.29	4.77	4.44	4.20	3.89	3.69	3.41
17	8.40	6.11	5.19	4.67	4.34	4.10	3.79	3.59	3.31
18	8.29	6.01	5.09	4.58	4.25	4.01	3.71	3.51	3.23
19	8.18	5.93	5.01	4.50	4.17	3.94	3.63	3.43	3.15
20	8.10	5.85	4.94	4.43	4.10	3.87	3.56	3.37	3.09
21	8.02	5.78	4.87	4.37	4.04	3.81	3.51	3.31	3.03
22	7.95	5.72	4.82	4.31	3.99	3.76	3.45	3.26	2.98
23	7.88	5.66	4.76	4.26	3.94	3.71	3.41	3.21	2.93
24	7.82	5.61	4.72	4.22	3.90	3.67	3.36	3.17	2.89
25	7.77	5.57	4.68	4.18	3.85	3.63	3.32	3.13	2.85
26	7.72	5.53	4.64	4.14	3.82	3.59	3.29	3.09	2.81
27	7.68	5.49	4.60	4.11	3.78	3.56	3.26	3.06	2.78
28	7.64	5.45	4.57	4.07	3.75	3.53	3.23	3.03	2.75
29	7.60	5.42	4.54	4.04	3.73	3.50	3.20	3.00	2.73
30	7.56	5.39	4.51	4.02	3.70	3.47	3.17	2.98	2.70
40	7.31	5.18	4.31	3.83	3.51	3.29	2.99	2.80	2.52
50	7.17	5.06	4.20	3.72	3.41	3.19	2.89	2.70	2.42
60	7.08	4.98	4.13	3.65	3.34	3.12	2.82	2.63	2.35
70	7.01	4.92	4.07	3.60	3.29	3.07	2.78	2.59	2.31
80	6.96	4.88	4.04	3.56	3.26	3.04	2.74	2.55	2.27
90	6.93	4.85	4.01	3.53	3.23	3.01	2.72	2.52	2.24
100	6.90	4.82	3.98	3.51	3.21	2.99	2.69	2.50	2.22
125	6.84	4.78	3.94	3.47	3.17	2.95	2.66	2.47	2.19
150	6.81	4.75	3.91	3.45	3.14	2.92	2.63	2.44	2.16
200	6.76	4.71	3.88	3.41	3.11	2.89	2.60	2.41	2.13
∞	6.63	4.61	3.78	3.32	3.02	2.80	2.51	2.32	2.04

Note: Table entry gives F^c corresponding to Pr $(F_{n,d} \geq F^c)$ = .01.
Source: Computed using Fortran subroutines from the IMSL Library.

TABLE A.7 Durbin–Watson Statistic—Significance Points for d_l and d_u (For One-Tailed Tests, α = .05)

	$k = 1$		$k = 2$		$k = 3$		$k = 4$		$k = 5$	
n	d_l	d_u	d_l	d_u	d_l	d_u	d_l	d_u	d_l	d_u
15	1.08	1.36	0.95	1.54	0.82	1.75	0.69	1.97	0.56	2.21
16	1.10	1.37	0.98	1.54	0.86	1.73	0.74	1.93	0.62	2.15
17	1.13	1.38	1.02	1.54	0.90	1.71	0.78	1.90	0.67	2.10
18	1.16	1.39	1.05	1.53	0.93	1.69	0.82	1.87	0.71	2.06
19	1.18	1.40	1.08	1.53	0.97	1.68	0.86	1.85	0.75	2.02
20	1.20	1.41	1.10	1.54	1.00	1.68	0.90	1.83	0.79	1.99
21	1.22	1.42	1.13	1.54	1.03	1.67	0.93	1.81	0.83	1.96
22	1.24	1.43	1.15	1.54	1.05	1.66	0.96	1.80	0.86	1.94
23	1.26	1.44	1.17	1.54	1.08	1.66	0.99	1.79	0.90	1.92
24	1.27	1.45	1.19	1.55	1.10	1.66	1.01	1.78	0.93	1.90
25	1.29	1.45	1.21	1.55	1.12	1.66	1.04	1.77	0.95	1.89
26	1.30	1.46	1.22	1.55	1.14	1.65	1.06	1.76	0.98	1.88
27	1.32	1.47	1.24	1.56	1.16	1.65	1.08	1.76	1.01	1.86
28	1.33	1.48	1.26	1.56	1.18	1.65	1.10	1.75	1.03	1.85
29	1.34	1.48	1.27	1.56	1.20	1.65	1.12	1.74	1.05	1.84
30	1.35	1.49	1.28	1.57	1.21	1.65	1.14	1.74	1.07	1.83
31	1.36	1.50	1.30	1.57	1.23	1.65	1.16	1.74	1.09	1.83
32	1.37	1.50	1.31	1.57	1.24	1.65	1.18	1.73	1.11	1.82
33	1.38	1.51	1.32	1.58	1.26	1.65	1.19	1.73	1.13	1.81
34	1.39	1.51	1.33	1.58	1.27	1.65	1.21	1.73	1.15	1.81
35	1.40	1.52	1.34	1.58	1.28	1.65	1.22	1.73	1.16	1.80
36	1.41	1.52	1.35	1.59	1.29	1.65	1.24	1.73	1.18	1.80
37	1.42	1.53	1.36	1.59	1.31	1.66	1.25	1.72	1.19	1.80
38	1.43	1.54	1.37	1.59	1.32	1.66	1.26	1.72	1.21	1.79
39	1.43	1.54	1.38	1.60	1.33	1.66	1.27	1.72	1.22	1.79
40	1.44	1.54	1.39	1.60	1.34	1.66	1.29	1.72	1.23	1.79
45	1.48	1.57	1.43	1.62	1.38	1.67	1.34	1.72	1.29	1.78
50	1.50	1.59	1.46	1.63	1.42	1.67	1.38	1.72	1.34	1.77
55	1.53	1.60	1.49	1.64	1.45	1.68	1.41	1.72	1.38	1.77
60	1.55	1.62	1.51	1.65	1.48	1.69	1.44	1.73	1.41	1.77
65	1.57	1.63	1.54	1.66	1.50	1.70	1.47	1.73	1.44	1.77
70	1.58	1.64	1.55	1.67	1.52	1.70	1.49	1.74	1.46	1.77
75	1.60	1.65	1.57	1.68	1.54	1.71	1.51	1.74	1.49	1.77
80	1.61	1.66	1.59	1.69	1.56	1.72	1.53	1.74	1.51	1.77
85	1.62	1.67	1.60	1.70	1.57	1.72	1.55	1.75	1.52	1.77
90	1.63	1.68	1.61	1.70	1.59	1.73	1.57	1.75	1.54	1.78
95	1.64	1.69	1.62	1.71	1.60	1.73	1.58	1.75	1.56	1.78
100	1.65	1.69	1.63	1.72	1.61	1.74	1.59	1.76	1.57	1.78

Note: n = number of observations, k = number of regressors.

Source: J. Durbin and G. S. Watson, "Testing for Serial Correlation in Least Squares Regression. II," *Biometrika* 38 (1951), p. 173. Reprinted with permission of the Biometrika Trustees.

TABLE A.8 Durbin–Watson Statistic—Significance Points for d_l and d_u (For Two-Tailed Tests, α = .05)

	$k = 1$		$k = 2$		$k = 3$		$k = 4$		$k = 5$	
n	d_l	d_u	d_l	d_u	d_l	d_u	d_l	d_u	d_l	d_u
15	0.95	1.23	0.83	1.40	0.71	1.61	0.59	1.84	0.48	2.09
16	0.98	1.24	0.86	1.40	0.75	1.59	0.64	1.80	0.53	2.03
17	1.01	1.25	0.90	1.40	0.79	1.58	0.68	1.77	0.57	1.98
18	1.03	1.26	0.93	1.40	0.82	1.56	0.72	1.74	0.62	1.93
19	1.06	1.28	0.96	1.41	0.86	1.55	0.76	1.72	0.66	1.90
20	1.08	1.28	0.99	1.41	0.89	1.55	0.79	1.70	0.70	1.87
21	1.10	1.30	1.01	1.41	0.92	1.54	0.83	1.69	0.73	1.84
22	1.12	1.31	1.04	1.42	0.95	1.54	0.86	1.68	0.77	1.82
23	1.14	1.32	1.06	1.42	0.97	1.54	0.89	1.67	0.80	1.80
24	1.16	1.33	1.08	1.43	1.00	1.54	0.91	1.66	0.83	1.79
25	1.18	1.34	1.10	1.43	1.02	1.54	0.94	1.65	0.86	1.77
26	1.19	1.35	1.12	1.44	1.04	1.54	0.96	1.65	0.88	1.76
27	1.21	1.36	1.13	1.44	1.06	1.54	0.99	1.64	0.91	1.75
28	1.22	1.37	1.15	1.45	1.08	1.54	1.01	1.64	0.93	1.74
29	1.24	1.38	1.17	1.45	1.10	1.54	1.03	1.63	0.96	1.73
30	1.25	1.38	1.18	1.46	1.12	1.54	1.05	1.63	0.98	1.73
31	1.26	1.39	1.20	1.47	1.13	1.55	1.07	1.63	1.00	1.72
32	1.27	1.40	1.21	1.47	1.15	1.55	1.08	1.63	1.02	1.71
33	1.28	1.41	1.22	1.48	1.16	1.55	1.10	1.63	1.04	1.71
34	1.29	1.41	1.24	1.48	1.17	1.55	1.12	1.63	1.06	1.70
35	1.30	1.42	1.25	1.48	1.19	1.55	1.13	1.63	1.07	1.70
36	1.31	1.43	1.26	1.49	1.20	1.56	1.15	1.63	1.09	1.70
37	1.32	1.43	1.27	1.49	1.21	1.56	1.16	1.62	1.10	1.70
38	1.33	1.44	1.28	1.50	1.23	1.56	1.17	1.62	1.12	1.70
39	1.34	1.44	1.29	1.50	1.24	1.56	1.19	1.63	1.13	1.69
40	1.35	1.45	1.30	1.51	1.25	1.57	1.20	1.63	1.15	1.69
45	1.39	1.48	1.34	1.53	1.30	1.58	1.25	1.63	1.21	1.69
50	1.42	1.50	1.38	1.54	1.34	1.59	1.30	1.64	1.26	1.69
55	1.45	1.52	1.41	1.56	1.37	1.60	1.33	1.64	1.30	1.69
60	1.47	1.54	1.44	1.57	1.40	1.61	1.37	1.65	1.33	1.69
65	1.49	1.55	1.46	1.59	1.43	1.62	1.40	1.66	1.36	1.69
70	1.51	1.57	1.48	1.60	1.45	1.63	1.42	1.66	1.39	1.70
75	1.53	1.58	1.50	1.61	1.47	1.64	1.45	1.67	1.42	1.70
80	1.54	1.59	1.52	1.62	1.49	1.65	1.47	1.67	1.44	1.70
85	1.56	1.60	1.53	1.63	1.51	1.65	1.49	1.68	1.46	1.71
90	1.57	1.61	1.55	1.64	1.53	1.66	1.50	1.69	1.48	1.71
95	1.58	1.62	1.56	1.65	1.54	1.67	1.52	1.69	1.50	1.71
100	1.59	1.63	1.57	1.65	1.55	1.67	1.53	1.70	1.51	1.72

Note: n = number of observations, k = number of regressors.

Source: J. Durbin and G. S. Watson, "Testing for Serial Correlation in Least Squares Regression. II," *Biometrika* 38 (1951), p. 174. Reprinted with permission of the Biometrika Trustees.

Answers to Selected Problems

Chapter 1: Introduction

1.2 $\hat{C} = 0.568 + (0.907)(1000) = 907.568$.

1.5 $Q = \beta_0 + \beta_1 P + u$. This is formally similar to the demand model. In constrast to the demand model, however, we expect that $\beta_1 > 0$; the sign of β_0 is not clear.

1.7 No. The monthly survey and estimation of the unemployment rate are primarily based on general statistical techniques.

1.11 Moving from A to B, $\Delta y = 5.0 - 2.5 = 2.5$ and $\Delta x = 8 - 3 = 5$; thus the slope $= \Delta y/\Delta x = 0.5$.

1.12 Since the slope $= 0.5$, $\Delta y = 0.5\Delta x$. If $\Delta x = 4$, $\Delta y = 2$; if $\Delta x = -1$, $\Delta y = -0.5$.

1.14 When $x = 2$, $y = 4$; the slope equals $(2)(2) = 4$ and the elasticity equals $(4)(2)/(4) = 2$. When $x = 4$, $y = 16$; the slope equals $(2)(4) = 8$ and the elasticity equals $(8)(4)/(16) = 2$ again.

1.16 Using initial values of y and x as bases, the arc elasticity equals $(\Delta y/y)/(\Delta x/x) = (12/4)/(2/2) = 3$. Using average values as bases, the arc elasticity equals $(12/10)/(2/3) = 1.8$.

Chapter 2: Economic Data

2.2 The answer depends on the nature of the actual data. Most likely, the number of plants and the number of product lines are clearly discrete, and the number of employees and the amount of output can be considered practically continuous.

2.3 The objective in taking the sample is to make an estimate of the unemployment rate in the whole nation. If an unrepresentative sample is taken, such as from only a prosperous region or from only white-collar workers, the resulting estimate will be a distorted (or biased) estimate.

2.7 For this problem it is assumed that people correctly report their total wealth, but incorrectly report the savings account component. The failure to report any account leads to measured $S = 0$ for some individuals. The misreporting leads to measured S being greater than true S at low levels of W (hence low S), and oppositely for high W. The resulting pattern of data points has an apparent slope that is flatter than β_1, leading to an underestimate of it.

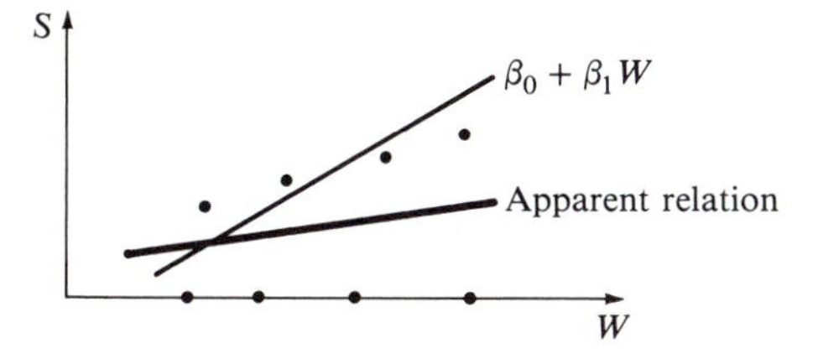

2.11 Predictions of future consumption are made along the line labeled "Continued process," which is an estimate of the behavior described in Equation (1.1). If the process changes, future values of consumption will lie around the line labeled "New process," and the model previously estimated gives distorted projections. Correct predictions would need to be based on knowledge of the new process.

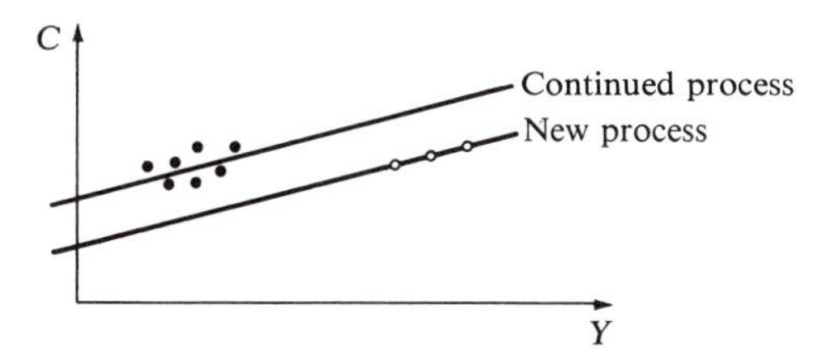

2.14 The calculations are made on the basis of Equation (2.2). For 1968, $CPI_{i-1} = 100$, and $RINF2_i$ is given in Table 2.3 as 4.200; hence $CPI_i = 104.20$. Similarly, for 1969, $CPI_i = 109.7997$, or about 109.80.

2.17 The four successive values of the absolute growth in X are: $\Delta X = 5, 7, 9, 11$. The five values of $\ln X$ increase over time, but the successive differences are $\Delta \ln X = 0.41, 0.38, 0.34, 0.30$. Hence the absolute growth is increasing as time passes, but the rate of growth decreases as time passes.

2.19 When $r = -20$ percent, the graphs are:

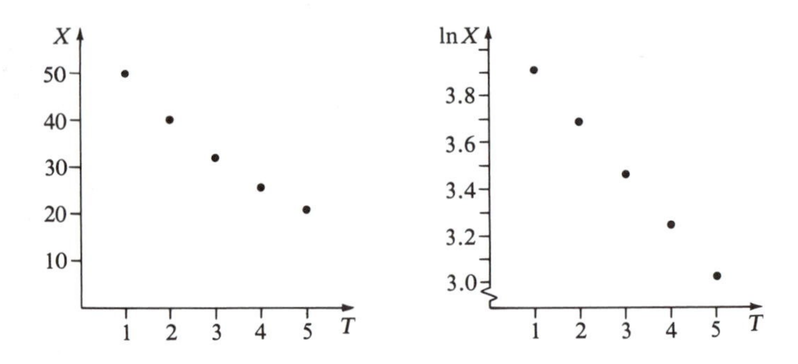

2.22 If a number is doubled, its logarithm is increased by adding ln 2 (= 0.693). That is, if $y = 2x$, $\ln y = \ln 2 + \ln x$.

2.23 If $Q = AP^b$, $\ln Q = \ln A + b \ln P$.

2.27 If $\ln Y = a + Z \ln b$, $Y = e^a b^Z$.

2.28 Based on Equation (2.26), $\ln 42 = \ln 10 + 4 \ln(1 + r)$. Solving, this yields $\ln(1 + r) = 0.359$, so $r \approx 0.43$. [Note that $\ln(1 + r) \approx r$ is a poor approximation because r is so large.]

Chapter 3: Descriptive Statistics

3.1 The mean is 5.38; the median is 5.5.

3.4 No, the median cannot be determined this way.

3.8 With $\overline{X} = 5.38$ and $S_X = 0.914$, the one-standard-deviation interval runs from 4.466 to 6.294 and includes 60 percent of the observations. The two-standard-deviation interval runs from 3.552 to 7.208 and includes 100 percent of the observations.

3.11 $\Sigma\, d_i^2 = \Sigma\, X_i^2 - (1/n)(\Sigma\, X_i)^2 = 296.96 - (1/10)(53.8)^2 = 7.516$. Passing through the data once, we find $\Sigma\, X_i = 53.8$ and $\Sigma\, X_i^2 = 296.96$. Hence $\Sigma\, d_i^2 = 7.516$. These intermediate terms lead to a mean of 5.38, a variance of 0.835, and a standard deviation of 0.914. These are the same as in Problem 3.7, of course.

3.12 $\overline{Y} \approx 44$, $S_Y \approx 18$.

3.17 (a) $W_i = 24 + 12X_i$.
(b) $\overline{W} = 24 + 12\overline{X} = 240$; $S_W = 12S_X = 106.8$.

3.18 The standard deviation of income is increased by 20 percent, but the standard deviation of the logarithm of income is not affected.

3.21 Covariance $= -0.788$; correlation $= -.872$.

3.23 $S_{YZ} = bS_{YX}$.

Chapter 4: Frequency Distributions

4.3 The one-standard-deviation interval runs from 1.25 to 3.39 and contains 54 percent of the observations; the two-standard-deviation interval runs from 0.18 to 4.46 and contains 100 percent of the observations. (The uniform distribution assumption is not applied to discrete variables.)

4.6 Consider two low values of X with a given difference ΔX and two high values of X with the same ΔX between them. The transformation $Y_i = \ln X_i$ yields a set of four Y values such that the difference between the lower-valued pair is greater than that between the higher-valued pair (see Figure 2.3). If the original X has equal differences between all adjacent X values, the difference between adjacent Y values decreases as Y increases.

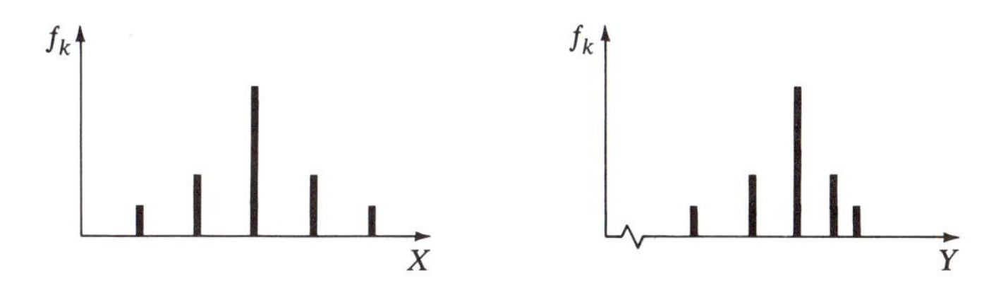

4.9 The frequency table is constructed like Table 4.2, with seven classes. The absolute frequencies are 3, 5, 30, 38, 14, 9, 1. The indicated class boundaries have a round number as the midpoint (3, 6, 9, etc.), and no observations can fall on a boundary.

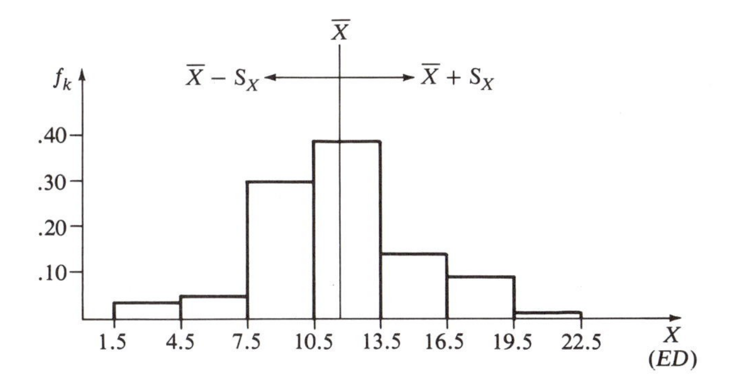

4.10 The mean is 11.58, and the standard deviation is 3.49. These statistics are not expected to be exactly equal to those calculated from the raw data on *ED* because in the computations we use the class marks instead of the actual values of *ED* for each observation. (In the raw data, the mean is 11.58 and the standard deviation is 3.44; hence, in this case the means computed from raw and grouped data are identical, but the standard deviations are slightly different.)

4.13 Inspection of the frequency table shows that the median lies in the fourth class. Its value is determined to be 11.45, based on the uniform distribution assumption. (Note that the lower boundary of the fourth class is 10.5.)

Chapter 5: Simple Regression: Theory

5.2 Yes. If all the disturbances are positive, all the actual Y_i values lie above the true regression line. This is an unlikely occurrence, however, except in very small samples.

5.6

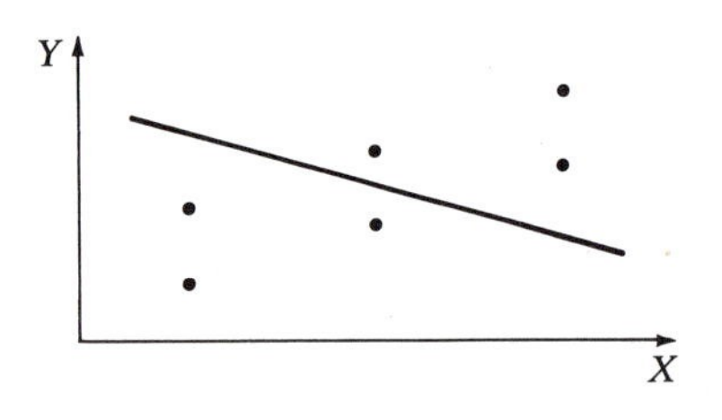

5.11 The slope is flatter for observations 1–5 than for 26–30. Both slopes serve as estimates of the same β_1 value in the theoretical model. Since they are different, at least one of them must contain some error.

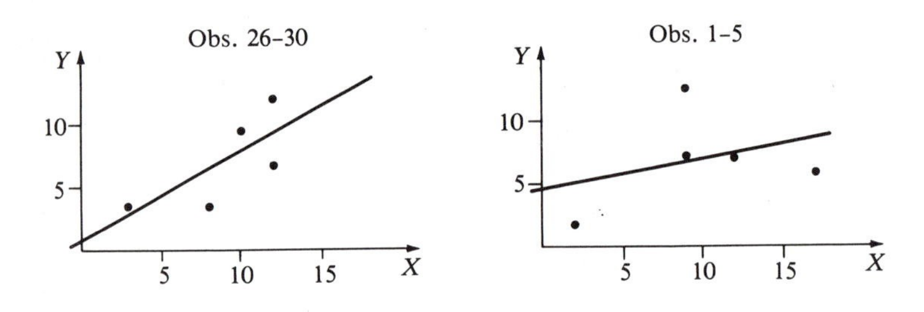

5.13 $R^2 = .507$, $SER = 2.992$.

5.18 $\hat{Y}_0^* = 763.9$ and $\hat{Y}_1^* = 701.8$. (Note that the new earnings variable is 1000 times greater than the original one.)

5.23

$\Sigma e_i^2 = \Sigma (Y_i - \hat{\beta}_1 X_i)^2$ see Equation (5.44)

$(d/d\hat{\beta}_1)[\Sigma (Y_i - \hat{\beta}_1 X_i)^2] = 0$ see (5.46)

$-2 \Sigma (Y_i - \hat{\beta}_1 X_i) X_i = 0$ see (5.48)

$\Sigma Y_i X_i = \hat{\beta}_1 \Sigma X_i^2$ see (5.50)

$\hat{\beta}_1 = \Sigma Y_i X_i / \Sigma X_i^2$ see (5.53)

(Note that R^2 as defined by (5.26) is not a useful measure of goodness of fit in this case.)

Chapter 6: Simple Regression: Application

6.1 The disturbance equals -2.

6.5 The rest of the variation is accounted to the residuals; we can say that this is due to the disturbances.

6.6 The 100 percent inflation means that all measured earnings in 1980 are twice as large as they would be in the 1963 data. Thus, an additional year of schooling increases predicted earnings by (2)(0.797) = 1.594 thousand dollars [see Equation (5.31)].

6.10 For 1974, $APC = CON/DPI = 763.6/858.4 = 0.890$. From Equation (6.14), for $T = 19$, predicted $APC = 0.907$. The error of fit is -0.017, which is somewhat larger (in absolute value) than *SER*.

6.14 *DPI* was 865.3 billion dollars in 1973 and 858.4 in 1974. Based on Equation (6.22), predicted *CON* for 1974 is 10.913 + (0.923)(865.3) = 809.6 billion dollars. Based on Equation (6.10), predicted *CON* for 1974 is 0.568 + (0.907)(858.4) = 779.1 billion dollars.

6.18 $Y = 8.60, 12.85, 15.29, 16.92$.

6.20 The increase in Y from 12.85 to 15.29 amounts to 18.99 percent. Since X increases by 100 percent, the ratio of proportionate changes is 0.1899. This is less than the point elasticity, 0.25.

6.22 If $Q = AP^b$, b is the elasticity of Q with respect to P (i.e., the price elasticity of demand). Such an elasticity is usually negative, corresponding to a downward-sloping demand curve as in Figure 6.3a. The elasticity b can be estimated by regressing $\ln Q$ on $\ln P$, as specified in Equation (6.30).

6.24 $\ln(1 + \text{pc of } Y) = (0.25)(0.693) = 0.1733$. Taking antilogs yields $(1 + \text{pc of } Y) = 1.189$, so the predicted increase in Y is 18.9 percent. This is identical to the change in Problem 6.20 (except for differences in computational rounding).

Chapter 7: Multiple Regression: Theory and Application

7.3 If income is fixed, the relation between predicted demand and price can be graphed as a straight line in a two-dimensional graph. For example, when $Y = Y_1$, predicted demand is a function only of price: this is a *ceteris paribus* relation. The vertical intercept depends on the regression constant, the amount of income, and its coefficient. When $Y = Y_2$ the intercept takes on a new value, and a new *ceteris paribus* relation between demand and price is established.

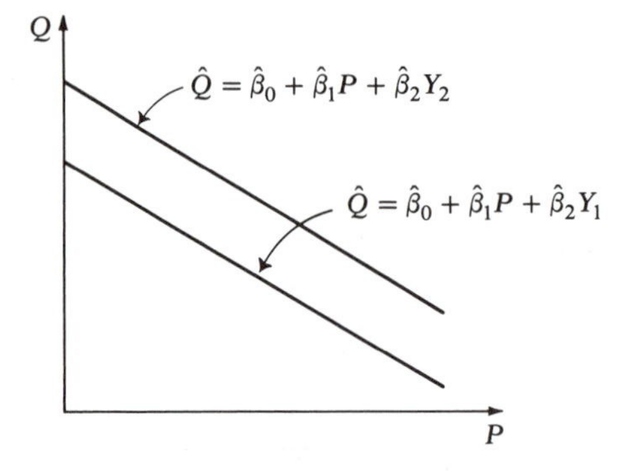

7.5 The first man has greater predicted earnings. Comparing the first with the second, $\Delta EARNS = 0.978 \Delta ED + 0.124 \Delta EXP = (0.978)(4) + (0.124)(-4) = 3.416$ thousand dollars.

7.12 For a black man with 16 years of schooling, predicted *EARNS* is $-0.778 + 0.762(16) + 1.926(1) = 9.488$ thousand dollars. For a white man with 12 years of schooling, predicted *EARNS* is $-0.778 + 0.762(12) = 8.366$ thousand dollars.

7.15 Let $M = 1$ if married, 0 if not; let $D = 1$ if divorced, 0 if not. The model is: $EARNS_i = \beta_0 + \beta_1 ED_i + \beta_2 M_i + \beta_3 D_i + u_i$.

7.24 Let Q = output, L = labor input, and K = capital input. The model is: $\ln Q_i = \beta_0 + \beta_1 \ln L_i + \beta_2 \ln K_i + u_i$.

7.25 Extending the notation of Problem 7.24, let T be a time trend. The model is: $\ln Q_i = \beta_0 + \beta_1 \ln L_i + \beta_2 \ln K_i + \beta_3 T_i + u_i$. As in a simple semilog specification, the coefficient on T equals $\ln (1 + g)$, where g is the rate of technical progress.

Chapter 8: Probability Theory

8.2 (a) Letting the first digit be the outcome on the first flip, and similarly for the second, the 16 outcomes are: 11, 12, 13, 14, 21, 22, 23, 24, 31, 32, 33, 34, 41, 42, 43, 44.
(b) Since each is equally likely, $\Pr(e_i) = 1/16$.
(c) In a three-flip activity, there are $4^3 = 64$ outcomes.

8.6 (a) Since each is equally likely, $\Pr(e_i) = 1/10$.
(b) Pr (5) = 1/10; Pr (6) = 2/10; Pr (7) = 3/10; Pr (8) = 4/10.
(c) Letting A be "getting a number less than or equal to 7," $\Pr(A) = 6/10$.

8.7 (a) Yes. (b) No. (c) No.

8.8 (a) $E_1 = \{e_2, e_3, e_4, e_5, e_6, e_7\}$; $\Pr(E_1) = 6/8$.
(b) $E_2 = \{e_1, e_2, e_3, e_5, e_6, e_7\}$; $\Pr(E_2) = 6/8$.
(c) $E_3 = \{e_2, e_4, e_5, e_6, e_7, e_8\}$; $\Pr(E_3) = 6/8$.

8.10 (a) Pr $(E_1) = 3/8 + 3/8 - 0 = 6/8$.
(b) Pr $(E_2) = 4/8 + 3/8 - 1/8 = 6/8$.
(c) Pr $(E_3) = 4/8 + 4/8 - 2/8 = 6/8$.

8.13 (a) 64 percent. (b) 32 percent. (c) 4 percent.

8.15 (a) The game ends with either 7 or 8. In this reduced sample space, Pr $(8) = 5/11$.
(b) The probability of establishing 8 as the point is 5/36, and then the probability of winning with 8 is 5/11. The total probability is $(5/36)(5/11) = 25/393$, or about 6.4 percent.

Chapter 9: Random Variables and Probability Distributions

9.3 No, because $\Sigma\, p(Y_k) = 21/20 \neq 1$.

9.5 $\mu_x = 7.0$, $\sigma_x = 1.0$.

9.6 $p(X) = 1/6$, $X = 1, 2, 3, 4, 5, 6$. $\mu_x = 3.5$, $\sigma_x = 1.71$.

9.11 No. If the sample size is not a multiple of 1000, then it is impossible for the relative frequency for $X = 2$ to be exactly .001. Also, even if n is a multiple of 1000, it is not necessary that $X = 2$ a specific number of times.

9.15 Since the number of offers has a binomial distribution, $E[X] = nP = 1$. Thus "getting fewer than the expected number" corresponds only to $X = 0$; $p(0) = 0.328$.

9.19 From Figure 9.5a it can be determined that $p(0.2) = .8$ and $p(0.8) = .2$. These density values give the relative likelihoods of these X values occurring. If we construct equal-width small intervals around each X value (such as 0.2 ± 0.001 and 0.8 ± 0.001), the probability of X occurring in the first interval ("X about 0.2") will be four times the probability of X occurring in the second ("X about 0.8") because the areas in the two thin rectangles will be in that ratio.

9.24 $\mu_Y = 2.5$, $\sigma_Y = 0.50$. The relations between these values and those from Problem 9.5 are in accord with Equations (9.21) and (9.23), of course.

9.26 In Problem 9.6 it was determined that $\mu_X = 3.5$ and $\sigma_X = 1.71$ for the toss of a single die. Let X_1 be the outcome of the first die and X_2 be the outcome of the second. Letting $W = X_1 + X_2$ and recognizing independence, we apply Equations (9.30) and (9.31) and find that $\mu_W = 7.0$ and $\sigma_W = 2.42$.

Chapter 10: The Normal and *t* Distributions

10.2 (a) $Z^c = 2.05$; (b) $Z^c = 1.28$; (c) $Z^c = 1.04$;
(d) $Z^c = 2.326$; (e) $Z^c = 1.645$; (f) $Z^c = 1.44$.

10.5 (a) Since there is 20 percent probability in each tail, the corresponding $Z^c = 0.84$; thus the X values are 87.4 and 112.6.
(b) For all normal distributions this probability is .382.

10.10 Pr ($Z \geq 1.00$) = .159, as determined from Table A.1. In Table A.2, looking down the column for $t^* = 1.00$ we see that Pr ($t \geq 1.00$) decreases as df increases. Thus, as df increases the distribution becomes more compact about its mean (σ decreases). Also, we see that Pr ($t \geq 1.00$) approaches .159 as df increases, illustrating that the shape approaches the standard normal.

10.11 Using Table A.3: (a) $t^c = 2.518$; (b) $t^c = 1.721$; (c) $t^c = 1.323$; (d) $t^c = 2.831$; (e) $t^c = 2.080$; (f) $t^c = 1.721$.

10.16 For the chi-square family, the graph of the probability distribution (i.e., density function) shifts to the right as df increases.

10.17 Using Table A.5, we find that $F^c = 2.71$.

Chapter 11: Sampling Theory in Regression

11.2 Pr ($u_3 > 0$) = .5. In the normal regression model, each disturbance is independent.

11.3 Let the disturbance be v. If $E[v_i] = \delta$, then $u_i = v_i - \delta$ conforms to Equation (11.2). The original model $Y_i = \beta_0 + \beta_1 X_i + v_i$ can be rewritten as $Y_i = \beta_0 + \delta + \beta_1 X_i + u_i$. The new intercept can be denoted by β_0', where $\beta_0' = \beta_0 + \delta$.

11.8 From sampling theory we know that $p(\hat{\beta}_1)$ is normal with $E[\hat{\beta}_1] = 10$ and $\sigma(\hat{\beta}_1) = 8$. Hence Pr ($\hat{\beta}_1 \leq 0$) = Pr ($Z \leq -1.25$) = .106, or 10.6 percent.

11.16 Using the definition of t in Equation (11.21), its value is $(0.797 - 0.0)/0.128 = 6.2$. Looking at Table A.2 we realize that there is practically no chance that a random variable having a t distribution would turn out to be this large; it would not occur often.

Chapter 12: Hypothesis Testing

12.1 (a), (b), and (d) can be only alternative hypotheses; (c) can be only a null.

12.4 Let β_2 denote the coefficient on *DRACE*. To judge whether a person's race affects his earnings, we set up H_0: $\beta_2 = 0$ and H_1: $\beta_2 \neq 0$. Since $t^* = -1.09$, we would conclude that race does not affect earnings (based on conventional levels of significance).

12.5 Let β_2 denote the coefficient on *EXP*. The appropriate hypotheses are H_0: $\beta_2 = 0$ and H_1: $\beta_2 \neq 0$. With df = 97, $t^c = 1.66$ for $\alpha = .10$ and $t^c = 1.98$ for $\alpha = .05$. Since $t^* = 2.07$, we would conclude that experience does affect earnings, at both levels of significance. (Note that experience is *not* significant at the 1 percent level.)

12.12 Let β_2 denote the coefficient on *LNRTB*. The appropriate hypotheses are H_0: $\beta_2 = 0$ and H_1: $\beta_2 < 0$. Since $t^* = -0.67$, we would conclude that the interest elasticity is not significantly negative (based on conventional levels of significance).

12.13 Using Table A.3, $t_1^c = 1.660$ and $t_2^c = 1.984$.

12.18 Let β_4 denote the coefficient on *LNMONTHS*. The appropriate hypotheses are H_0: $\beta_4 = 1$ and H_1: $\beta_4 \neq 1$. Since $t^* = -0.25$, the elasticity is not significantly different from 1 (based on conventional levels of significance).

Chapter 13: Estimation and Regression Problems

13.1 From Equation (12.8), $\hat{\beta}_1^* = 0.797$ and $s^*(\hat{\beta}_1) = 0.128$. With $t^c = 1.98$, the 95 percent confidence interval for β_1 is 0.797 ± 0.253, or 0.544 to 1.050.

13.3 No. A 100 percent confidence interval for the marginal propensity to consume would be $-\infty$ to ∞. The interval definitely includes the true value of the parameter, but it is of no help in tax analysis.

13.6 (a) The confidence interval contains β_j in about 95 percent of the samples.

(b) No. We have 95 percent confidence that it contains β_j, but we cannot determine whether it does or not.

13.8 Based on Equations (13.31) and (13.32), $\hat{Y}_p = 1814.57$ billion dollars, and $s_p = 15.36$. The 95 percent confidence interval is 1814.57 ± 31.78, or 1782.79 to 1846.35 billion dollars. (Note that the confidence interval is substantially wider than in the example in Section 13.2, where income was about one trillion dollars.)

13.10 If we know the true values of the parameters in the regression model, our predictions are not affected by estimation error. For the *p*th observation, as for every other one, the actual Y is viewed as the outcome of a normal random variable with mean $\beta_0 + \beta_1 X_p$ and standard deviation equal to σ_u. The 90 percent confidence interval is $(\beta_0 + \beta_1 X_p) \pm (1.645)(\sigma_u)$, where 1.645 is Z^c such that $\Pr(|Z| \geq Z^c) = .10$.

Chapter 14: Autocorrelation and Heteroscedasticity

14.2 $E[u_2] = \rho u_1^* = (-.9)(20.0) = -18.0$. Taking $\epsilon_2 = 1.0$ as a possible outcome, $u_2^* = -17.0$. Then, taking $\epsilon_3 = -1.0$ as a possible outcome, $u_3^* = (-.9)(-17.0) - 1.0 = 14.3$.

14.4 We cannot expect to make a conclusive determination, because the table gives us information about the bounds for a critical value of d, not the critical value itself. In all cases in the table, the value of d_l increases with the number of observations, suggesting that the sampling distribution of d becomes more compact. However, when $k = 3$, 4, or 5, the value of d_u decreases as n increases up to a point, after which d_u starts to increase with n.

14.6 Standard practice is to report the original regression, not the corrected one. When there is no autocorrelation ($\rho = 0$), the Hildreth–Lu estimators are likely to have larger mean square errors.

14.7 Measurement error in the dependent variable is one of the factors incorporated into the disturbance. If this has decreased over time, the true disturbance is heteroscedastic.

14.10 The disturbance v_i cannot be normally distributed because the dependent variable is limited to being nonnegative in all these cases.

14.11 The original model is $Y_i = \beta_0 + \beta_1 X_i + u_i$, with $\sigma(u_i) = \sigma\sqrt{X_i}$. Dividing through by $\sqrt{X_i}$ yields

$$\left[\frac{Y_i}{\sqrt{X_i}}\right] = \beta_0 \left[\frac{1}{\sqrt{X_i}}\right] + \beta_1[\sqrt{X_i}] + \epsilon_i, \qquad \text{where} \quad \epsilon_i = \left(\frac{1}{\sqrt{X_i}}\right) u_i$$

Since $\sigma(u_i) = \sigma\sqrt{X_i}$, $\sigma(\epsilon_i) = \sigma$. Thus the disturbance in this specification is homoscedastic, and β_0 and β_1 can be estimated by OLS. (Note that there is no intercept in this specification, so a special form of OLS is required.)

Chapter 15: More Testing and Specification

15.2 The appropriate hypotheses are given by Equation (15.16). In the test, the number of restrictions is $r = 3$ and there are $k = 7$ regressors in the unrestricted form. Letting $\alpha = .05$, the critical value for $F_{3,92}$ is $F^c \approx 2.71$. In Table 15.1, $SSR_R = 17.7241$ and $SSR_U = 17.0904$. This leads to $F^* = 1.137$, which is not in the critical region. We conclude that the region of residence does not significantly affect earnings.

15.5 The unrestricted model is given by Equation (15.18):

$$Y_i = \beta_0 + \beta_1 LABINC_i + \beta_2 PROPINC_i + u_i$$

The appropriate hypotheses are H_0: $\beta_2 = 0.9\beta_1$ and H_1: $\beta_2 \neq 0.9\beta_1$. The restricted model incorporates the statement of the null hypothesis:

$$\begin{aligned} Y_i &= \beta_0 + \beta_1 LABINC_i + 0.9\beta_i PROPINC_i + u_i \\ &= \beta_0 + \beta_1[LABINC_i + 0.9PROPINC_i] + u_i \\ &= \beta_0 + \beta_1 ADJINC_i + u_i \end{aligned}$$

where *ADJINC* is a new regessor representing adjusted income. In the F test, the number of restrictions is $r = 1$.

15.7 We see in Table 15.2 that, for example, the estimated coefficient on *LNGNP* (for $i = 1$ to 15) is 0.292 in regressions (2) and (4), but that the standard errors are different (0.029 versus 0.034). This difference arises because regression (4) estimates the standard errors on the presumption that the disturbances are homoscedastic (i.e., that σ_u is the same for all observations), whereas the separate estimation of regressions (2) and (3) allows for different variability in the disturbances for the two periods. Indeed, we find that $SER = 0.0191$ in the earlier period and $SER = 0.0278$ in the latter.

15.9 For each observation, the disturbance $u_i = Y_i - E[Y_i]$. For observation 1, suppose that $E[Y_i] = 0.25$; the outcome of u_1 is either -0.25 or 0.75. For observation 2, suppose that $E[Y_i] = 0.5$; the outcome of u_2 is either -0.5 or 0.5. The requirement that $E[u_i] = 0$ fixes the probability values.

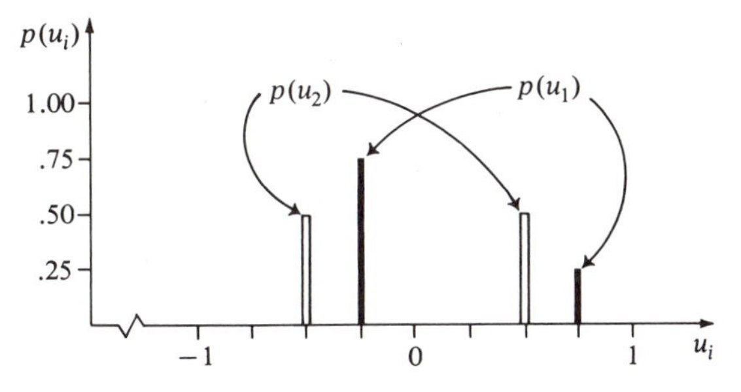

15.10 The disturbances in Figure 15.3a take on many different values and appear to be continuous. They might be approximately normal with a fairly small standard deviation, but they are not exactly normal because Y is bounded between 0 and 1.

15.11 Although the data points in Figure 15.3a seem to snake around a straight line, there is no basis for believing that the disturbances are autocorrelated. Instead, the systematic relation seems to be nonlinear (also see Figure 14.3).

Chapter 16: Regression and Time Series

16.2

$$\Delta CON_i = 0.886\Delta DPI_i + 0.0116\Delta DPILAG1_i + 0.0100\Delta DPILAG2_i$$

Period i: $\Delta CON = 0.886$
Period $i + 1$: $\Delta CON = 0.886 + 0.0116 = 0.8976$
Period $i + 2$: $\Delta CON = 0.886 + 0.0116 + 0.0100 = 0.9076$
Period $i + 3$: $\Delta CON = 0.886 + 0.0116 + 0.0100 = 0.9076$
Period $i + 4$: $\Delta CON = 0.886 + 0.0116 + 0.0100 = 0.9076$

16.3 If a simple regression of Y on X is estimated when Equation (16.3) is correct, the estimated regression suffers from an exclusion-of-variables type of misspecification. If X is increasing over time and if β_1, β_2, and β_3 are all positive, then the expected value of the slope coefficient in the simple regression will be greater than β_1. The slope in the simple regression captures the combined effects on Y of the current and past values of X.

16.6 With Y_i^* fixed, Y_i approaches Y_i^* over time.

16.7 If X grows at a constant absolute amount, so does Y_i^* [ignoring the disturbance in Equation (16.13)]. With Y_i^* growing at a constant amount, Y_i approaches Y_i^* over time.

16.9 In the presence of a lagged dependent variable, the appropriate test is based on Durbin's h statistic, calculated according to Equation (14.11). From (16.11) we have $n = 23$ and $V = (0.136)^2$. Given $d = 1.25$, we compute $h \approx 2.37$. In large samples Durbin's h has a standard normal distribution, so the one-tailed critical value is $Z^c = 1.645$. We conclude that positive autocorrelation is present. (Note that when $d < 2$, $h > 0$; thus the critical region for h is in the right-hand tail.) In this case, d is in the indeterminate region.

16.13 The dummy variable coefficients measure differences from the excluded category. If *Q4* is excluded, the coefficient on *Q1* would be -3.47; that on *Q2* would be -1.53; and that on *Q3* would be -1.56.

16.15 Compared with the first quarter, third quarter *NPE* is predicted to be 1.91 higher because of a seasonal effect and (2)(0.108) = 0.216 higher because of a trend effect. The total prediction is 2.126 higher.

Chapter 17: Simultaneous-Equation Models

17.1

$$WAGE_i = \left(\frac{\gamma_0 - \beta_0}{\beta_1 - \gamma_1}\right) + \left(\frac{\gamma_2}{\beta_1 - \gamma_1}\right)OTHINC_i + \left(\frac{v_i - u_i}{\beta_1 - \gamma_1}\right)$$

$$HOURS_i = \left(\frac{\beta_1\gamma_0 - \beta_0\gamma_1}{\beta_1 - \gamma_1}\right) + \left(\frac{\beta_1\gamma_2}{\beta_1 - \gamma_1}\right)OTHINC_i + \left(\frac{\beta_1 v_i - \gamma_1 u_i}{\beta_1 - \gamma_1}\right)$$

17.3 This is a special case of a simultaneous-equation model. Because X depends on Z but not on Y or u, the model can be thought of as a sequence of two single-equation models. Once X is determined, it serves as an "exogenous" variable in the first equation. The "reduced form" is

$$X_i = \gamma_0 + \gamma_1 Z_i + v_i$$

$$Y_i = \beta_0 + \beta_1\gamma_0 + \beta_1\gamma_1 Z_i + \beta_1 v_i + u_i$$

17.5 By making A endogenous and GNP exogenous, the model becomes recursive. The "reduced form" is

$$CON_i = \beta_0 + \beta_1 GNP_i + \beta_2 RTB_i + \beta_3 CON_{i-1} + u_i$$

$$A_i = -\beta_0 - (\beta_1 - 1)GNP_i - \beta_2 RTB_i - \beta_3 CON_{i-1} - u_i$$

17.9 No. The "endogenous" variable X on the right-hand side of the first equation is not related to the disturbance u, as can be seen in the "reduced form" given above. Thus there is no OLS bias in this case. The second structural equation, with no endogenous variables on the right-hand side, stands as a regular single-equation model.

17.10 Since the $\hat{\pi}$'s are OLS estimators, they are linear combinations of the Y_j's [as used in Equation (17.16)]. Each equation of the reduced form shows that each Y_j depends on the disturbances of the system. Hence the $\hat{\pi}$'s are linear combinations of the disturbances.

17.11 Equation (17.3) is just identified: two predetermined variables are excluded while two endogenous variables appear on the right-hand side. Equation (17.5) is over-identified: two predetermined variables are excluded while one endogenous variable appears on the right-hand side.

17.14 Considering the coefficients on *OTHINC* in the reduced form, the ratio of that in the *HOURS* equation to that in the *WAGE* equation is simply β_1. Suppose that we estimate the reduced form equations. The ratio of the two estimated coefficients on *OTHINC* is the ILS estimate of β_1.

Chapter 18: Inference for the Mean and Variance

18.2 X has a binomial distribution with $E[X] = 50$ and $\sigma(X) = 5$.
(a) $E[\overline{X}] = 50$, $\sigma(\overline{X}) = 1.25$.
(b) Treating $\overline{X}$ as approximately normally distributed, $\Pr(\overline{X} \leq 48) = \Pr(Z \leq -1.6) = .055$.

18.5 The possible values of $\overline{X}$ are -1, 0, 1, occurring with probabilities .25, .50, .25, respectively.

18.7 A two-tailed test is called for. With $n = 400$, $t^* = (103 - 100)/(15/20) = 4.0$ and $t^c \approx 1.97$. The conclusion of the test now would be to reject the null hypothesis that $E[X] = 100$.

18.8 With $n = 81$, $s(\overline{X}) = 15/9 = 1.67$ and $t^c = 1.99$. The 95 percent confidence interval is $103.0 \pm (1.99)(1.67)$, or 103.0 ± 3.32, which ranges from 99.68 to 106.32. (Since the confidence interval includes the value $\overline{X} = 100$, the null hypothesis H_0: $\mu_X = 100$ would not be rejected.)

18.11 (a) Since $\sigma_X^2 = 1$, $E[S_X^2] = 1$.
(b) With $(n - 1) = 9$ and $\sigma_X^2 = 1$, $\chi^2 = 9S_X^2/1$ has a chi-square distribution with df = 9. Pr $(S_X^2 > 2)$ = Pr $(\chi^2 > 18)$. In Table A.4 we see that with df = 9, Pr $(\chi^2 \geq 16.93) = .05$ and Pr $(\chi^2 \geq 19.03) = .025$. Hence Pr $(\chi^2 \geq 18)$ is probably between 3 and 4 percent.

18.13 With df = 15, $(\chi^2)_l^c = 6.26$ and $(\chi^2)_r^c = 27.50$. Based on Equation (18.27), the 95 percent confidence interval extends from 30.0 to 131.8.

Chapter 19: Chi-Square Tests and Analysis of Variance

19.1 With 600 workers, $(\chi^2)^* = 7.50$. With $\alpha = .05$, $(\chi^2)^c = 5.99$, as before. Thus the null hypothesis is rejected, and we conclude that the data contradict the theory.

19.4 Based on the standard normal distribution, the predicted frequencies are: 16 less than -1, 34 between -1 and 0, 34 between 0 and 1, and 16 greater than 1. Based on Equation (19.1), $(\chi^2)^* = 6.34$. There are four categories, so df = 3. With $\alpha = .05$, $(\chi^2)^c = 7.82$, and we do not reject the null hypothesis. We conclude that the observations might have been taken from a standard normal variable.

19.6 In the table, $(\chi^2)^* = 124.7$ and df = 4. With $\alpha = .05$, $(\chi^2)^c = 9.49$ and we reject the null hypothesis. We conclude that income is related to education.

19.9 The basic F test in the first regression corresponds to a one-way ANOVA by occupation. For $F_{2,97}$, $F^c \approx 3.09$ at $\alpha = .05$. Based on Equation (15.12), $F^* = 26.12$, and we conclude that there are significant differences by occupation.

19.10 In a two-way ANOVA, we test for differences by sex by comparing the second regression (unrestricted) with the first (restricted). For $F_{1,96}$, $F^c \approx 3.94$ at $\alpha = .05$. Based on Equation (15.9), $F^* = 3.05$, and we conclude that the differences by sex are not significant. (Note that $F^c \approx 2.8$ at the 10 percent level of significance, so the differences would be significant.)

Bibliography

The first two sections list standard texts in the fields of mathematical statistics and econometrics. Succeeding sections give references for selected chapters.

Intermediate Mathematical Statistics

Freund, John E., and Walpole, Ronald E. *Mathematical Statistics*. 4th ed. Englewood Cliffs, N.J.: Prentice-Hall, 1987.

Hoel, Paul G. *Introduction to Mathematical Statistics*. 5th ed. New York: Wiley, 1984.

Hogg, Robert V., and Craig, Allen T. *Introduction to Mathematical Statistics*. 4th ed. New York: Macmillan, 1978.

Mood, Alexander M., Graybill, Franklin A., and Boes, Duane C. *Introduction to The Theory of Statistics*. 3rd ed. New York: McGraw-Hill, 1974.

Intermediate and Advanced Econometrics

Goldberger, Arthur S. *Econometric Theory*. New York: Wiley, 1964.

Gujarati, Damodar. *Basic Econometrics*. New York: McGraw-Hill, 1978.

Hanushek, Eric A., and Jackson, John E. *Statistical Methods for Social Scientists*. New York: Academic Press, 1977.

Intriligator, Michael D. *Econometric Models, Techniques, and Applications*. Englewood Cliffs, N.J.: Prentice-Hall, 1978.

Johnston, J. *Econometric Methods*. 3rd ed. New York: McGraw-Hill, 1984.

Kelejian, Harry H., and Oates, Wallace E. *Introduction to Econometrics*. 2nd ed. New York: Harper & Row, 1981.

Kennedy, Peter. *A Guide to Econometrics*. 2nd ed. Cambridge, Mass.: MIT Press, 1986. (A useful commentary; not a text.)

Kmenta, Jan. *Elements of Econometrics*. 2nd ed. New York: Macmillan, 1986.

Maddala, G. S. *Econometrics*. New York: McGraw-Hill, 1977.

Pindyck, Robert S., and Rubinfeld, Daniel L. *Econometric Models and Economic Forecasts*. 2nd ed. New York: McGraw-Hill, 1981.

Theil, Henri. *Principles of Econometrics*. New York: Wiley, 1971.

Wonnacott, Thomas H., and Wonnacott, Ronald J. *Econometrics*. 2nd ed. New York: Wiley, 1979.

Chapter 2: Economic Data

Ferber, Robert, et al. "Validation of a National Survey of Consumer Characteristics: Savings Accounts." *Review of Economics and Statistics* 51 (November 1969), 436–444.

Projector, Dorothy S., and Weiss, Gertrude S. *Survey of Financial Characteristics of Consumers*. Washington: Board of Governors of the Federal Reserve System, 1966. (This describes and analyzes the survey from which the cross-section data set was taken.)

Projector, Dorothy S. *Survey of Changes in Family Finances*. Washington: Board of Governors of the Federal Reserve System, 1968.

Economic Report of the President (Transmitted to the Congress February 1982). Washington: U.S. Government Printing Office, 1982. (This annual report includes a large appendix of statistical tables, from which the time-series data set was taken.)

Chapter 7: Multiple Regression: Theory and Application

Goldfeld, Stephen M. "The Demand for Money Revisited." *Brookings Papers on Economic Activity* (3:1973), 577–638.

Gordon, Robert J. "Wage-Price Controls and the Shifting Phillips Curve." *Brookings Papers on Economic Activity* (2:1972), 385–421.

Laidler, David E. W. *The Demand for Money: Theories and Evidence*. 2nd ed. New York: Harper & Row, 1977.

Mincer, Jacob. *Schooling, Experience, and Earnings*. New York: National Bureau of Economic Research, 1974.

Perry, George L. *Unemployment, Money Wage Rates, and Inflation*. Cambridge, Mass.: MIT Press, 1966.

Chapter 14: Autocorrelation and Heteroscedasticity

Breusch, T. S., and Pagan, A. R. "A Simple Test for Heteroscedasticity and Random Coefficient Variation." *Econometrica* 47 (September 1979), 1287–1294.

Cochrane, D., and Orcutt, G. H. "Application of Least Squares Regression to Relationships Containing Autocorrelated Error Terms." *Journal of the American Statistical Association* 44 (March 1949), 32–61.

Durbin, J., and Watson, G. S. "Testing for Serial Correlation in Least Squares Regression. II." *Biometrika* 38 (June 1951), 159–178.

Durbin, J. "Testing for Serial Correlation in Least-Squares Regression When Some of the Regressors Are Lagged Dependent Variables." *Econometrica* 38 (May 1970), 410–421.

Glejser, H. "A New Test for Heteroskedasticity." *Journal of the American Statistical Association* 64 (March 1969), 316–323.

Hildreth, Clifford, and Lu, John Y. "Demand Relations with Autocorrelated Disturbances." Michigan State University Agricultural Experiment Station, *Technical Bulletin* 276 (November 1960).

Park, R. E. "Estimation with Heteroscedastic Error Terms." *Econometrica* 34 (October 1966), 888.

Chapter 15: More Testing and Specification

Chow, Gregory C. "Tests of Equality Between Sets of Coefficients in Two Linear Regressions." *Econometrica* 28 (July 1960), 591–605.

Fisher, Franklin M. "Tests of Equality Between Sets of Coefficients in Two Linear Regressions: An Expository Note." *Econometrica* 38 (March 1970), 361–366.

Chapter 16: Regression and Time Series

Griliches, Zvi. "Distributed Lags: A Survey." *Econometrica* 35 (January 1967), 16–49.

Goldberger, Arthur S. "The Interpretation and Estimation of Cobb-Douglas Functions." *Econometrica* 36 (July-October 1968), 464–472. (For making predictions with a logarithmic regressand.)

Chapter 17: Simultaneous-Equation Models

Fisher, Franklin M. *The Identification Problem in Econometrics*. New York: McGraw-Hill, 1966.

Kuh, Edwin, and Schmalensee, Richard L. *An Introduction to Applied Macroeconomics*. Amsterdam: North-Holland, 1973. (Builds and analyzes a small macroeconomic model.)

Suits, Daniel B. ''Forecasting and Analysis with an Econometric Model.'' *American Economic Review* 52 (March 1962), 104–132.

Working, E. J. ''What Do Statistical 'Demand Curves' Show?'' *Quarterly Journal of Economics* 41 (February 1927), 212–235.

Index